Airport Engineering
Planning and Design

Airport Engineering
Planning and Design

Subhash C. Saxena

B.Sc. Engg. (Civil), Ranchi; M.A.Sc. (Transportation), Toronto;
F. ASCE, F. ZwIE, F. IE, M. CI (USA), M. TDI (USA), M. IRC

Director (Technical) & Chief Consultant (Traffic & Transportation), TES, Jaipur
Formerly : Professor & Dean, Faculty of Engineering,
National University of Science & Technology, Bulawayo, Zimbabwe
Head, Civil Engineering, University of Zimbabwe, Bulawayo, Zimbabwe
Professor, University of Dar es Salaam, Tanzania; University of Malaya, Malaysia;
University of Sulaimania, Iraq; Regional College of Engineering & Technology, Surat; and
Lecturer, Indian Institute of Technology, Roorkee

Author of *A Textbook of Railway Engineering*

CBSPD

CBS Publishers & Distributors Pvt Ltd

New Delhi • Bengaluru • Chennai • Kochi • Kolkata • Lucknow • Mumbai
Hyderabad • Jharkhand • Nagpur • Patna • Pune • Uttarakhand

Airport Engineering
Planning and Design

ISBN: 978-81-239-1550-0

First Edition: 2008
Reprint: 2010, 2011, 2012, 2013, 2015, 2017, 2020, 2023

Published by **Satish Kumar Jain** and produced by **Varun Jain** for
CBS Publishers & Distributors Pvt Ltd
4819/XI Prahlad Street, 24 Ansari Road, Daryaganj, New Delhi 110 002, India
Ph: 011-23289259, 23266861   Website: www.cbspd.com
                                    e-mail: delhi@cbspd.com

Corporate Office: 204 FIE, Industrial Area, Patparganj, Delhi 110 092, India
Ph: 011-4934 4934   Fax: 011-4934 4935   e-mail: publishing@cbspd.com;
                                             publicity@cbspd.com

Branches

- **Bengaluru:** Seema House 2975, 17th Cross, KR Road, Banasankari 2nd Stage, Bengaluru 560 070, Karnataka, India
  Ph: +91-80-26771678/79   Fax: +91-80-26771680   e-mail: bangalore@cbspd.com
- **Chennai:** 7, Subbaraya Street, Shenoy Nagar, Chennai 600 030, Tamil Nadu, India
  Ph: +91-44-26680620, 26681266   Fax: +91-44-42032115   e-mail: chennai@cbspd.com
- **Kochi:** 42/1325, 1326, Power House Road, Opp KSEB, Power House, Ernakulum Kochi 682 018, Kerala, India
  Ph: +91-484-4059061-65,67   Fax: +91-484-4059065   e-mail: kochi@cbspd.com
- **Kolkata:** 147, Hind Ceramics Compound, 1st Floor, Nilgunj Road, Belghoria, Kolkata-700056, West Bengal, India
  Ph: +033-25633055, 033-25633056   e-mail: kolkata@cbspd.com
- **Lucknow:** Basement, Khushnuma Complex, 7 Meerabai Marg (Behind Jawahar Bhawan),Lucknow-226001, UP, India
  Ph: +91-522-4000032   e-mail: tiwari.lucknow@cbspd.com
- **Mumbai:** PWD Shed, Gala no 25/26, Ramchandra Bhatt Marg, Next to JJ Hospital Gate no. 2, Opp. Union Bank of India Noorbaug, Mumbai-400009, Maharashtra, India
  Ph: 022-66661880/89   e-mail: mumbai@cbspd.com

Representatives

• Hyderabad	0-9885175004	• Jharkhand	0-9811541605	• Nagpur	0-9421945513
• Patna	0-9334159340	• Pune	0-9923910676	• Uttarakhand	0-9716462459

Printed at Chaman Enterprises, Daryaganj, Delhi, India

To

My Grandchildren

ARJUN, ANVITA & AADI

Preface

The growth of economy of any country depends upon the development of transportation. In country like India, having second largest system of railways in the world and fully developed highway transportation, there is growing demand for air transportation. The steady rise in consumer income is having favourable impact on air travel. With more and more cheap fare commercial airlines like Air Deccan, Spice Jet, Kingfisher and many others coming up, internal airtravel will continue to increase and people will have more time for leisure. This potential, together with phenomenal increase in international travel, has created severe congestion at the airports in the large cities, needing additional facilities for rapidly growing short-haul domestic markets.

In the years to come, it is evident that substantial financial and technical resources shall be required for investment in airways, airport improvements and development of new airports, in all growing cities around the world. Books on airport engineering are rare and those written long time ago are outdated in content and concepts. There is, therefore, a need for a book based on recent developments and standards for airport planning and design.

International Civil Aviation Organisation (ICAO), Canada is the international agency which provides guidelines for development of airports throughout the world. It provides international design standards and recommends practices which are applicable to nearly all airports serving international air travel. Author is deeply indebted to ICAO, Canada and Federal Aviation Administration, USA, for not only allowing him to use their literature freely for writing this book, but also mailing their publications regularly for several years. Material has been drawn freely from the publications of these two agencies in producing this book providing latest information.

The text presented in this book introduces the subject of airport engineering meant for graduate and postgraduate students of Civil and Transportation Engineering as well as practising engineers. It deals with basic concepts of planning and designing of airports and is based on recommendations and standards of ICAO and FAA available till beginning of this century. However, it should be realised that techniques, regulations and standards regarding airport planning and design change from time to time.

Various chapters, after presenting the introductory overview, deal with airport planning, aircraft characteristics and design of various elements of an airport, including the terminal building.

Number of solved, unsolved and objective type examples are included in the book for better understanding and practising the theory for the students. This book contains all the new and updated material drawn from published work of many organisations and individuals, including International Civil Aviation Organisation (ICAO), Canada, Federal Aviation Agency (FAA), USA and Airport Authority of India. The author is deeply indebted to all of them.

Valuable suggestions and comments from the readers in India and abroad, for further improvement of the book, shall be gratefully acknowledged.

For the encouragement, environment and impetus provided to complete this book, the author is thankful to his wife Aruna, son and daughter-in-law Sundeep and Ritu. For proof-reading and providing idea for cover page design, my special thanks to daughter Shivani, son-in-law Vishesh and Mr. Ajay Saxena of Airport Authority of India, New Delhi.

Subhash C. Saxena

About the Author

Prof. Subhash C. Saxena is well-known in Civil Engineering field due to his research, teaching and academic achievements in India and abroad. Besides over 65 research and technical papers, his books "Railway Engineering (1973)", "Tunnel Engineering" (1972), "Traffic Planning and Design (1989)" are most popular books among engineering students and practising engineers.

Prof. Saxena obtained Bachelor of Science (Civil Engineering) degree from Ranchi University, Ranchi in 1961 and Master of Applied Sciences (Transportation Engineering) degree from University of Toronto, Toronto, Canada in 1964, with 'A' grade in all technical subjects. His research thesis was entitled "Progressive Maintenance of Roads in Developing Countries."

He worked for two years as a Research (Pool) Officer with Central Road Research Institute, New Delhi (1965-67), where he started several research projects for cutting down the maintenance costs for roads and bridges, involving laying experimental quality controlled stretches of roads on national highways in Haryana and UP.

In 1967, Prof. Saxena joined Roorkee University (now Indian Institute of Technology, Roorkee) as a lecturer in Civil Engineering. He introduced a new area of research through 'Radioisotopes' in Highway Engineering for determining bitumen content of bituminous mixes, soil moisture and density, for quick quality control through non-destructive technique. He was selected as Professor of Civil Engineering at Sardar Vallabh Bhai Regional Engineering College, Surat (Gujarat) in 1967.

At S.V.R. College, Surat, Professor Saxena, as Head of Transportation Engineering, conducted several traffic and transportation projects for improving traffic conditions, safety and efficiency on Surat roads. He organised a National Seminar on "Traffic Problems of Surat" to highlight the research work in this area. While at Surat he completed and published his two most popular books, "Text Book of Railway Engineering" and "Tunnel Engineering (1972-73). During this period he was Chairman of the Debating and Literary Society of the college and warden of students hostels. Government of India sent him as an expert to Govt. of Tanzania in 1974 for two years to start a new Technical College in Dar es Salaam, Tanzania.

A Recruitment Board from Iraq selected him as a Professor of Civil Engineering for University of Sulaimania, Sulaimania, Iraq in 1978, where he developed curriculum and laboratory facilities of the Civil Engineering Department and started two new postgraduate programmes leading to 'Diploma' and 'Masters' degree in Transportation Engineering. Besides guiding several postgraduate research projects, he guided a project on Planning and Design of an airport for Erbil city in Iraq.

In 1981 he joined University of Malaya, Kuala Lumpur, Malaysia as the head of Transportation. He initiated the idea of constructing university roads by the students using their technical knowledge in field project. One such hilly road was completed by students during summer vacation under his supervision and guidance. Several research projects in traffic engineering were completed for Malaysian roads and papers published in international journals and conferences.

In 1984, Prof. Saxena joined University of Dar es Salaam, Dar Es Salaam, Tanzania. In recognition of his academic and published work he was offered additional salary by the Swiss Development Corporation paid in Switzerland. He was named best teacher of the department in the evaluation by the students, and was faculty coordinator of University Teaching and Learning Improvement Programme. Several professional development programmes were organised by him for improving academic and research standards. He was Chairman/Member of several academic/technical committees of the department, faculty and the university. His book on "Traffic Planning and Design" was written and published in 1989.

In 1989, Professor Saxena joined University of Zimbabwe, Harare, and was posted as Head, Civil Engineering department at university campus at Bulawayo, Zimbabwe. His contributions in establishment of a new university of science and technology, at Bulawayo, were significant. National University of Science & Technology, Bulawayo, came into existence in 1990, and he took over as the founder Dean of the Faculty of Engineering. He was responsible for development of curriculum, staff, laboratories and research facilities for the three departments of the new university: Civil and Water Engineering, Industrial Engineering and Electronic Engineering, In 1993, Prof. Sexena was invited to travel to Russia and Czech Republic, as a member of the American delegation to examine current airport projects and plan to improve and better utilise the airports of Russia and the Czech Republic. As the Professor and Dean of the faculty, Prof. Saxena chaired/participated in all the academic and administrative committees of the university including Council, Senate, academic board, faculty board, appointments and selection committees etc.

In 1996, Professor Saxena returned to India, and started providing educational and technical consultancies to various organisations and consulting firms. As the Principal, he established a new Engineering College, Maharishi Arvind Institute of Engineering & Technology at Jaipur. As Traffic and Transportation Planner, he developed a 20 year traffic and transportation plan for Jaipur city incorporating the stage construction of suggested improvement measures on immediate, short and long term basis. As a Traffic and Pavement Design Engineer, he has provided consultancy services on several projects such as improvement of Rajasthan State Highways, improvement and widening of National Highways, NH-8, NH-12, NH-26, NH-205, NH-47, NH-4, NH-77 under the nationwide programme of National Highway Authority of India (NHAI), for highway improvements.

At present, Prof. Saxena is Technical Director of Theme Engineering Services, Jaipur; and Chief Consultant for Traffic and Transportation Projects.

All these years, while teaching in India and abroad, Prof. Saxena felt the need of a simple and comprehensive book on Airport Engineering for use of students and practising engineers. He therefore, approached the Director of International Civil Aviation Organisation (ICAO), Canada and Federal Aviation Agency (FAA), USA, to supply him with their literature regularly and permission to reproduce it in his book. Having that permission and receiving published material for over 10 years, Prof. Saxena decided to write this book on "Airport Engineering—Planning and Design", to meet the demands of airport development in fast growing domestic and international air travel.

Contents

List of Figures

List of Tables

Common Abbreviations

ACN	Airport Classification Number
AEP	Airport Entry Permit
ARP	Airport Reference Point
ARSA	Airport Radar Service Area
ARSR	Air Route Surveillance Radar
ARTC	Air Route Traffic Control
ASR	Airport Surveillance Radar
ASDA	Accelerate Stop Distance Available
ATC	Air Traffic Control
BCAS	Bureau of Civil Aviation Security
CBR	California Bearing Ratio
CoSCA	Commission of Security Civil Aviation
CRCP	Continuously Reinforced Concrete Pavement
CTOL	Conventional Takeoff and Landing
DGCA	Director General of Civil Aviation
EMP	Environment Management Plan
ESWL	Equivalent Single Wheel Load
FATO	Final Approach and Take Off
FARA	Final Approach Reference Area
FAI	Federal Aeronautics International
FAA	Federal Aviation Agency
GPS	Global Positioning System
HAPI	Heliport Approach Path Indicator
IFR	Instrument Flight Rules
ICAN	International Commission for Air Navigation
ICAO	International Civil Aviation Organisation

ICAA	International Civil Airports Association
IAAI	International Airport Authority of India
ITP	Inclusive Tourist Package
ITU	International Telecommunication Union
ILS	Instrument Landing System
IATA	International Air Transport Association
LDA	Landing Distance Available
LCN	Load Classification Number
LTA	Lighter Than Air
LoRAN	Long Range Aerial Navigation
MLS	Microwave Landing System
NAA	National Airport Authority
NAAI	National Airport Authority of India
NDT	Non Destructive Testing
NIAMAR	National Institute of Aviation Management and Research
ONGC	Oil and Natural Gas Commission
PAR	Precision Approach Radar
PCN	Pavement Classification Number
PAPI	Precision Approach Path Indicator
PDSO	Private Domestic Scheduled Operator
PMS	Pavement Management System
REIL	Runway End Identifier Lights
RESA	Runway End Safety Area
RVR	Runway Visible Range
STOL	Short Take Off and Landing
TACAN	Tactical Air Navigation
TLOF	Touchdown and Lift Off
TORA	Take Off Runway Available
TODA	Take Off Distance Available
UHF	Ultra High Frequency
VASI	Visual Approach Slope Indicator
VFR	Visual Flight Rule
VHF	Very High Frequency
VTOL	Vertical Take off and landing
WEAA	Western European Airport Association
WMO	World Meteorological Organisation

Conversion Factors

F.P.S.		Metric
1 ton (2000 lbs)	=	0.9072 tonnes = 907.2 kg
1 ton (2240 lbs)	=	1.0160 tonnes
1 statute mile (5280 feet)	=	1.6090 km = 1609 m
1 nautical mile (6080 ft)	=	1.8531 km = 1853 m
1 ton-mile (2000 lbs-mile)	=	14600 tonne–kilometers
1 ton-mile (2240 mile)	=	1.6352 tonne-kilometer
1 ft. (feet)	=	0.3048 m
1 inch	=	2.54 cm = 0.254 m
1 lb (pound)	=	0.4536 kg
Degree Fahrenheit	=	Degree celcius = $(^{\circ}F - 32)\dfrac{5}{9}$
1 gallon (US liquid)	=	0.003785 m^3
1 acre	=	4046.8 m^2
1 mile^2 (statute)	=	$2{,}589{,}988 \text{ m}^2$
1 ft^2	=	0.0929 m^2
1 ft/second	=	0.3048 m/s

CHAPTER 1

Historical Developments of Aviation and Aircrafts

1.0 INTRODUCTION

The growth of air transportation is one of the most remarkable technological developments of last century. Air travel is expected to increase enormously in the decades ahead, although the rate of growth will be quite different for each country. Rise in income level favour air travel. As income grows people are able to spend more on air transportation and enjoy more leisure time by air travel - short as well as long hauls.

Air transportation provides for movement of people and goods and makes possible the creation and expansion of various business and industries, often opening up vast opportunities for exploiting raw materials and markets in otherwise inaccessible areas, thus opening new vistas of economic development. It also plays the role of uniting people of different countries and giving them the benefits of civilisation and peace. Air transportation contributes in :

- Economic growth of the country : As per ICAO (International Civil Aviation Organisation), every $ 100 spent on air transport generates benefits worth $ 325 for economy.
- Faster development and distribution of products and services : It increases country's/industries global reach.
- Opportunities for tourists for adventures during vacation : It boosts tourism.
- Bringing social changes by uniting people closer.
- Better understanding of inter regional problems.
- Widening and increasing business and thus industry grows. It enables to get to market quickly.
- Frequent exchange of information.
- Helps countries to push GDP growth and earn foreign exchange.
- Sharing and enjoying the cultures and traditions of distance lands.
- Savings of time and energy due to high speed of air transportation.
- Connecting otherwise inaccessible regions. Helps in providing relief operations in emergencies like tsunami or earthquakes.
- Airports generate several jobs and increase value of land around them.

The development of a safe and efficient air transportation system is essential for urban development and a vital feature of modern cities, and urban life. The future challenges will be to accommodate the expansion in air travel in the most rational way and correct anticipation of demand for services, choosing the right type of facility to provide the required service, to preserve the environment, and meet the various social objectives. Improvements in the existing airports and development of new airports will have to be undertaken. Civil and transportation engineers will be making major contributions to provide the facilities to meet this growth by system and master planning, design, construction and maintenance of airports including airport parking and internal circulation system.

The tremendous changes in operating characteristics of aircrafts require enormous research and efforts to develop navigational aids, runaway lights and control devices so that modern all jet aircrafts could land safely, pavements of adequate thickness to support their impacts and pressure, and procedures to permit them to maneuver easily on and off runways and around terminals.

In India, development of airports, has not been commensurate with the growth of air traffic. The new airlines coming up and continuance of the present pattern of growth of air traffic, the airports are exhausting their capacity, despite maximum utilisation of its facilities and possible expansions. New airports will have to be planned and developed for many cities in India.

1.1 HISTORY OF AVIATION

The steam age, beginning in the early nineteenth century, freed man from dependence on human and animal transport and vastly increased his mobility over land and sea. Railways and automobiles provided man with rapid transport over land. Wright brothers introduced flights at the beginning of twentieth century allowed man to navigate through the air from point to point over the earth's surface, avoiding or minimising geographical barriers that impede other forms of transport.

Following the Wright brothers' flights at Kitty Hawk in 1903 development of aircrafts was quite rapid. During the period 1914–1918 technical development of aircrafts and engines in terms of dependability and load carrying capacity accelerated, partly due to World War I. Europe remained the leader in commercial aviation. Germany had regularly scheduled passenger service by "Zeppelins" even before World War I.

After the close of world war in 1919 and 1920, scheduled airline services started in Europe between Paris and London and between Amsterdam and London. In USA first scheduled flight started on January 1, 1914 linking Tampa and St. Petersburg a distance of 34 kms. In India, regular air flight started in 1929 between Karachi and Delhi, by Imperial Airways Service, and in 1933 the Indian Trans-Continental Airways Limited started foreign flights. First airmail route in USA was established in 1918 between New York and Washington, followed by New York and Chicago in 1919, extended to San Francisco in 1920.

The official airline guide was first published in January 1927, listing regularly scheduled services to and fro 128 cities served by 28 airlines in USA. The first longest flight of 4500 kms was undertaken from New York to Paris on May 20–21, 1927. A series of record breaking flights for distance, endurance and speed followed which added confidence in the dependability of air flights and stimulated public acceptance and interest in air transportation. Transoceanic passenger service was inaugurated in 1937 with passenger service by flying boat began in 1939.

In America, the first of the famous high wing corrugated metal monoplanes known as "Tin Goose" was produced by Ford Motor Company in 1927. Though noisy, these aircrafts carried passengers and represented a large step forward in passenger comforts by starting stewardess service. Other aircrafts of this period were the Curtiss Condor, Fokker Trimotor, Lockheed Orion, Stinson Trimoter and the Boeing B–80 and 247. They were, then superseded by Douglas DC–2 in 1934 and Douglas DC-3 in 1936.

During these early days, largely the business people used to travel by air who cared more for savings in time and comforts of journey to justify expenses. It was in late 1940s when cabin aircrafts such as the DC–6, Convair 240 and Lockheed Constellation were introduced into service, making possible greater speeds and more comfort, higher altitude flights, enhancing the demand for civil air transport.

During the World War II, all available air transport technology was diverted to the war effort. War accelerated the technology and new jet technology enhanced the demand for air travel in 1960 and beyond. Development of jet and rocket engines provided jump in speed and payload capacity, enabling improvements in aircrafts and travel. Regular commercial jet service was started by the British Overseas Airways Company with its comet in 1952. Full scale commercial jet service was offered in late 1950s by number of airlines using the Boeing 707.

With these advances in technology and development of bigger and heavier aircrafts, the important technical feature of runway length and strength requirements for different type of aircrafts attracted the attention. The required runway lengths grew rapidly as heavier, faster aircrafts come into service following the DC–3. This trend reached a peak with the introduction of the 707–120 and then stabilised with the 707–320 at about 3.5 times the length required for a DC–3 or at about 3500 m for the heaviest long distance aircraft.

Early transport aircrafts were designed for operation on unpaved strips. When the first permanent commercial passenger airline service started in the USA in April, 1927, between Boston and New York, landing and take offs at both ends of the journey were made on packed cider or dirt land fields. As airline industry grew, speed, payloads and runway length requirements needed paved runway surfaces and markings for pilot orientation as well as lighting for night operation. Also the airline hanger and terminal where passengers gathered for flight boarding, were introduced, usually a separate building for each airline. As the scheduled airline passenger services grew, needs for better terminal facilities finally consolidated in "Union" or single terminal structure. Supplying of airport landing fields and terminal facilities at a bargain rental then commenced.

In the beginning of airline passenger service, convenient public transport between airport and city was seldom available. The early airlines used to provide "airline limo" "bus" service usually between airport and one or more hotels in the town. The primary means of public transportation to airport was limosine or charter bus service. Today airports are served by several types of road transport and there is trend towards providing direct rail connections between city centres and airports, like London, Zurich, Amsterdam, Frankfort, Boston, Chicago, New York etc.

The air transportation has grown rapidly since 1950 because of the introduction of jet transport services. These aircrafts gave rise to increased noise problem and complaints from airport neighbours, requiring extensive construction of new or expanded airports.

The airplanes development, brought with it international problems. The coordination of techniques and laws, the dissemination of technical and economic information, the need for safety and regularity in air transport, establishing uniform navigational aids and weather reporting systems. The need for establishment of standards for rules of the air, for air traffic control, for personnel licensing, for the design of airports and many other aspects of safety and efficiency required combined and coordinated efforts of all the nations.

There was the question of commercial rights – what arrangements would be made for airlines of one country to fly into and through the territories of another.

Fifty two nations met in Chicago in November, 1944 and considered the problems of international civil aviation. A permanent body called 'Internation Civil Aviation Organisation' (ICAO) was created on April 4, 1947, with its headquarter in Montreal, Canada. It is a specialised agency of the United Nations.

1.2 INTERNATIONAL CIVIL AVIATION ORGANISATION (ICAO)

The aims and objectives of ICAO are to develop the principles and techniques of international air navigation and to foster the planning and development of international air transport so as to :

  (i) ensure the safe and orderly growth of international civil aviation throughout the world.

  (ii) encourage the arts of aircraft design and operation for peaceful purposes.

  (iii) promote the development of airways, airports and air navigation facilities for international civil aviation.

  (iv) meet the needs of the people of the world for safe, regular, efficient and economical air transport.

  (v) prevent economic waste caused by unreasonable competition.

  (vi) ensure that the rights of contracting nations are fully respected and each has a fair opportunity to operate international airlines.

  (vii) avoid discrimination between contracting nations.

  (viii) promote safety to flight in international air navigation.

  (ix) promote generally the development of all aspects of international civil aviation.

ICAO works in close cooperation with other specialist institutions such as World Meteorological Organisation (WMO), International Telecommunication Union (ITU), International Civil Airports Association (ICAA), Western European Airport Association (WEAA), International Air Transport Association (IATA), Agency for Air Navigation Safety (ASECNA), Federal Aeronautics International (FAI), Federal Aviation Agency (FAA) etc. In order to achieve its objectives ICAO develops "Standards" and "Recommended practices" for personnel licensing, rules of the air, meteorology, general aviation, aircraft registration, airworthiness, certificates, aeronautical telecommunications, air traffic services, search and rescue, aircraft accident inquiries, aerodromes, aircraft noise etc. It also outlines the following Rights (Freedom) of Air :

First freedom gives right to overfly a country.

Second freedom gives aircraft the right to land for non-commercial reasons (technical stop).

Third freedom gives aircraft the right to disembark passengers, mail and freight coming from a country of which the aircraft has the nationality.

Fourth freedom gives aircraft the right to embark passengers, mail and freight for the purpose of carrying them to the country of which the aircraft has the nationality.

The **fifth** freedom gives the airline of a country the right to embark or disembark on the territory of another country passengers, mail and freight bound for or coming from a third country. There are also subtle combinations between Third, Fourth and Fifth freedom. Other freedoms are exchanged by countries through bilateral arrangements.

ICAO publications are the international source for standardisation, regulatory and guidance material in the air industry. Its standards and recommendations are applicable to nearly all airport serving international traffic.

ICAO has distributed air transport in following six geographical regions :

(i) Asia and Pacific

(ii) Europe

(iii) North America

(iv) Latin America and Carribean

(v) Africa, and

(vi) Middle East.

Federal Aviation Agency (FAA) was created in USA as an independent agency, directly responsible to US president. In 1958, agency and its functions were transferred to Department of Transportation. In 1967, name of Federal Aviation Agency was changed to Federal Aviation Administration.

ICAO also provides assistance to developing countries by provision of international experts, instructors and training. India has been a member of International Civil Aviation Organisation and is also on the Council of ICAO since its inception.

In India, National Airport Authority (NAA) was created through an act of parliament on June 1, 1986, for proper planning, monitoring, managing and coordinating all aspects of civil aviation and ensure safety of operation of national and international airports in India. Director General of Civil Aviation (DGCA) is responsible for management and other bilateral issues like tariffs and schedules. As the growth and demand for air traffic increased and jet crafts came in use, the congestion, delays and other problems at international airports began to increase. A separate body International Airports Authority of India (IAAI) was established in 1972 for planning, developing, constructing expanding and maintaining the international airports in India. IAAI is responsible for safety, security and environmental controls at airports.

1.3 CHALLENGES OF AIR TRANSPORTATION

The present examples of the major issues that are impacting the evolution of the airport, its planning and its management are summarised below.

1.3.1 Problem of growth of traffic

The rapid air traffic growth and inadequate support facilities is one of the challenges for airport authorities.

The extent of air traffic growth was in most cases unplanned because it resulted from sudden technological breakthroughs. Huge air traffic loads occurred so quickly that there was no time to plan and expand the facilities. This resulted in congestions and delays. Finding a solution to what appears to be a never ending demand is a big challenge.

1.3.2 Environmental problem

Airports do have negative environmental impacts. The impact of changing technology on the environment was not fully appreciated in the beginning. Failure to provide solutions to environmental disbenefits caused serious constraints on efforts by airlines and airports to supply facilities to accommodate future growth in traffic.

Potential pollution of the local environment adjoining airports must be recognised early and dealt with effectively to proclude or minimise adverse impacts. The smells and visible smoke of burned fuel and pollution of local water by inadequately protected surface drainage system may add other dimensions to the unfavourable imprints of airports on their environs.

1.3.3 Noise problem

The noise produced by the jet aircrafts has four primary effects on people :

    (i) Annoyance is caused by jet sound of an enlarged percentage of high frequency tones. Medical tests have shown that the average human ear can accept about 30 DB more of low frequency than of high frequency sound at the same level of discomfort.

    (ii) Masking of speech is more likely to occur as a result of the higher frequencies in jet sources that obliterate consonant sound.

    Remote airports may be relatively noise-complaint free but the access time and costs will reduce the demand for travel. Noise pollution correction at the source is expensive, but if used, will expansion and operational constraints be alleviated? Improved flight path control, steep approaches and quieter propulsion systems are being tried to reduce noise pollution.

    (iii) Interference with TV and radio is two fold. The masking effect plus overall intrusion of the jet noise is a distraction. In the case of TV there is also the interference of the flyover causing image flutter on the picture tube.

    (iv) Frequency and tone also exert distracting forces on the hearer. When repeated flyovers persist over a period of time, the hearer's tolerance drops markedly.

1.3.4 Planning problems

The site of the new airports must be as close as possible to the communities they serve to reduce the ground access time, yet must not be so close as to involve low flight and noise over populated areas. The planning must be modular and facilities flexible to preserve service life in the face of threatened obsolescence arising from the new technological developments.

1.3.5 Problems of airport users

Airport users are irritated when facilities provided for their comfort and convenience are costly or do not meet their needs. Excess costs, long distance, travel time to and from airports, processing procedures, are the usual causes of dissatisfaction.

1.3.6 Problem of safety

Aviation's rapid growth is putting fresh impetus to ensure air safety, but the safety hurdles are many, ranging from inadequate satellite and reliever airports to ageing and often obsolete air traffic control systems.

1.3.7 Problem of airport access

There is a general frustration by the fact that constant improvement, in the speed of air travel have not been matched by the time taken to reach an airport. Since introduction of jet aircraft, it is now quite common for the trip to and from the airport to take as much time as the trip in the air or even more for shorter flights. Something must, therefore, be done to reduce travel time to airport otherwise the quantity of local air travellers may drop. One of the obvious solution is to establish a separate route to the airport to avoid the congestion and delays that occur on city roads during peak hours, to permit high speed travel to airports. Some countries like UK and USA have used railroads to provide faster access to airports. The alternative which is most desirable from economic and speed point of view should be found through a cost-effectiveness analysis.

1.4 AIRPORT TYPES

The world's first airports were no more than natural turfed and levelled grounds, often surrounded by trees and farm buildings. Flight were achieved on grass runways no more than few hundred meters long. First aerodrome was built in UK in 1910.

The first modern airport was built in Germany, before the second world war. Todays paved runways have lengths over 2000 m and large airports may occupy a space of several square kilometers.

An airport is an area of land or water that is used or intended to be used for the landing and takeoff of aircraft and includes its buildings and facilities. Within this broad definition airports can be classed in different categories :

(a) **Based on functions :**
 (i) *Heliports* : Meant for helicopters.
 (ii) *STOL ports* : Mean for short take off and landing.
 (iii) *Conventional* : Meant for normal commercial aircraft.
 (iv) *Sea plane bases* (Aqua ports or seadromes) : Those which float on the sea.

(b) **Based on usage or ownership :**
 (i) *Public* : Owned by the government.
 (ii) *Private* : Owned by individuals.
 (iii) *Military* : Used for military purposes only.
 (iv) *Joint use* : Used by both civil and military aviation.

(c) **Based on facilities available :**
 (i) Number and type of runway.
 (ii) Navigational aids.

(d) **Based on type of aircraft designed to serve :**
 (i) *Basic utility (BU) airports* : Accommodates most single engine and light twin-engine aircrafts.
 (ii) *General utility (GU) airport* : Additionally accommodates medium twin-engine aircraft under 5670 kg.
 (iii) *Basic transport (BT) airport* : Can handle business jets under 27,200 kg.
 (iv) *General transport (GT) airport* : Can accommodate almost anything that flies.

1.5 ROLE OF AN AIRPORT

The basic purpose of a commercial airport facility is to provide an exchange point for people and goods arriving and departing in a variety of air and land vehicles. These vehicles have dissimilar operating characteristics.

Outbound or departing passengers arrive at the airport terminal, then proceed through the terminal building and are then 'compressed' into an airplane. Inbound or arriving passengers conversely go through an 'expansion' process as they scatter from arriving aircraft to connecting flights or to various local destinations in a variety of vehicles.

This interchange must be provided in a safe, convenient and pleasant manner for the air traveller and accompanying greeters : Airport user expect a variety of aminities and necessities.

Because of differing operating characteristics and spatial requirements of air and land vehicles, it is convenient to divide an airport into two parts : Airside or "airplane" part of the airport and landside or "people" part of the airport.

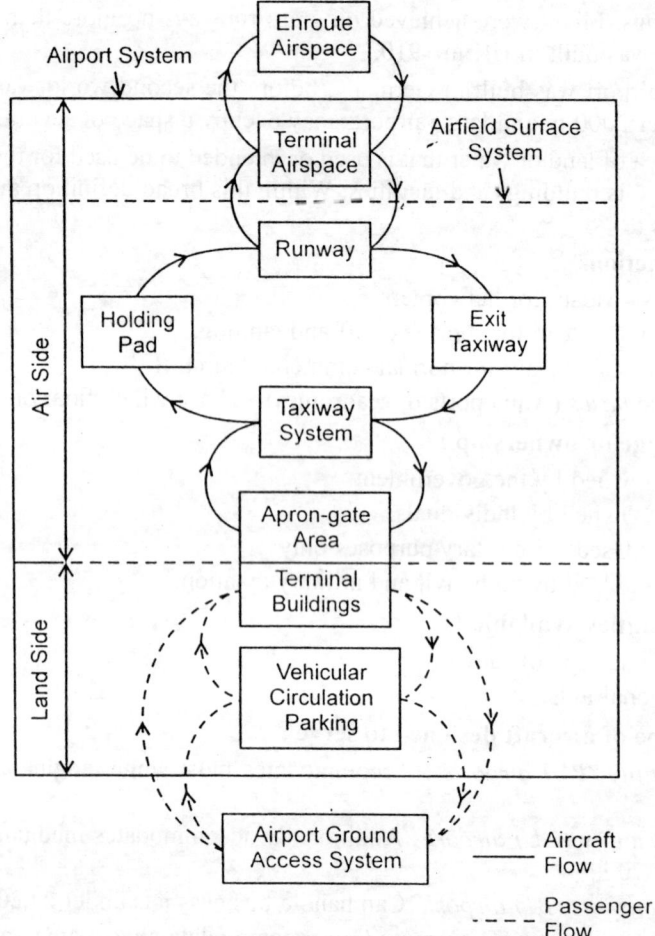

Fig. 1.1. Landside and airside of an airport.

Airside part of airport

In this part aircrafts land, take off, are serviced and separated or united with loads they have carried or will carry. It involves interfaces with supplier operators, automative traffic control and emergency rescue services.

The principal tasks at airside is the safe and efficient operation of the aeroplane and all its facilities. They include maintaining the flight schedules, servicing and maintaining of planes, cleaning, fueling, repair services, air traffic control, meteorological activities, maintenance and operation of runways and their approaches, departing zones, taxi ways, aprons, field areas, fencing, lighting, navigational and guidance aids, signing, obstruction marking, service roads, hangers and other airside buildings and equipments. The airside tasks of the airport are highly technical and operational in nature.

Landside part of airport

In the landside part passengers, cargo and mail arrive and depart by various ground transport means or local transport. This part consists of holding facilities (for taxi and buses) and short term and long term car parks, complexes of terminal buildings and services for the air passangers. Passengers, cargo and mail are processed in the terminal building. Today's international airports are characterised by area of car parks, complex and multistoried terminal buildings providing service facilities. The main tasks involved in landside are maintenance of buildings and land, asthetic and environmental considerations, efficient day to day operation, coordination with airlines and public relations to serve best the tenants and patrons. It involves dealing with business firms that provide services, shops and other facilities in the terminal building and also with government agencies such as immigration, public health, customs, agriculture and animal husbandry controls. Adequate provisions for handling passengers and cargo, provisions for anti-hijacking and anti-sabotage security in the passenger and baggage handling process. Regulation of landside traffic through police are also necessary.

1.6 DEPARTURE AND ARRIVAL PROCESS FOR AN INTERNATIONAL FLIGHT

To familiarise the readers who might have not undertaken an air journey, the operation involved for the passenger terminal is described in following paragraphs.

A passenger can go to the airport terminal by bus, taxi or by ones own means. Some airports may be connected by local trains. The car can be parked in the parking area. Many airports have garages if the car has to be parked for long periods.

Passengers generally buy the tickets before hand through travel agencies or through one of the airlines offices in town. After possession of his ticket, he turns up with his baggage at the check-in counter of the particular airline on whose aircraft he has to fly or of the airline representing it in the terminal. The airline's groundstaff checks in the passenger, weighs his baggage, attaches a label indicating the destination airport and the flight number, and put it on a conveyor belt which takes it to the despatching room where it will be placed on transporters and then conveyed to the aircraft. The airline staff gives the passenger back his ticket and issues a boarding pass, which contains seat number, the departure gate number and departure time. The passenger then has to go through the immigration area where passport and visa formalities are verified through immigration officers.. After his passport has been checked, he is authorised to enter the transit zone of the terminal. There he will find duty free shops where on showing his ticket, he will be able to buy duty free goods. The transit lounge contains bars, restaurant, shopping centre and various entertainment zones.

Before departure, a blinking light on the departure display board (or an announcement over the public address system) warns the passenger that embarkation has started. He goes to the specific departure gate through the security hold region for his flight. An airport staff or sometimes an employee of the airline in question tears the counterfoil from his boarding card and authorises him to board the aircraft which is at the end of corridor of a telescoping loading bridge that has been brought up to the aircraft door. At airports which donot have facility of loading bridges, passengers are taken to steps brought up to the aircraft door by bus services facilitated by that particular airlines.

On his return, the passenger will again go through the immigration and finally to customs where he will find his baggage which has been brought there by conveyor belts (there are several other systems of baggage transport). Finally he leaves the terminal to take land transportation to final destination.

The above departure and arrival process does not apply to domestic flights, as the passengers bound for a flight to town within the country are not subjected to immigration, custom or health formalities.

There are some long distance passengers in transit who remain on board their aircraft after landing or who disembark but stay in the transit lounge owing to the duration of the stop or a change of aircraft.

1.7 DEVELOPMENT OF AIRCRAFTS AND THEIR OPERATING CHARACTERISTICS

The development of aircrafts from Zeppelin in 1918 to ultrasophisticated machines capable of flying faster than the speed of sound is a fascinating story. At the start of aviation, aircrafts were equipped with piston engines (with propellers). Since the second world war these engines have been gradually replaced by turbine engines i.e. turboprops (with propellers) and jets (without propellers). Jet engined aircrafts are known as 'Jet Aircrafts' or 'Jumbo Jets'. Depending on whether an aircraft flies at a speed under or over that of sound, they are called "Subsonic" or "Supersonic" aircrafts.

The evolution in aircraft industry continues and the present Boeing and Concorde aircrafts are not the ultimate designs. There will certainly be both larger and faster aircrafts in future through improved technology. Quiet propulsion shall be an important achievement. Higher engine temperatures will compensate for efficiency losses due to noise suppression. Composite structures can reduce overall structural weight by 20 per cent. Powered lift will permit approach lift coefficients of 4 to 5 and field lengths less than 610 metres. Supercritical aerodynamics will enable aircrafts to cruise efficiently at mach number close to 1. Advanced supersonic arrow-wing configurations will produce cruise life-drag ratios approaching 10 at mach 3. Actively cooled structures may make possible hypersonic cruise with aluminium or titanium aircraft. Control configured vehicles which depend on reliable SAS may improve efficiency by as much as 10 per cent. Various advances in integrated avionics will reduce pilot workload, relieve terminal congestion, improve all-weather operation and increase comfort and safety.

New technology in the form of advanced airfoils and flap systems, rotary combustion engines, low cost gas turbines, improved control and navigation systems offer opportunities for significantly increasing the overall utilisation of general aviation aircraft.

This is a very brief overview of some of the new civil aircraft which might appear on the scene in the years ahead.

1.7.1 Trends in size and weight

Aircraft size and weights are crucial determinants in the design and construction of runways, taxiways, apron and terminal loading arrangement. In the early days of flight, aircraft size was often simply expressed in terms of wing span. 'Spruce Goose' all plywood flying boat made by M.R. Howard Hughes was one of the largest aircraft of those times having a wing span of 91 m.

With today's high-lift air foils and superpowerful jet engines, huge wingspans are no longer the measure of either size or weight. Boeing 747 can carry more than 500 passengers with only two-thirds the wing span of the 'Spruce Goose'. The present aircraft size and weight have to be defined by length, height, wheel base and gross weight. Double decker aircrafts are likely to be introduced soon.

From the revenue generating point of view the best measure to size and weight, is probably the payload capacity. The trend has been sharply upward since the introduction of the Boeing 707 in the late 1950s. The Douglas aircraft of the 1930s and 1940s all had fewer than 100 seats. Not until 1970 with the introduction of Boeing 747, were more than 200 seats available on a single aircraft as shown in Fig. 1.2. Conceptual studies are underway for aircraft, that can have capacity of 700 to 1000 passengers.

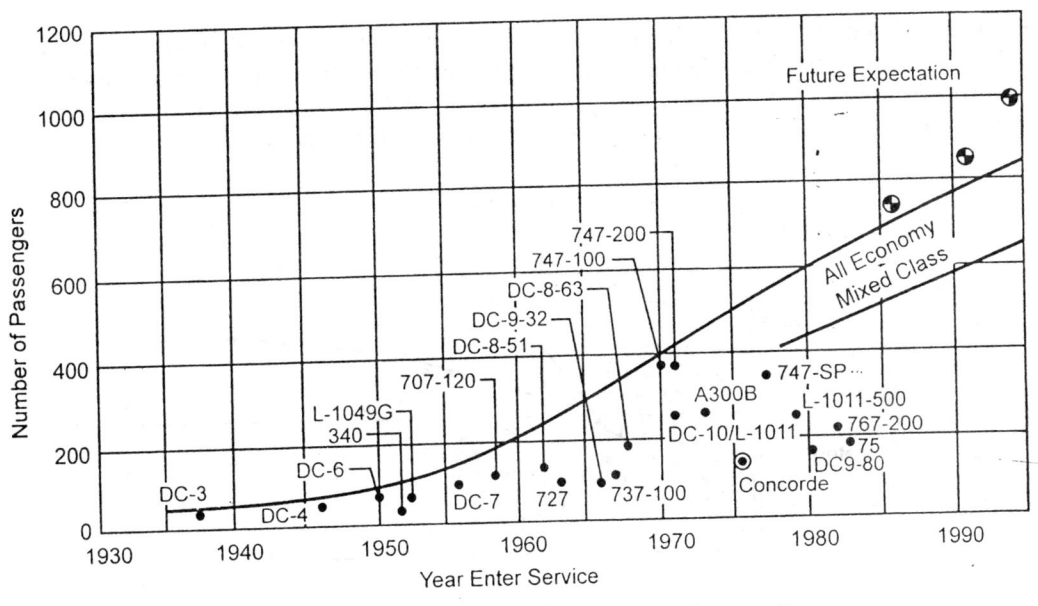

Fig. 1.2. Growth trend in passenger aircraft capacity.

Pay loads have increased significantly from 18 tonnes in 1952 Douglas DC-6B to more than 118 tonnes in the 1972 Boeing 747-200F. With the pay loads increase, the gross weight of an aircraft is also increased. As shown in Fig. 1.3, a fully loaded Boeing 747-200 weighs just over 340 tonnes. Even with that weight distributed among 16 main gear tires, each tire carries more than 18,100 kg. Although aircraft manufacturers have through the use of a multiple landing gear, wide lateral and longitudinal

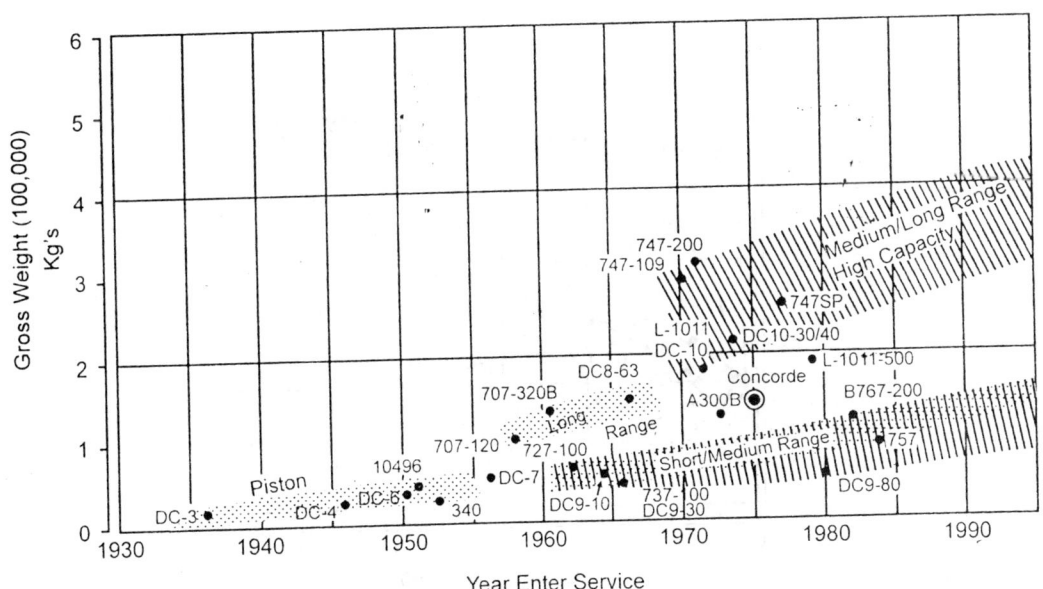

Fig. 1.3. Growth trend in gross weight of aircrafts (ENO).

wheel spacings, and large tires attempted to stay within existing pavement strengths, there will be a need for providing extra strengthening of pavements for future. With higher gross weights and the fuselages are becoming higher and longer as shown in Fig. 1.4 and so also wing spans have increased as shown in Fig. 1.5. The landing gear treads have become wider as shown in Fig. 1.6.

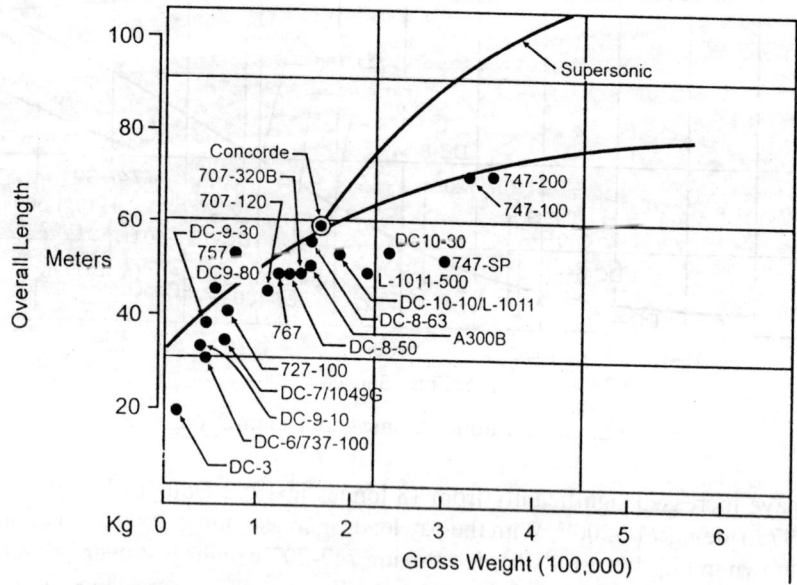

Fig. 1.4. Growth of overall length vs gross weight (ENO).

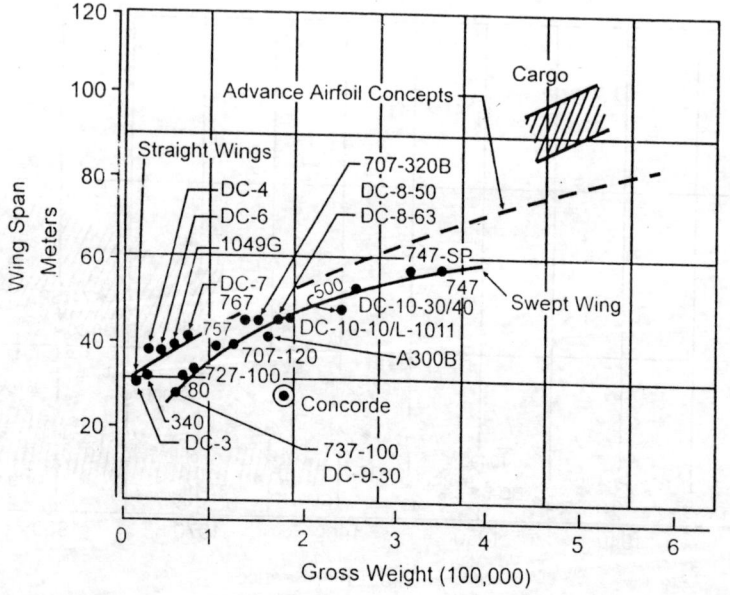

Fig. 1.5. Growth of wing span vs gross weight (ENO).

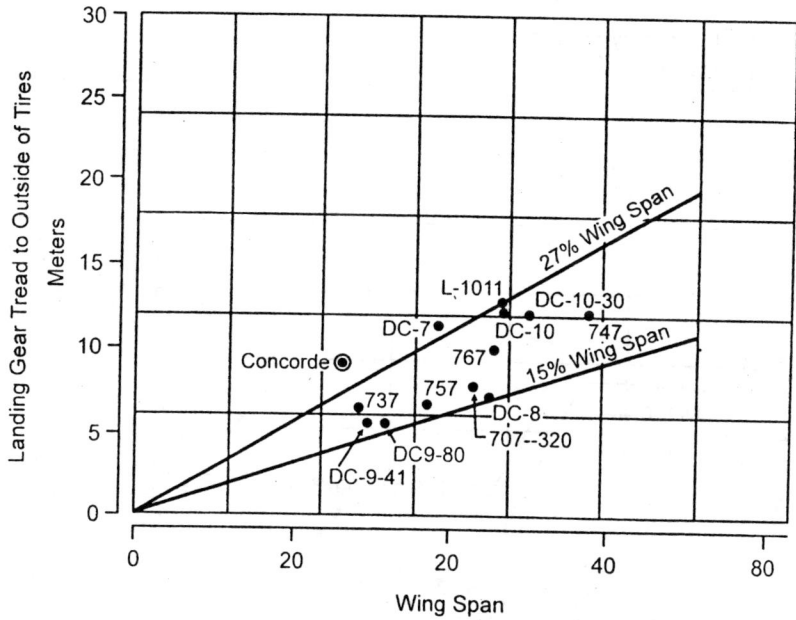

Fig. 1.6. Trends : landing gear vs wing span (ENO).

These larger aircrafts obviously require greater ramp and apron space for maneuvering. Thus the ramp requirements per aircrafts (based on the area encompassed by the wing span plus 7.6 m and the overall length plus 7.6 m) have also increased. As shown in Fig. 1.7, the mid size jets require between 1860 sq.m. and 3720 sq.m., while Boeing 747 requires about 5570 sq.m.

For serving lesser number of people smaller turbojet, turbo prop, and piston power short haul aircrafts are developing. During the 1970s, a number of small aircraft of about 40 to 80 seats were introduced. During the 1990s, new short haul aircrafts with capacity of about 200 seats were in use.

Designers of airports, must consider the gradually increasing size and weight of the aircrafts which may need longer runways, stronger pavements and other improvements in airport facilities.

1.7.2 Aircraft noise trends

The new aircrafts produce less noise than their predecessors, as shown in Fig. 1.8, however, it is still a persistent problem, annoying people living near the airports. As air travel increases noise may become a serious problem.

In spite of the use of noise-suppressing devices, as well as extensive research into quieting jet engines and the introduction of revised noise abatement aircraft operating procedures, the aircraft noise problem has yet to be completely solved.

1.8 DEVELOPMENT OF FREIGHT TRANSPORTATION BY AIR

Airmail had been carried on an experimental basis by army planes as early as 1911. The first scheduled commercial flight with cargo took place on January 1, 1914 from St. Petersburg to Tampa, Florida carrying a consignment of smoked hams.

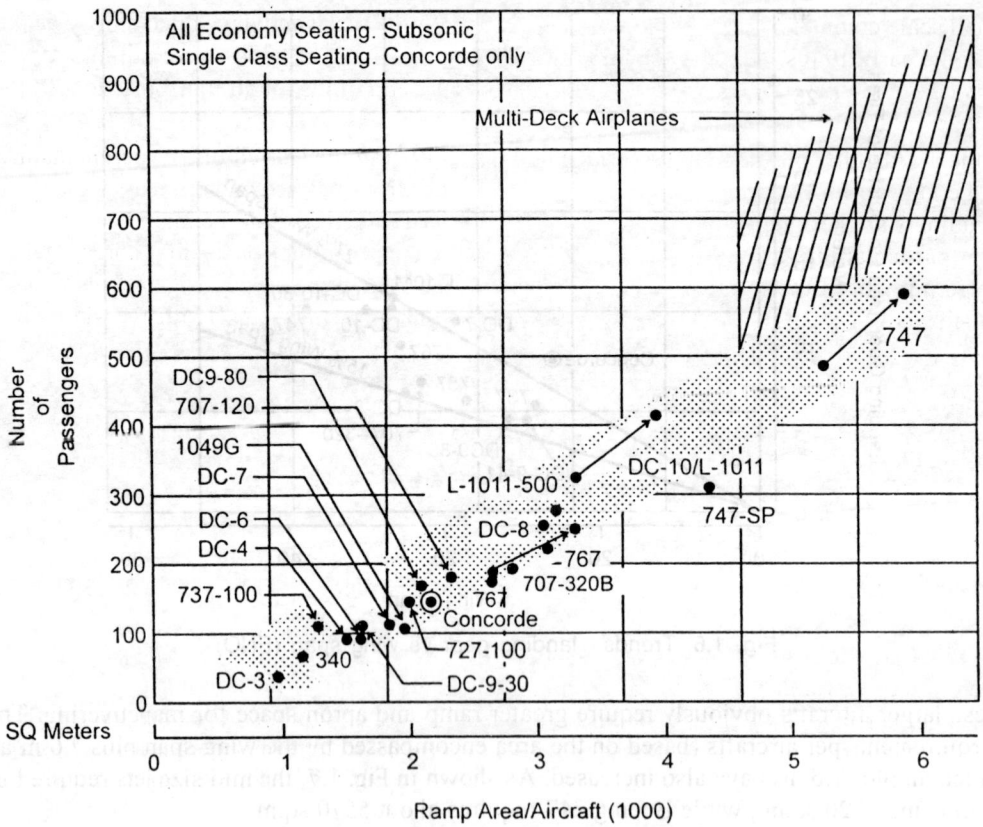

Fig. 1.7. Ramp area trend (ENO).

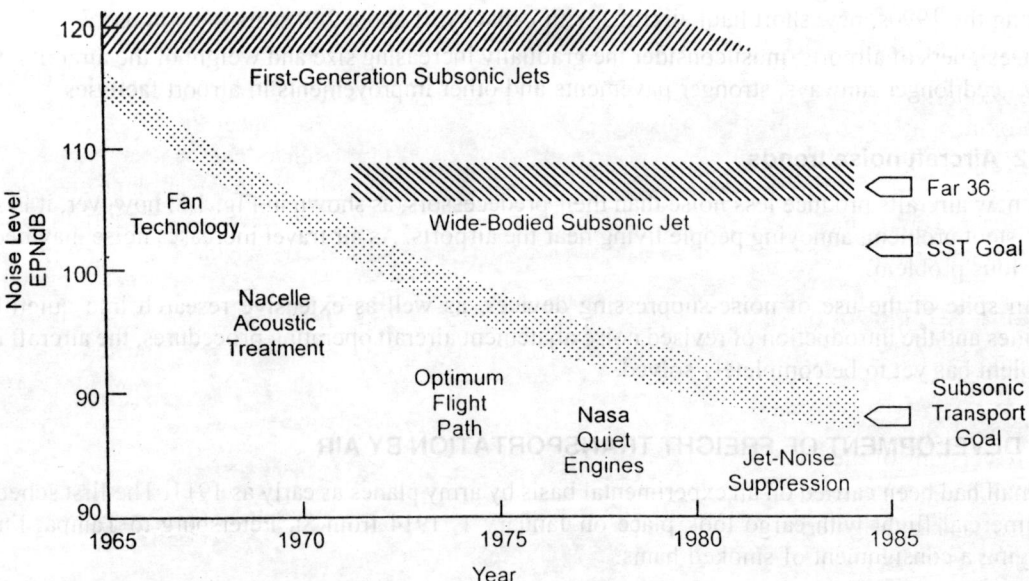

Fig. 1.8. Aircraft noise trends multiengine CTOL aircraft (ENO).

Small scale commercial air operations started after World War I and developed slowly between the wars. In the early 1940s, airplanes devoted wholly to the carriage of goods were put into service. The Warsaw convention, a treaty governing liability procedure and limits for air transportation was adopted in 1929.

In the 1970s, the inter-modal air-surface container programme started off. Experiments involving through container movements of air shipments combined with truck sea and rail cars were successful. Air containers are designed to make full use of the internal dimensions of the main deck or lower lobe space of particular aircraft. Air containers are not always compatible for loading abroad another aircraft type. Freight in the lower-lobe compartments of airfreighters and passenger airplanes is loaded and off-loaded with automated container handling equipment or by hand, with the aid of conveyor belts.

1.8.1 Freighter aircraft developments

Some developments in air surface intermodal carriage and transfer techniques are reviewed below.

Boeing 747F

The Boeing 747 freighter is the only plane able to accommodate $8 \times 8 \times 20$ ft (2.5 m $\times$ 2.5 m $\times$ 6.5 m) internodal container efficiently. This airplane allows containers to be positioned in an efficient side by side configuration. The main deck of the 747F accommodates $8 \times 8 \times 20$ ft internodal containers as well as other type of air containers.

There are convertable 747s that permit the main cabin interior of the airplane to be configured for all passenger carriage, for all cargo carriage or cargo-passenger combinations with the freight carried in the forward portion of the main deck and passengers in the aft portion. There is also a "Combi" 747 with permanent provisions for approtioning the main-deck cabin to passengers in the forward section and cargo in the aft. Cargo is loaded through a forward nose door in the convertible version and through an aft side door in the Combi. Inter modal containers ($8' \times 8' \times 20'$) can be accommodated in both types of 747s.

McDonnel Douglas DC-10 and lockheed L1011

The two large 3 engine airplanes McDonnel Douglas DC-10 and Lockheed 1011 or Tristar, can carry approximately 150,000 pound (70,000 kg) of cargo. In the cargo version of the DC-10, which has a large side-loading cargo door $8 \times 8 \times 10$ ft and $8 \times 8 \times 20$ ft containers can be accommodated. The placement of these large containers has to be in a single line fore-and-aft in the airplane on or close to the deck's centre line. The cargo lift capability of the DC-10 freighter makes it appropriate for transporting small air containers that can be stacked to the contour of the fuselage.

Boeing 727, McDonnell Douglas DC9 and DC8-63/73 and Canadair CL-44

The most popular old passenger plane used for cargo purposes is the Boeing 727. The 3 engine Boeing 727 is the most widely used, because of its availability and its small size. It is suited to the flow of express traffic on medium and short routes and is economical to operate with a two man cockpit crew. Early versions of the 2 engine DC-9 are used for the same reasons.

McDonnell Douglas DC8-73 is a 4 engine airplane modified with new fuel-efficient, quiet engines is also used as freighter. DC-8-63 is fitted with 'hush kits' to limit noise, rather than use new engines to be used as freighters, but they do not provide as great of fuel saving.

1.8.2 New airplane designs

Consolidated Freightways and Eastern Airlines have put together a pioneering concept of contracted capacity in the lower-lobe compartment of Eastern's A-300 airplanes. Although the A-300 is not compatible with some other airplane containers, it does have a large lower lobo capacity which makes this feasible.

The new advanced air freighter to transport cargo in inter-modal containers shall weigh 450,000 kgs (1 million pounds), the plane would be able to carry 177,000 kg (390,000 lbs) of cargo over a distance of 7300 kms (4600 miles). Container would be accommodated side by side in three rows as compared to two rows in the current 747 freighter.

1.8.3 Spanloader

Other airfreighter design include the spanloader, the twin fuselage and the flying flatbed. The spanloader, a delta wing airplane, eliminates fuselage weight and drag and reduces wing loads by spreading the cargo weight across the wing. It will carry a 350,000 kg (770,000 pound) pay load for 5600 kms (3500 miles). Intermodal containers will be loaded through the wing tips.

The twin fuselage airfreighter reduces structural aircraft weight per tonne pay load carried, achieving lower operating costs than an equivalent single fuselage airplane.

The flying flatbed eliminates the upper portion of the airfreighter's fuselage to allow loading and unloading of outsized freight, as well as various sizes of intermodal containers.

1.8.4 Lighter than air (LTA) aircraft

Many countries are researching and developing LTA vehicles for cargo carriage. These vehicles are planned to be used in intermodal transportation. LTA aircrafts can move containers, outsized pieces or bulk commodities in tanks or bags to and from interchange points of other modes, regardless of terrain. LTA's slow speed may affect its productivity, as compared to a jet. The advantages of LTA are :

- ability to hover and use small airfields.
- low noise characteristics of engines as compared to those of jet aircrafts.
- ability to carry containerised cargo and outsized loads.
- need for minimal support facility at airports served.
- in some instances unprepared fields can be used.
- fuel savings.
- Less investments.
- Improved door-to-door capability.

LTA vehicle should be used for short distance moving of very large pieces of cargo. Research is being conducted to determine the niche in which the LTA vehicle will furnish its optimum performance, as compared to competitive vehicles

1.9 DEVELOPMENT OF AIR TRANSPORTATION IN INDIA

1.9.1 Growing demand

Imperial Airways service started regular flights between Karachi–New Delhi in 1929. Tata Airways Limited started flights between Allahabad, Calcutta (now Kolkata) and Colombo in 1939. Tata Airlines changed its name to Air India Limited in July 1946.

First international service was started in June 1948 to London using Constellation–749 aircraft. In 1953, Air India (AI) was made to operate long distance international services and Indian Airlines (IA) were formed to fly domestic and neighbouring country flights. Jet services to London and New York using Boeing 707 aircraft were started in 1960. First Boeing 747 (Jumbo jet) was introduced in April 1971.

Several new low cost airlines started flying domestic flights at most competitive fare, in 2004-05. Jet Airways, Sahara Airways, Air Deccan, King Fisher Airlines, Spice Jet Airlines, Go-Air Airlines started functioning in 2004-05. Several other private airlines are planning to cover domestic routes at economical fares. Jet and Sahara Airways are also permitted for undertaking international flights along with Air India, to Sri Lanka, Nepal, UK and other countries. Pawan Hans Helicopter Ltd. provides helicopter services to ONGC in its off-shore operations and to inaccessible areas and difficult terrain.

The increased number of airlines and their increasing fleet size and flight frequencies has created competitive domestic services with fare reductions, in spite of zooming oil prices. The number of passengers flying routes has grown. A-320's offering a capacity of about 174 seats in a single class configuration is getting the popular carrier. With at least a dozen more airlines waiting in the wing, the passenger rush is certain to choke the undersized airports even further.

Boom in air connectivity has shown 21.7% growth in 2005 over 48.7 million passengers handled in 2004. The total number of passengers both international and domestic on Indian airports was 6 crores in financial year 2004-05, as against 4 crore in financial year 2003-04. The expected increase in air traffic is more than 25%. Airport capacity and number of passengers which were predicted for the year 2010, have already exceeded in 2005. On November 24, 2005, Delhi airport handled record number of 540 flights in a day. Number of passengers using Delhi airport has already exceeded 4000 per day. Yearly passengers handled by Delhi airport are 1 crore and 12 lacs, which is expected to grow to 4 crores 20 lacs in 2024.

Demands for larger capacity and more facilities at the airport are increasing at a fast rate. Mumbai and Delhi airports are the busiest, having only one runway. The new Shanghai airport in China is having 28 runways, while in Malaysia the airport terminal building is 7 storied, to house offices of 500 airline companies. Indian airport besides runway capacities are also lacking very much in accommodating new airline offices and their demands for handling passengers. Present growth rate places Indian airports among the fastest growing in the world, next to China.

Recently government of India has encouraged private participation in development of new airports with 26% share of the government and 74% private. New Greenfield airports based on this principle are under construction at Devanahalli, Bangalore and Shamshabad, Hyderabad and also planned for Chennai and Kolkata. National Airport Authority of India (NAAI) manages 126 airports which include 13 international, 86 domestic airports. Location of different airports in India is shown in Fig. 1.9, 1.10 and 1.11.

The International airports are at New Delhi, Amritsar, Mumbai, Ahmedabad, Pune, Kolkata, Guwahati, Kochi, Trivandrum, Chennai, Bangalore, Hyderabad and Goa. On December 29, 2005, Jaipur has been given the status of 14th International airport.

Modernisation work are being planned for the airports at Ahmedabad, Amritsar, Goa, Guwahati, Jaipur, Lucknow, Madurai, Mangalore, Trivandrum and Udaipur. For modernisation of airports following allotments have been made by the Government of India in the budget :

Delhi airport	Rs. 6500 crores
Mumbai airport	Rs. 6300 crores
Hyderabad and Bangalore airports	Rs. 3596 crores
Other 23 small airports	Rs. 5000 crores

Fig. 1.9. International airport in India (AAI).

The new terminal at recently upgraded international airport of Jaipur will cost Rs. 56.77 crores. 10 aeroplanes shall be capable of landing at one time at this airport, which will have runway length of 12,000 feet (4000 m) so that Boeing 747 aircrafts can easily land. Central Government shall provide money for its total development to international airport standards in two stages. Works worth 115 crore in the first phase are already in progress. In the second phase 100 crores are allotted to this airport.

Fig. 1.10. Domestic airports in India (AAI).

1.9.2 Airport Organisation in India

The aviation organisation has three main functions – regulatory, infrastructural and operational. The Government has ended the monopoly of Indian Airlines and Air India on the scheduled operations by repealing the Air Corporation Act, 1953. Private airlines are now operating on domestic network offering a wide choice of flights and fares to passengers. An effort is now being that new civil aviation policy gives more emphasis to privatisation in civil aviation sector.

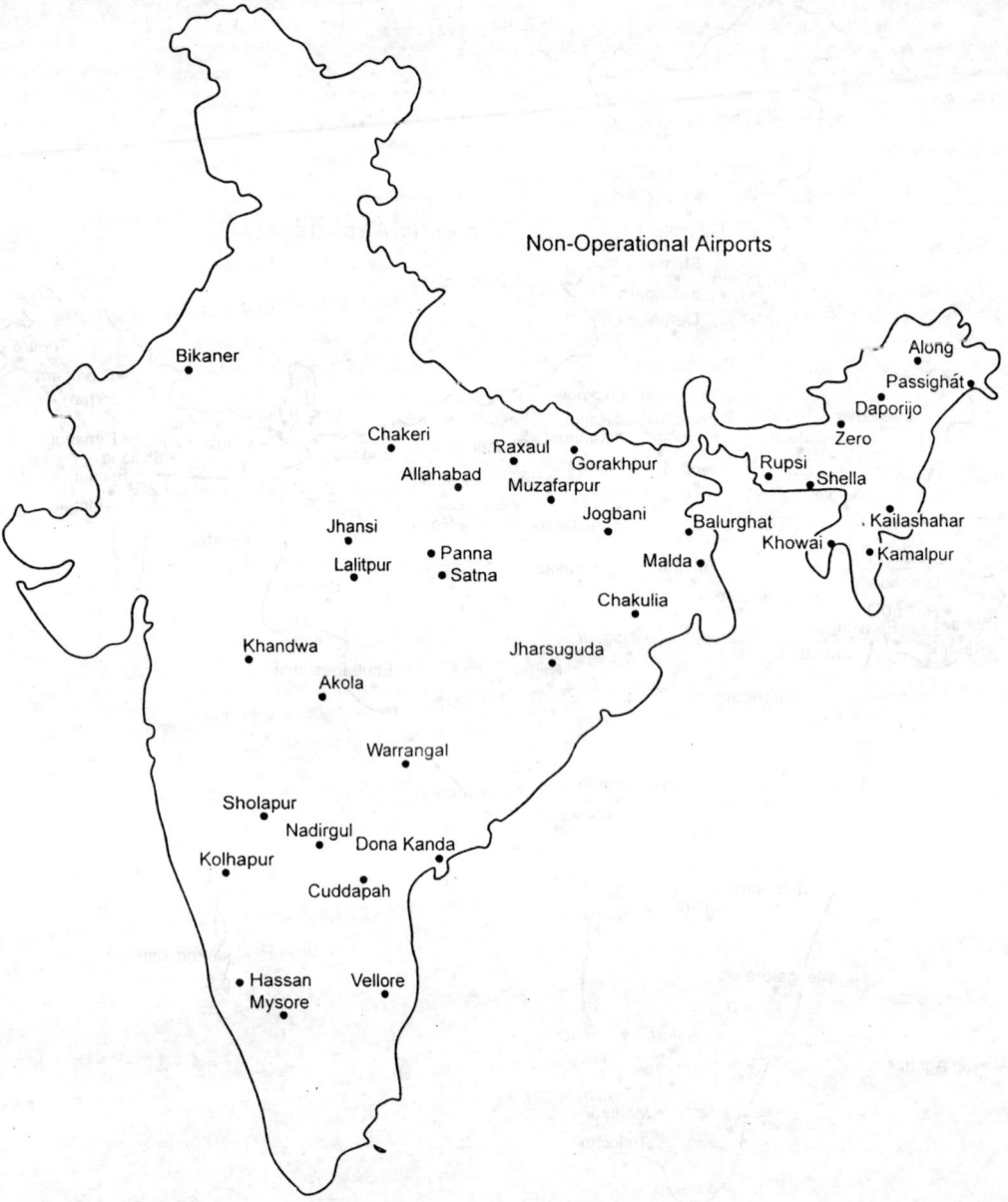

Fig. 1.11. Non-operational airports in India (AAI).

The Ministry of Civil Aviation is responsible for the formulation of national policies and programmes for development and regulation of civil aviation and for devising and implementing schemes for orderly growth and expansion of civil air transport. Its functions include :

  (i) overseeing the provisions of airport facilities.

  (ii) provision and monitoring of air traffic services.

(iii) ensuring of aviation safety.

(iv) security and carriage of passenger and goods by air.

The Ministry administers its control through various organisations, details of which are available on their website (*www.civilaviation.nic.in*).

Cargo

The Government of India introduced in April 1999 an "Open Sky Policy" for cargo. Under this policy foreign airlines or association of exporters can bring any frighters to the country for upliftment of cargo, permitting market forces to determine cargo tariff, with IATA rates as the floor rates.

On May 31, 2004, India has bilateral Air Services Agreement with 100 countries, permitting Private Domestic Scheduled Operators (PDSO) to commence operations to SAARC countries. Inclusive Tourist Package (ITP) charter flights are also permitted now.

1.9.3 Directorate General of Civil Aviation (DGCA)

The Directorate General of Civil Aviation (DGCA) is the regulatory body in the field of Civil Aviation. Its functions and responsibilities are :

   (i) Regulation of air transport services to/from and within India.

  (ii) Licensing of pilots, aircraft maintenance engineers.

 (iii) Monitoring of flight crew standards.

 (iv) Registration of civil aircrafts.

  (v) Coordination of the work relating to ICAO.

 (vi) Investigation of minor air accidents and rendering technical assistance to courts and committees.

 (vii) Supervision of training activities of flying/gliding clubs.

(viii) Licensing of aerodromes and air carriers.

 (ix) Rendering advice to the government on matters pertaining to air transport including bilateral air services agreements with foreign countries.

  (x) Development of light aircraft, gliders and winches.

 (xi) Laying down airworthiness of such aircrafts.

 (xii) Processing amendments to the aircrafts act, 1934 and aircraft rates 1937 and other acts relating to aviation.

(xiii) Type certification of air craft.

1.9.4 Research and development directorate

Research and Development Directorate of DGCA is responsible for :

  (i) Type certification of civil aircrafts, engines, propellers and type approval of instruments, avionics and equipment.

 (ii) Approval of modifications and repair schemes of civil aircrafts.

(iii) Development of testing of indigenous aircraft material, parts and equipments.

 (iv) Scientific laboratory investigations of in-services failed aircraft/power plant/structural parts and components.

  (v) Airworthiness cum operational monitoring of recording of cockpit voice recorders and flight data recorders.

 (vi) Providing assistance to inquiry commissions on accidents to decode the recorders to establish cause of accidents.

(vii) Quality monitoring and control, testing of fuel/oil samples of aircrafts, including those used by VIPs.

(viii) Economics of aircraft operation and performance evaluation.

(ix) Development of airworthiness/design codes/CARs etc.

(x) Approval of design organisations involved in designing aircraft/components and their renewals.

(xi) Human resource development on airworthiness engineering.

1.9.5 Bureau of Civil Aviation Security (BCAS)

BCAS is headed by Director General of Police, designated as Commission of Security Civil Aviation (COSCA). BCAS has its headquarter in Delhi and four regional offices in Delhi, Kolkata, Mumbai and Chennai airports each under a Regional Deputy Commissioner of Security. BCAS defines and allocates the tasks for national civil aviation safety, and issue of Airport Entry Permits (AEP) to personnel working at airport.

1.9.6 Airport Authority of India (AAI)

The Airport Authority of India was formed in April 1995. It is responsible for :

(i) Providing safe, efficient air traffic services, communication and navigational aids at airports.

(ii) Plan, develop, construct and maintain runways, taxiways, apron, terminal buildings etc.

(iii) Provide air safety services.

(iv) Arrange search and rescue facilities in coordination with other agencies.

(v) Performing other functions as mentioned in AAI act.

AAI controls and manages the entire Indian space extending beyond the territorial limits of the country, as accepted by the International Civil Aviation Organisation (ICAO). The authority has a Civil Aviation Training College at Allahabad for imparting training on various operations like Air Traffic Control, Radars, Communication etc. AAI maintains the National Institute of Aviation Management and Research (NIAMAR) at Delhi, which conducts management training programmes and refresher courses.

1.10 AIR INDIA, INDIAN AIRLINES AND PAWAN HANS HELICOPTERS LTD.

Air India owns a fleet of 18 aircrafts. Four B747-200, two B747-300, Six B747-400 and eight A310-300. It has leased another eleven A310-300 and one B747-400. Air India, operates 189 flights per week serving 28 international and 13 domestic stations. During 2003-04, Air India carried approximately 3.8 million passengers on its scheduled flights.

Indian Airlines name has been recently changed to "Indian".. It is the major domestic air carrier of the country. It also operates to 17 international stations. It has a fleet of 62 aircrafts. Many more low cost airlines are now operating on domestic flights.

Since its inception in 1985, the Pawan Hans Helicopters Ltd. (PHHL) has operated number of helicopters by offering wide range of services to its clients through a well balanced fleet of 31 helicopters consisting of Robinson (R-44), Bell 20624, Bel 407, Dauphin SA365N, Dauphin A5365N3 and Mi-172. Its corporate office is located in Delhi with Regional offices at Delhi and Mumbai. It carries out ONGC off-shore tasks, provide services to Government of Punjab, operates regular flights to Kedarnath Shrine from Augustmuni.

Some statistical economic facts

It is estimated that $ 325 worth of activity is generated for every 100 dollars planned and invested in a country's civil aviation sector. 1.5 million new jobs can be created by modernisation of India's airports. Hundred direct jobs in air transport create 610 new indirect jobs. $ 22.5 billion is the foreign exchange earnings directly facilitated by India's civil aviation. This can be doubled if airports are of global standards. Rs. 5000 per minute is the fuel burnt by an aircraft when it hovers over Delhi and Mumbai airports for 30-45 minutes due to landing congestion.

Table 1.1 shows growing statistics of Indian aviation industry.

Table 1.1. Growing statistics of Indian Aviation Industry

	Year 2000	Year 2005	Year 2010 (estimated)
(a) Passengers	16 crore	20 crore	6 crore
(b) Cargo	1.6 lac ton	3 lac ton	8 lac ton
(c) Aircrafts	–	220	640
(d) Daily takeoff and landing	–	Mumbai 500	1000
		Delhi 450	900
(e) Pilots	–	3000	7500
(f) Increase in domestic traffic	–	24%	–
(g) Increase in international traffic	–	16%	–

Main Indian airline companies

Several new private airlines are now in competition to fly on Indian skies at low fares. This will increase air traffic and it will be big challenge for the existing airports to provide facilities to meet this growing demand. Some of the domestic airlines at present are shown in Table 1.2.

Table 1.2. Domestic airlines in India

Airline	Year of start	Head office	Aeroplanes	Destination
Air India	1932	Mumbai	42	44
Jet Airways	1993	Mumbai	51	45
Indian (Indian Airlines)	1953	New Delhi	56	56
Air Sahara	1991	New Delhi	26	26
Air Deccan	2003	Bangalore	22	57
Spice jet	2004	New Delhi	6	11
King Fisher	2005	Bangalore	7	13
Go Air	2005	Mumbai	9	8
Paramount Airways	2005	Coimbatore	1	4
Air India Express	2005	Coimbatore	1	4

CHAPTER 2

Airport Terminology

Wide variety of terms are in use throughout the world to describe facilities, procedures and concepts for airport operation, planning and design. A glossary of aeronautical terms is presented in this chapter to familiarise the readers with a particular technical terminology. Most of these terms have been used in this book, and conform with the terms and definitions adopted by International Civil Aviation Organisation.

Aerodrome : A defined area on land or water (including any building, installations and equipment) intended to be used either wholly or in part for the arrival, departure and surface movement of aircraft.

Aeroplane : A power driven heavier than air aircraft which derives its life in air through aerodynamic reactions on its surface.

Aircraft : Any machine which can derive support in the atmosphere from the reactions of the air and is intended for aerial navigation. It may be heavier or lighter than air.

Airfield : The landing area including taxiways, runways and holding aprons.

Aileron : A movable flap to serve as a lateral control in the flight of an aeroplane.

Airport : Any aerodrome at which a permanent custom and immigration facilities are provided.

Aircraft maintenance area : All the ground space and facilities provided for aircraft maintenance. It includes aprons, hangers, buildings, workshops, vehicle parks and roads associated therewith.

Aircraft stand : A designated area on an apron intended to be used for parking of aircraft.

Aerodrome beacon : Aeronautical beacon used to indicate the location of an aerodrome from the air.

Aerodrome identification sign : A sign placed on an aerodrome to aid in identifying the aerodrome from the air.

Aeronautical beacon : An aeronautical ground light visible at all azimuths, either continuously or intermettiently to designate a particular point on the surface of the earth.

Aeronautical ground light : Any light specially provided as an aid to air navigation, other than a light displayed on an aircraft.

Aeroplane reference field length : The minimum field length required for takeoff at maximum certificated take-off mass, sea level, standard atmospheric conditions, still air and zero runway slope.

Aircraft classification number (ACN) : A number expressing the relative effect of an aircraft on a pavement for a specified standard subgrade strength.

Airport reference code : It is a coding system used to relate airport design criteria to the operational and physical characteristics of the aeroplanes intended to operate at the airport.

Airside : The movement area of an airport, adjacent terrain and buildings or portions thereof, access to which is controlled.

Airside waiting area : Space between the departures concourse and airside exists from the passenger building.

Aircraft movement : An aircraft take off or landing at an aerodrome. For aerodrome traffic purposes one landing and one take off is counted as two movements.

Airport capacity : The maximum number of aircraft operations that can take place in an hour. A landing or take off operation is counted as one operation.

Airport established elevation : The elevation of the highest point of the landing area (airport's usable runway) above mean sea level.

Airport pavement : It is the prepared and designed thickness laid over the subgrade, to carry the aircraft wheel loads.

Airport reference point : The point representing the designated geographical location of an airport. The latitude and longitude of the approximate centre of the airport.

Air ship : A power driven lighter than air aircraft.

Air strip : A long comparatively narrow strip suitable for landing and take off of aircrafts.

Airport layout plan : The plan of an airport showing the layout of existing and proposed airport facilities.

Aircraft traffic demand : The hourly demand is the sum of number of aircrafts desiring to land and the number of aircrafts desiring to depart, from the aerodrome during that hour.

Airway : The air route along which navigational aids are provided and maintained.

Amphibian : An aeroplane which is designed to take off and land on water as well as on the ground.

Approach area : The wide chearance areas of ground beyond both ends of runways.

Approach surface : The imaginery inclined plane representing the obstruction clearance line and located directly above the approach area.

Apron : A defined area on land intended to accommodate aircraft for purposes of loading or unloading passengers, mail or cargo, fuelling, parking, or maintenance etc.

Apron management service : A service provided to regulate the activities and the movement of aircraft and vehicles on an apron.

Arrival concourse : Space between baggage claim area on customs inspection and landside exists from the passenger building.

Aspect ratio : It is the relationship between the wing chord and the wing span. A short wing span and wide chord means a low aspect ratio.

Asphaltic concrete (Bitumenious concrete) : A graded mixture of aggregate and filler with asphalt or bitumen, placed hot or cold and rolled.

Baggage : Peronal properties of passengers or crew carried on an aircraft.

Baggage claim area : Space in which baggage is claimed.

Baggage container : A receptacle in which baggage is loaded for conveyance aircraft.

Baggage sorting area : Space in which departure baggage is sorted into flight loads.

Baggage storage area : Space in which baggage is stored pending transport to aircraft.

Barrette : Three or more aeronautical ground lights closely spaced in a transverse line so that from a distance they appear as a short bar of lights.

Beanfort scale : It is a scale to indicate wind velocities as follows :

Beanfort number	Description	Velocity at 10 m above flat open ground (knots)
0	Calm	1
1	Light air	1-3
2	Light breeze	4-6
3	Gentle breeze	7-10
4	Moderate breeze	11-16
5	Fresh breeze	17-21
6	Strong breeze	22-27
7	Moderate gale	28-33
8	Fresh gale	34-40
9	Strong gale	41-47
10	Whole gale	48-55
11	Storm	56-63
12	Hurricane	64-71

Blast fence : A barrier used to divert or dissipate jet blast or propeller wash.

Blast pads : Special shoulders provided at the take off ends of runway and along taxiway to protect them from erosion due to high velocity of the jet blast.

Boundary lights : Road lights to show the boundary of a landing area.

Boundary marker : Lines or markers used to indicate the boundary of a landing area.

Biplane : An aeroplane whose main supporting surface is divided into two parts one above the other.

Building restriction line : A line which identifies suitable building area locations on airports.

Base course (Base) : The layer or layers of specified or selected material of designed thickness placed on a sub-base or subgrade to support a surface course.

Bearing strength (bearing capacity, pavement strength) : The measure of the ability of a pavement to sustain the applied load.

Calm period : The duration of time in which wind velocity is calm generally less than 6.4 KPh.

Cargo : Freight other than passenger baggage and mail, which is carried by aircraft.

Cargo area : All the ground space and facilities provided for cargo handling. It includes aprons, cargo buildings and warehouses, vehicle parks and roads.

Cargo building : A building through which cargo passes between air and ground transport and in which processing facilities are located.

Cargo warehouse : A building in which cargo is stored pending transfer to air or ground transport.

Check in : The process of reporting to an airline for acceptance on a particular flight.

Check in concourse : The space between the passenger building landside entrance and the check in position.

Check in position : The location of facilities at which check in is carried out.

Clear way (CWY) : A defined rectangular area at the end of the landing strip (runway) in the direction of take off, selected or prepared as suitable area over which an aeroplane may make a portion of its initial climb to a specified height.

Conical surface : An imaginary surface which extends upwards and outwards from the periphery of the horizontal surface with a slope of 1 : 20 measured in a vertical plane.

Control area : The air space within which regulations of air traffic control are applicable.

Control zone : An air space within which rules additional to those governing flights in control area apply for the safety of air traffic.

Control tower : Situated usually on the top of the terminal building, it is a room from which air traffic within the airport is controlled, by directing and supervising the arriving and departing aircrafts.

Cross wind component : It is the component of wind at right angles to the direction of travel of an aircraft, when wind blows in an inclined direction to the direction of landing and take off.

CTOL : It represents conventional take off and landing.

Commercial air carrier : A carrier performing scheduled or non-scheduled air transport services or both, available to the public for the carriage of passengers, mail or cargo for remuneration.

California bearing ratio (CBR) : The bearing ratio of soil determined by comparing the penetration load of the soil to that of a standard material. The method covers evaluation of the relative quality of subgrade soils but is applicable to sub-base and some base-course materials.

Composite pavement : A pavement consisting of both flexible and rigid layers with or without separating granular layers.

Displaced threshold : A threshold not located at the extremity of a runway. The portion of pavement behind a displaced threshold may be available for take off in either direction and landings from the opposite direction.

Departure concourse : The space between the cheek in position and the airside waiting area.

Domestic flight : A flight having exclusively destinations within the country by an airline of that country.

Domestic scheduled airline : An airline, registered in a country which operates any scheduled services within the country but which does not operate scheduled international services.

Direct transport passengers : Passenger's stopping temporarily at the airport under consideration and departing on a flight having the same flight number as the flight on which they arrived.

Declared distances :

(a) *Take-off run available (TORA)* : The length of runway declared available and suitable for the ground run of an aeroplane taking off.

(b) *Take-off distance available (TODA)* : The length of the take off run available plus the length of the clearway, if provided.

(c) *Accelerate stop distance available (ASDA)* : The length of the take-off run available plus the length of the stopway if provided.

(d) *Landing distance available (LDA)* : The length of runway which is declared available and suitable for the ground run of an aeroplane landing.

Dependent parallel approaches : Simultaneous approaches to parallel or near parallel instrumental runways where radar separation minima between aircraft on adjacent extended runway centre lines are prescribed.

Elevator : A mechanical automatic device for climbing different floors of a building.

Expansibility : The ability to be physically extended to the limits of the site to provide additional space and extra capacity using either new or existing operating procedures.

Emigrants : These include emigrants, temporary emigrants, visitors on holiday, visitors on business and persons in transit.

(i) *Permanent emigrant :* Residents who leave country permanently.

(ii) *Temporary emigrant :* Residents who leave country with intention of comeback.

(iii) *Visitors on holiday :* Those persons who leave country for holiday trip.

(iv) *Visitors on business :* Those who leave country for a business trip.

(v) *Persons in transit :* Those who leave the country after a transit stop.

Effective intensity : The effective intensity of a flashing light is equal to the intensity of a fixed light of the same colour which will produce the same visual range under identical conditions of observation.

Flight stage : The operation of an aircraft from take off of its next landing, technical stops are excluded from commercial movements but included in non-commercial movements.

Fin : The fixed part of the tail unit whose purpose is to contribute to the transverse and directional stability.

Fixed light : A light having constant luminous intensity when observed from a fixed point.

Flight time : The total time from the moment an aircraft takes off to the movement it comes to rest at the end of the flight.

Frequency ranges :

Nomenclature	Description	Range of frequency
VLF	Very low frequency	30 kc/s
LF	Low frequency	30-300 kc/s
MF	Medium frequency	300-3000 kc/s
HF	High frequency	3000-30000 kc/s
VHF	Very high frequency	30-300 mc/s
UHF	Ultra high frequency	300-3000 mc/s
SHF	Super high frequency	3000-30000 mc/s
EHF	Extremely high frequency	30000-300000 mc/s

Fuselage : The part of the aircraft which forms the main body.

Flexibility : The ability to adapt to new and radically different technical and physical requirements and methods of operation, with consequent changes in the use and population of specific areas and also the ability to be gradually modified in accordance with evalutionary changes. It also means the ability to increase the operations capacity within existing physical limits.

Flexible pavement : A pavement structure that maintains intimate contact with and distributes loads to the subgrade and depends on aggregate interlock, particle friction and cohesion for stability.

Gate position : The space allotted to an aircraft parking at a loading apron.

General aviation activities : All civil aviation operations other than scheduled air services and non-scheduled air transport operations for remuneration or hire.

Hanger : Large shed erected at the airports for parking, servicing and repairing of aircrafts.

Helicopter : A heavier than air aircraft which is supported in atmosphere by the reaction of the air on one or more power driven rotors rotating about an approximately vertical axis.

Heliport : An area for landing and take off of helicopter.

Holding apron : The portion adjacent to the end of runways to allow check of aircraft instruments and engine operation prior to take off and to enable the aircrafts to wait for take off clearance.

Horizontal surface : It is the imaginary horizontal surface located at a level of 45 m above the airport established elevation. It is circular in plan.

Hazard beacon : An aeronautical beacon used to designate danger to air navigation.

Holding bay : A defined area where aircraft can be held, or bypassed, to facilitate efficient surface movement of aircraft.

Hazard to air navigation : An object which will have a substantial adverse effect upon the safe and efficient use of navigable airspace by aircraft, operation of air navigable facilities, or existing or potential airport capacity.

IFR flight : The flight in accordance with the instrument flight rules.

Instrumental landing system (ILS) : A navigational aid which facilitates landing with the help of radio beams. It provies lateral and vertical guidance to the aircrafts and also indicates the distance of the aircraft from the runway threshold.

Instrumental runway : The runway of an airport which is provided with instrumental landing facilities to land under conditions of poor visibility.

International air service : An air service which passes through the airspace over the territory of more than one country.

International airport : Any airport in the country which is an airport of entry and departure where formalities incident to customs, immigration, public health, agriculture quarantine and similar procedures are carried out.

International flight : A flight that contains one or more international flight stages.

International scheduled airline : An airline which operates any scheduled international air transport service regardless of the proportion of international services offered as compared with all other kinds of services offered.

International flight stage : A flight stage with one or both terminals in a territory of country, other than the country in which the airline is registered. Technical stops are not considered in classifying flight stages.

Immigrants : These include new immigrants, returning residents, visitors on holidays, visitors on business and persons in transit.

Immigration control : The immigration and/or police inspection of arrival passengers.

Identification beacon : An aeronautical beacon emitting a coded signal by means of which a particular point of reference can be identified.

Independent parallel departures : Simultaneous departures from parallel or near parallel instrument runways.

Inner transitional obstruction free zone : The airspace above the surface located on the outer edge of the runway obstruction free zone and the inner-approach obstruction free zone. It applies to precision instrument runway.

Landing area : That part of movement area intended for landing and take off of aircraft.

Landing strip : A long and narrow area, forming part of an airport, which is suitable for landing and take off of aircrafts.

Landside : That area of an airport and buildings to which non-travelling public has free access.

Landing direction indicator : A device to indicate visually the direction currently designated for landing and for take-off.

Light failure : A light shall be considered to have failed when for any reason the average intensity determined using the specified angles of beam elevation, toe in and spread falls below 50 per cent of the specified intensity of a new light.

Lighting system reliability : The probability that the complete installation operates within the specified tolerances and that the system is operationally usable.

Large airplane : An airplane of more than 5700 kg maximum certified take-off weight.

Mach number : This indicates the speed of an aircraft relative to the local speed of sound Mach (M) one means speed is equal to that of local speed of sound. Subsonic Mach number below 0.75, Transonic Mach number 0.75-1.2, Supersonic Mach number 1.2-5.0, Hypersonic Mach number above 5.00.

Manoeuvering area : That part of an aerodrome to be used for the take off, landing and taxiing of aircraft excluding aprons.

Movement area : That part of an aerodrome to be used for take-off, landing and taxiing of aircraft and for the surface movement of aircraft – the aprons.

Mono plane : An aeroplane which is supported on a single wing extending equally on each side of the body.

Marking : A symbol or group of symbols displayed on the surface of the movement area in order to convey aeronautical information.

Marker : An object displayed above ground level in order to indicate an obstacle or delineate a boundary.

Non-commercial movements : Landing and take off by an aircraft operating for the purpose and solely for the benefit of the owner or for the purpose of positioning for a scheduled flight or for the carriage of stores not for hire or reward. Included in this category are flights made by Government and military aircrafts and other unscheduled flights.

Non-precision approach runway : An instrument runway served by visual aids and a non-visual aid providing at least directional guidance adequate for a straight in approach. It has no existing or planned precision instrument approach procedure.

Non-instrument runway : A runway intended for the operation of aircraft using visual approach procedures.

Near-parallel runways : Non-interesting runways whose extended centre line have an angle of convergence/divergence of 15° or less.

Obstacle : All fixed (whether temporary or permanent) and mobile objects or parts thereof, that are located on an area intended for the surface movement of aircraft or that extend above a defined surface intended to protect aircraft in flight.

Off-airport processing facilities : A passenger or cargo transport link terminal at an urban population centre at which processing facilities are provided.

Obstacle free zone (OFZ) : The airspace above the inner approach surface, inner transitional surfaces and balked landing surface and that portion of the strip bounded by these surfaces which is not penetrated by any fixed obstacle other than a low-mass and frangibly mounted one required for air navigation purposes.

Object : It includes, but is not limited to above ground structures, Navaids, people, equipment, vehicles, natural growth, terrain and parked aircraft.

Object free area : A two dimensional ground area surrounding runways, taxiways and taxilanes which is clear of object except for objects whose location is fixed by function.

Inner approach obstacle free zone : The airspace above a surface centered on the extended runway centreline. It applies to runways with an approach lighting system.

Overlay : An additional surface that maintains intimate contact with and distributes load to the subgrade and depends on aggregate interlock, particle friction and cohesion for stability.

Passenger amenities : Facilities provided for passengers which are not essential for passenger processing.

Passenger area : All the ground space and facilities provided for passenger processing, including aprons, passenger buildings, vehicle parks and roads.

Passenger building : A building through which passengers pass between air and ground transport and in which processing facilities and amenities are located.

Passenger loading bridge : A mechanically operated adjustable ramp to provide direct passenger access between aircraft and buildings.

Passenger processing : The reception and control of passengers during their transfer between air and ground transport.

Passport control : The immigration and/or police inspection of passengers.

Passengers disembarked : Passengers finishing their trip at the airport under consideration or passengers who will continue their trip on a route with a different flight number.

Passengers embarked : Passengers starting their journey at the airport under consideration or connecting transit passengers who continue their journey on a route with a flight number different from the flight on which they arrived.

Pier : A corridor at, above or below ground level to connect aircraft stands to a passenger buildings.

Port health control : The medical inspection of documents and/or passengers, baggage, cargo.

Peak aircraft traffic demand : The aircraft traffic demand at an airport which will be reached in the most busy hour, averaged over two consecutive hours.

Primary runway(s) : Runway(s) used in preference to others whenever conditions permit.

Pavement classification number (PCN) : A number expressing the bearing strength of a pavement for unrestricted operations.

Precision approach runway : Precision instrument runway served by instrument landing system and visual aids intended for operations.

Pavement structure (pavement) : The combination of sub-base, base course and surface course placed on a subgrade to support the traffic load and distribute it to the subgrade.

Relocated threshold : The portional of pavement behind a relocated threshold is not available for take-off or landing. It may be available for taxiing of aircraft.

Rudder : An air control in the aircraft which helps the pilot to turn the nose of the aeroplane in any particular direction. It can move to and fro about a vertical axis through about 30°.

Runway : A long and narrow defined rectangular strip which is prepared for landing and take off of aircrafts at an airport.

Runway threshold : The beginning of the runway that is available for landing.

Runway end safety area (RESA) : An area symmetrical about the extended runway centre line and adjacent to the end of the strip primarily intended to reduce the risk of damage to an aeroplane undershooting or overrunning the runway.

Runway strip : A defined area including the runway and stopway if provided, intended to reduce the risk of damage to aircraft running off the runway and to protect aircraft flying over it during take-off or landing operations.

Runway visual range (RVR) : The range over which the pilot of an aircraft on the centre line of a runway can see the runway surface marking or the lights delineating the runway or identifying its centre line.

Runway blast pad : A surface adjacent to the ends of runway provided to reduce the erosive effect of jet blast and propeller wash.

Runway protection zone : An area off the runway end, used to enhance the protection of people and property on the ground.

Rigid pavement : A pavement structure that distributes load to the subgrade having as its surface course a portland cement concrete slab of relatively high bending resistance.

Saturation of an airport : Reached when the aircraft traffic demand equals or exceeds the corresponding airport capacity.

Service rate : The maximum aircraft movement rate which could be reached at an airport.

Sustainable capacity : The highest movement rate of an airport which could be continuously maintained for three hours or more under defined conditions.

Sea plane : An aeroplane which can take off and land on water.

Standard atmosphere : An imaginary atmosphere when air is perfectly dry, the temperature and pressure at sea level is 15°C and 760 mm of mercury respectively.

STOL : Short take off and landing.

STOL port : It is an area used for landing and take off of STOL aircrafts.

Streamline : To give a body such a shape that its head resistance to motion through the air is lessened.

Stopway : Rectangular area at the end of the runway in the direction of take off in which an aircraft can be stopped after an interrupted take off.

Subsonic aircraft : Aircraft which travels at speed less than the speed of sound.

Supersonic aircraft : Aircraft which travels at speed greater than the speed of sound.

Surveillance radar : Provides an overall picture of the surrounding atmosphere. The information about the aircraft within its range is received on its scope.

Segregated parallel operations : Simultaneous operations on parallel or near parallel instrument runways in which one runway is used exclusively for approaches and the other runway is used exclusively for departures.

Signal area : An area on an aerodrome used for the display of ground signals.

Shoulder : An area adjacent to the edge of paved runways, taxiways or aprons providing a transition between the pavement and the adjacent surface, support for aircraft running off the pavement, enhanced drainage and blast protection.

Small-airplane : An airplane of 5700 kg or less maximum certified take off weight.

Sub-base course (sub base) : The layer or layers of specified selected material of designed thickness placed on a subgrade to support a base course.

Sub grade (formation) : The upper part of the soil, natural or constructed, which supports the loads transmitted by the pavement.

Surface course (wearing course) : The top course of a pavement structure.

Tail : The rear portion of an aircraft to which rudder, elevator and fin are usually attached.

Take off runway : A runway intended for take off only.

Taxiway : A defined path over which an aircraft may taxi to and from the runway and loading apron, and to provide link between one part of the aerodrome and another.

Terminal area : Portion of the airport other than the landing area. It includes terminal building, facilities for loading and unloading the passengers, cargo and mail, apron for parking of aircrafts, vehicle parking area, cargo storage buildings and hangers.

Terminal building : It provides space for airline operations, facilities for passengers, offices for airline operations, facilities for passengers, offices for airport management and other non-aeronautical functions.

Transition surface : An imaginary inclined plane with a slope of 1:7 measured upward and outward in a vertical plane at right angles to the centre line of runway.

Transfer passengers/baggage : Passengers making direct connections between two different flights.

Transit passengers : Passengers departing from an airport on the same flight as that on which they arrived.

Transport link : Any form of transport system provided exclusively for operation between airport and urban population centre.

Transporter : Any vehicle used to convey passengers between aircraft and passenger building.

Traffic : For air transport purpose, traffic means the carriage of passengers, freight and mail.

Traffic units : It is the sum of embarking and disembarking passengers and freight loaded and off loaded (100 kg of freight is equivalent to 1 passenger).

Theoretical airport capacity : The maximum movement rate which could be reached with the mix of aircraft and of take offs and landings under defined conditions for that airport. Minimum separation being maintained between air aircrafts.

Threshold : The beginning of that portion of the runway usable for landing. When the threshold is located at a point other than at the beginning of the pavement, it is called as either a displaced or a relocated threshold depending on how the pavement behind the threshold may be used.

Taxi-holding position : A designated position at which taxiing aircraft and vehicles may be required to hold in order to provide adequate clearance from a runway.

Taxiway intersection : A junction of two or more taxiways.

Taxiway strip : An area including a taxiway intended to protect an aircraft operating on the taxiway and to reduce the risk of damage to an aircraft accidently running off the taxiway.

Touchdown zone : The portion of a runway, beyond the threshold, where it is intended landing aeroplanes first contact the runway.

Taxi lane : The portion of the aircraft parking area used for access between taxiway and aircraft parking positions.

Taxiway safety area : A defined surface alongside the taxiway prepared or suitable for reducing the risk of damage to an airplane unintentionally departing the taxiway.

Usability factor : The percentage of time during which the use of a runway or system of runways is not restricted because of the cross-wind component.

VFR flight : A flight in accordance with the Visual Flight Rules.

Visibility : The distance at which conspicuous points can be seen and identified.

Visual runway : A runway without an existing or planned straight-in instrument approach procedure.

Wind rose diagram : A diagram showing direction, duration and intensity of wind over a certain period.

Zoning : Restrictions to the development of the area surrounding the airport so that no structure protrudes above the obstruction clearance line.

CHAPTER 3

Planning of Airports

3.0 INTRODUCTION

Air travel is expected to increase enormously in the decades ahead. The rate of growth shall vary from country to country depending upon the level of people's income. Existing facilities at airports can rarely cope with the increase in activity and demands of the future. Programme of expansion and improvements in existing airport, development of new airports and preparing master plans, will have to be undertaken according to the rate of growth of air travel. The future challenge will be to accommodate the expansion in air travel in most rational way. Correct anticipations of demand for service, choosing right type of facilities to provide required service and to preserve the environment and meeting the various social objectives.

Airport planning has to be intensive and imaginative, requiring an approach which examines broad aspects of the problem as a system, with the ability to use analytical techniques to evaluate efficiently the enormous range of alternative designs, operating policies and procedures of implementation. The demand and flow of traffic and its relationship between different facilities are affected by a wide variety of inter-dependent factors. Airport planning is the evolution of a compromise between the conflicting features of the best plan for each of the individual facilities, so that planning of all the individual facilities contributes and combines into the most efficient total plan and provides the greatest degree of flexibility and expansibility for future development. We need to develp plans for the airport as part of a system, through the developing systems approach to airport planning.

3.1 OUTLINE OF TECHNICAL PLANNING PROCESS

The air transportation system like other systems is planned at several levels : The national, state and local.

Airport planners need to understand the peculiarities of the country or region to comprehend the nature of practical problems like policy issues of national planning, airport location and use of technology etc. It is, therefore, essential that initial planning is done considering the needs for airports, based on the nation as a whole. Large urban areas are most significant in the national air transportation system, and identifying priority development requirements. A systematic approach to planning for nation's airport needs is to be adopted. An overall planning of airports for all the regions of a country is called "Regional" planning.

The regional airport system plan is a representation of the aviation facilities required to meet the immediate and future transportation needs of the country and their timely and orderly development. It recommends the general location and characteristics of new airports and nature of expansion for existing ones. It relates airport system planning to the policy and coordinative transportation planning for the area, land use planning and urban environment, to establish a viable integrated network of airports at the national level.

3.1.1 Objectives of regional planning

The regional planning of airports is done to provide for the orderly and timely development of a system of airports adequate to meet the present and future air transportation needs of the country to :

- Develop a balanced regional multi-modal transportation system plan, with coordinated airport facilities.
- Protect and enhance the environment through the location and expansion of aviation facilities in such a way that impairment of the ecology and the intrusion of acceptable levels of noise and air pollution into the community are avoided.
- Provide a framework for individual airport development programme consistent with short, intermediate and long term airport system requirements.
- Accomplish the coordination with state airport system plan and regional airport plans, and serve as a basis for coordinating air navigation facilities, air space used and air traffic control procedures.
- Develop fiscal plans and establish appropriate priorities for airport financing, in short and long term governmental budgeting.
- Optimise the use of land and airspace and preserve these existing airport facilities which are consistent with overall objectives of the long term planning.

Generally an effective planning organisation (Civil Aviation Organisation in India) does the job of technical planning, cooperating and coordinating between states, other concerned organisations and political entities, to participate in the planning of air facilities. The organisational arrangements for the airport system planning process may vary from country to country and region to region.

An effective organisation for regional airport planning should be capable of :

- (a) *Establishing policies* : Short and long term policies must respond as nearly as possible to the desires, attitudes and long range air transportation requirements of the region. General policies should relate to and be integrated with comprehensive planning policy.
- (b) *Bringing coordination* : Aviation and non-aviation agencies, airport operators etc. should be brought together for active participation in planning.
- (c) *Technical ability* : The technical planning process begins with inventory step and continues up to the selection of most suitable alternative and replaced during continuing planning process.

The following basic information is collected for the regional planning :

(i) The planning period

The long-range time period for the plan is generally taken as 20 years. It includes the need and location for airports, heliports, STOL ports during this period. As it takes 8 to 10 years, for a major airport to be established after its need has been identified, a 20 year planning period which is also used in highway system planning seems to be alright. A shorter period of 2–5 years may be used for actual development of facilities and cost estimates and budgeting.

(ii) The geography and topography of the area.

The geographical limit of the planning areas must be established for data collection, forecasting and potential site selection. The political boundaries of the area are important to determine jurisdictional involvements. Each region or area may be different from topography point of view. In certain difficult terrain air transportation may be the only accessible mode.

(iii) The population served

The total population likely to be served by an airport now and in future is to be arrived at. Richer people use more of the air travel.

(iv) Types of airports

There may be some already existing airports in the planning area which might need upgrading. There may be need for more airport to accommodate scheduled airline operation with conventional aircrafts. VTOL (vertical takeoff and landing aircrafts) and STOL (short takeoff and landing aircrafts) ports should be an important future element of airports system plan. Facilities for sea plane or float airports may also be considered. Need for military airport for potential joint civil and military use should be considered.

(v) Ownership and operation of airports

An airport may be operated by Central Government, State Government or private individuals. Public ownership provide better stability, regular financial aid and certain domain powers. Private owner may be given incentives and encouragements to invest in airport systems on full or at least on interim basis. Alternatively an authority may be created at the outset, grouping all the major airports under one organisational structure such as "National Airport Authority".

3.1.2 Considerations in regional planning of airports

The various considerations involved in regional airport system planning are outlined below :

1. Study organisation

- Establishing policies
- Organisation and coordination
- Procedures

2. Inventories

- Airports and their operation :
  (a) Civil – air carrier
         – general aviation
             • publicly owned
             • privately owned
         – VTOL and STOL
         – special purpose

(b) Military,
(c) Expansion possibilities
- Airspace and navaids
- Aeronautical activity
  (a) Air carrier
    (i) operations
    (ii) origin and destination
    (iii) passengers
    (iv) cargo
  (b) General aviation and military operations and ownership distribution
- Environmental factors
- Comprehensive landuse plans
- Transportation plans
- Socio-economic factors
- Applicable laws, ordinances, regulation
- Financial resources

3. Forecasting

- Scheduled aviation :
  (a) (i) operations
      (ii) passengers
      (iii) cargo/mail
      (iv) origin and destination
  (b) Air Taxi
  (c) VTOL and STOL
- General aviation
  (a) operations
  (b) based aircrafts
  (c) distribution of owners
- Military
  (a) operation at military fields
  (b) operation at civil airports

4. Considerations of new technology

- Aircrafts
- Air navigation
- Ground access

5. System requirements

- Aircraft operational requirements

- Capacity/demand analysis
  - (i) airfield
  - (ii) terminal
  - (iii) air space
  - (iv) ground access

6. Alternate systems

- Ranking of alternatives
- Selection of particular system

7. Plan implementation

- Political considerations
- Environmental considerations
- Timing of project
- Financing
- Land acquisition and land use control

3.2 MASTER PLANNING FOR AN AIRPORT

The rapid growth of air transportation is exceeding the capacity of existing airports and there is an urgent need for preparing master plans for the expansion of existing airports wherever possible and development of new ones. The most efficient plan for the airport as a whole is that which provides the required capacity for aircraft, passenger, cargo and vehicle movements with maximum passanger, operator and staff convenience at lowest capital and operating cost.

3.2.1 The master plan

A generally accepted definition states that "An airport master plan presents the planner's conception of the ultimate development of a specific airport. It effectively presents the research and logic from which the plan was evolved and artfully displays the plan in a graphic and written report. Master plans are applied to the modernisation and expansion of existing airports and to the construction of new airports, regardless of their size or functional role. Development in this definition includes both aviation and non-aviation areas, and adjacent land uses.

A master plan should establish schedule of priorities and phasing for various elements as per traffic demands and available finances.

3.2.2 Goals of a master plan

An airport master plan serves the following purposes :
- Guides in development of physical facilities and development of land uses for surrounding areas.
- Determines the environmental effects of aerodrome construction and operation.
- Establishes airport access requirements.
- Used to provide short and long term policy guidance.
- Helps to develop work programmes, schedules and budgets.

- Assists in securing financial aid.
- Establishes coordination, monitoring procedures, data management and public monitoring system.
- In economic planning it provides a basis for benefits cost analysis and assessment of impact on area economy.

For physical planning it :
- Helps in development of air space and air traffic controls, terminal complex, support and service facilities, ground access systems and overall land use patterns.
- Determines community attitude and opinion.
- From financial point of view it determines the funding source and constraints, feasibility study of various development alternatives.
- It provides useful data on :
  - (i) Master planning work programmes
  - (ii) Inventory on existing conditions
  - (iii) Forecast of future air traffic demand
  - (iv) Time phased development plan
  - (v) Existing and future constraints
  - (vi) Alternative plans for comparative analysis

Master plan should be thoroughly evaluated and modified every 5 years or so in view of changes in economic, operationa, environmental and financial conditions. Successful expansion of existing airports and the development of new airports results from guidelines provided in master plan. From a master plan, actual implementation programmes are developed. Master planning is specialised phase of airport development which is necessary before design can begin.

3.3 PREPLANNING FOR AN AIRPORT PROJECT

For a master plan to be useful to airport authorities, a proper planning and pre-planning has to be done. The preplanning considerations include the following :

(a) *Preplanning coordination* : The airport master plan will be of interest to a diversity of people, organisations, aerodrome users, planning agencies, conservation groups, land transportation, officials, airlines and others with aviation interests. It is important that master planning team coordinate its efforts and seek advice of those interested groups in arriving at a well integrated and implementable plan.

(b) *Collection of information* : For sound master planning forecasting, preparation and collection of useful data on the usage of airport and its component is essential. The data to be collected covers physical facilities, utilisation of facilities, volume and composition of traffic, cost of transportation, financial status of people, airlines using the airport, Government policies and regulations etc. Reliable data is collected from concerned sources such as government departments, statistical offices, banks, financial institutions, airlines etc.

(c) *Goals and programming* : The specific goals of the master planning process as described earlier should be well understood during the pre-planning stage.

(d) *Economic and financial consideration* : The financial burden of major airport expansion or development of a new airport can be formidable. In order to know financial implications of such a project, early determination of economic feasibility is important. Preliminary broad cost estimates, for each element of airports should be worked out. This will form the basis for establishing financial resources, and determination of finances available and provisions for the future. A benefit cost analysis may be necessary to justify the project at the pre-planning stage.

(e) *Planning team, organisation and procedure* : The planning team should consist of selected people such as statistician, economists, financier, operational research scientists, architects, civil, mechanical, electrical and traffic engineers, pilots, air traffic controllers, airport managers etc. The advice of other experts, specialists and interested agencies should be sought as necessary, from time to time.

The organisation for planning and development of a master plan may vary from country to country depending upon the nature of airport, type of transport organisation and governmental responsibilities. An effective organisation shall make policies acceptable to users, coordinate well with concerned agencies, and assure that technical process is sound and effective master plan can be established for implementation.

The master plan should indicate the ultimate overall size, taking care of aircrafts, passengers, cargo and ground vehicle needs.

3.4 STEPS IN TECHNICAL PLANNING PROCESS

After deciding over the policy matters and the pre-planning efforts, the technical planning process can begin involving the following steps :

(a) Inventory of existing facilities.

(b) Forecast of aviation demand for future facility requirements.

(c) Development of alternative systems to meet future demands.

(d) Evaluation of alternative systems.

(e) Implementation and continuing planning process following the same general procedures established earlier.

A. Inventory of existing facilities

The planning period for a master plan is generally taken as 20 years. At the first step the planner should determine the existing aviation facilities and activities, and go through the process of data collection or inventories. The scope of data collection depends upon the area under study, geological and political factors, amount, type and quality of data already available. The following data is required for planning and projecting :

(i) Area to be served, population of the area and economic characteristics.

(ii) Type of airports (civil, military, VTOL, STOL etc.) and their existing operating characteristics and possibilities of future expansion.

(iii) Airspace and the navigation aids used. The dimensions and configrations of the control zones, flight rules etc. in the area.

(iv) General aeronautical activity : Existing air traffic, statistics and historical background, number of passengers, amount of cargo and mail, local origin and destination operation of general, commercial, military and other aircrafts, and their trends, subsonic and supersonic jets.

(v) Environmental factors, current noise and pollution levels, and their impact, location of parks, wild life, recreation areas.

(vi) Socio-economic factors such as population growth, income, economic status and type of business. Forecasts of population and its distribution, economic activity, gross national product, employment, per capita income, occupation, education, industrial activity etc..

(vii) Land use activity in the area : Industrial, agricultural etc., current land uses and estimate of land value today and in future.

(viii) Applicable laws, ordinances and regulations.

(ix) General transportation plan (comprehensive plan) for the area showing existing and planned landuses, utilities, highways, railways, school, hospitals etc. Available data on volume, origin-destination data, travel by mode and purpose, etc.

(x) Financial resources : History of financing, community attitude, bond financing etc.

(xi) Limitations or restrictions : Adverse terrain, existence of natural or manmade obstructions, densely populated areas and other environmental considerations.

(xii) Community attitude : Determination of the attitude of the community towards airport development is very important. A positive attitude helps in development.

These data may be collected from concerned departments or organisations. It may be necessary to conduct special studies for collecting the inventory data. There may be lot of such data easily available which might have been collected during the regional planning of airports on national basis. Availability of such information should be ascertained prior to conducting traffic, socio-economic and other studies for data collection. Use of surveys and questionnaires is made to supplement the available data. Fig. 3.1 shows an outline for technical planning process.

3.5 AVIATION FORECASTING

Forecasts of aeronautical activity is the basis for planning of facilities for future requirements. Rapid changes will continue to be a salient feature of air transport. Our inability to predict the course of rapidly changing pattern of air transportation and of the cost of providing facilities for it, implies that we need to adopt a flexible planning process. Forecasting is not an exact science, but an analytical approach to empirical data that improves with experience.

Air travel is expensive which people use more as they become richer. As national economies improve the air transport grows faster while the period of stagnation and recession might hamper the growth. With significant fluctuations in the rate of growth, planning problems become horrendous. It becomes most difficult to build right facilities at right time. A rapid spurt in traffic can easily swamp a facility with congestion, confusion and associated costs while a slump can leave a large investment idle for several years, thus wasting cost of construction and maintenance. We should, therefore, prepare contingency plan for the unexpected and plan the airport development such that it can be altered to suit the circumstances. In forecasting there is a need to know :

(a) Type of civil airport users : Air carrier, scheduled air taxi, general aviation and military services if applicable.

(b) The type and volume of operational activity : Aircraft operation, passengers, cargo, based aircrafts etc.

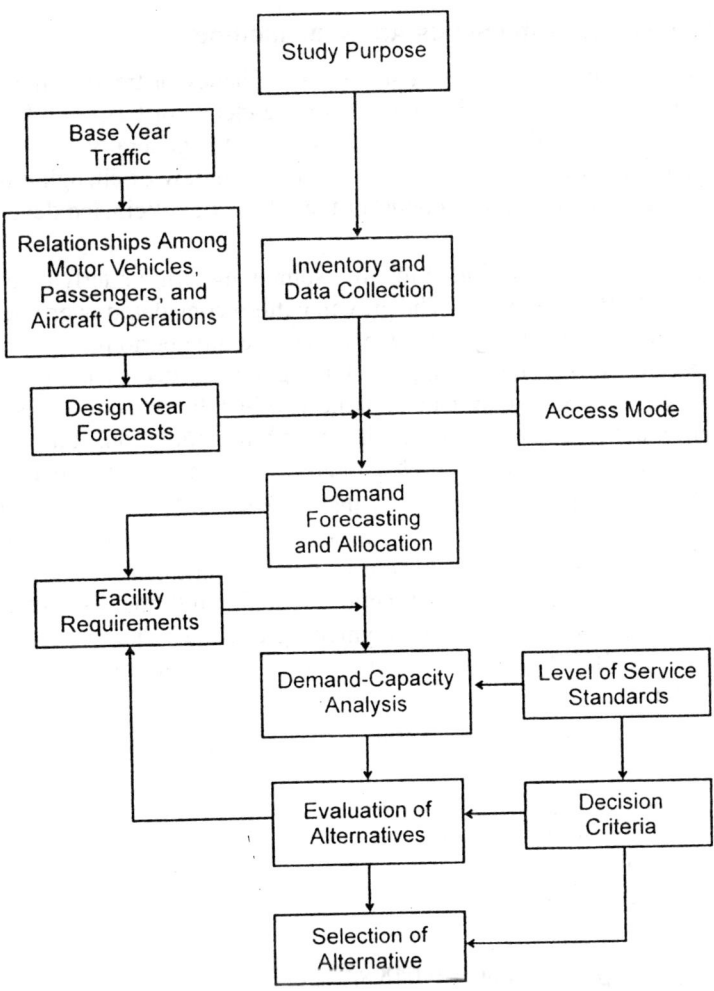

Fig. 3.1. Outline of Technical Planning Process (FAA).

(c) The aircraft mix : Supersonic, business and pleasure aircrafts, future vertical/short take off and landing aircrafts etc.

3.6 METHODS OF TRAFFIC FORECASTING

The principal methods of forecasting can be divided into broad categories : quantitative or scientific, and qualitative or judmental or a combination of these.

1. Trend extrapolation method (Time-series analysis)
2. Mechanical (Comparative method)
3. Category analysis method
4. Market research method
5. Analytical (Technological) method
6. Intutive (Delphi) method
7. Econometric method

3.6.1 Trend Extrapolations (Time Series Analysis) method

The basis of extrapolation procedures is that some past tendency or trend in the variable, being forecasted reflects what is going to happen in future. If this tendency or trend can be measured, forecasts may be derived by extrapolating the relation into the future. It is generally assumed that observed rate of growth will continue for sometime and then may change inview of changing factors at regional or national basis, requiring adjustments to account for the effects of anticipated developments or changes in the environment.

Trend extrapolation consists of trying to identify some long term underlying growth pattern which fits the behaviour of air traffic in the past. The growth pattern considered overtime is usually a straight line meaning a constant absolute change between successive time periods or asymptotic implying that development proceeds towards some limiting level at a gradually decreasing rate. The chosen growth factor is then projected. Fitting to data can be done graphically or using statistical techniques. For example aviation growth may be say 0.8 times the growth in population. Ratios are often used when total for larger area projected and analyst wishes to allocate them to lower areas.

This method of trend extrapolation assumes that a factor influencing air traffic in the past will continue to influence in the same way in future, which may not be true. It is a simple and inexpensive method to carry out, and relatively obvious about its assumptions. Correlations and mathematical expressions establish amount of air travel as a function of community factors such as population. Average income per capita, national product, employment, gross sales etc. and aviation factors such as airport quality, registered pilot in the country, flight hours, based aircrafts and system characteristics such as degree of isolation.

One model used in Tanzania was of the form :

$$L_n y_i = L_n a + x_i L_n b$$

where
- y_i = number of passengers or weight of cargo and mail in kgs
- x_i = Number of years
- a, b = are constants

Passengers :

Disembelled	$L_n y_i = 11.556 + 0.048 \, x_i$
Embelled	$L_n y_i = 11.368 + .049 \, x_i$
Transit	$L_n y_i = 10.903 - 0.068 \, x_i$
Cargo : Offloaded	$L_n y_i = 14.747 - .022 \, x_i$
Loaded	$L_n y_i = 14.487 - .015 \, x_i$

3.6.2 Mechanical (Comparative) Method

Mechanical method based on analogy, which could be used for forecasting traffic at a particular airport is called comparative method. In simplicity this method assumes that if two cities located in the same or different countries, have similar socio-economic conditions, then even air traffic growth rates in the two cities will be the same. Thus, if the growth of airtraffic is known for one city, the same growth rate will be assigned to the other city with similar conditions.

3.6.3 Category Analysis Method

A more insignificant version of the trend extrapolation method is to divide the travellers into a multitude of categories. These are defined by the characteristics of the users that appear to be important in determining the propensity to air travel, their age (younger generation is more likely to fly), the purpose of trip and so on. This method then prepares a forecast by projecting the trend of each category and summing up the results to obtain a total for all travellers. The detailed analysis on aggregated approach is likely to provide more accurate results but its execution requires statistics on individual behaviour which have to be obtained through extensive and costly surveys. The data collection must be associated with hundred of geographical zones sorted with multiple categories which depend upon the profession, age, income and other characteristics of the travellers. This huge data then rquires processing through computers and manipulated to fit mathematical models which define air travel in number of factors.

It has the advantage of being more easily relatable to cause and effect. Forecasts are usually more accurate and useful. It is, however, an expensive method.

A typical model developed for Baltimore region in USA was :

Trips from a zone = − 633.6 + 0.54936 × (Government employment in zone)
+ 0.06009 × (population) + 1.1678 × (per capita income)

3.6.4. Market Research Method

The market research method relates travel pattern of a given population segment to selected demographic and economic characteristics. In New York and New Jersey, this method was used based on a series of household surveys conducted over a period of 20 years. The result of these surveys indicate a strong correlation between travel pattern and such characteristic as income and occupation. One of the main purpose of such a survey is to determine whether a person is a 'flyer' or not and if an air traveller, how many trips are taken each year. Trends and growth rates are established for the number of trips per 1000 flyers in each cell. The air travel market was divided into large number of travel cells for personal and business travel. Personal travel cells are classified by age, occupation, education and income. Business travel cells are classed by industry, occupation and income. A typical matrix of personal travel was developed having 134 cells showing the population in each cell. Forecasts are made by first estimating population growth in each of the cells and then calculating the expected number of trips by multiplying the number of people in the population by percent of flyers and the trip per 1000 flyers. Estimates from each cell are then summed up and adjusted for elements not covered in the survey to produce totals for future years.

Market analysis method can be extremely useful in identifying population segment that generate most of the air activity and those that are good future potentials. The weakness of the method is that it does not consider service characteristics such as fare and trip time. The market analysis method for example, cannot relate changes in demand for air travel to changes in average fare level or introduction of new fares established to attract certain market.

3.6.5 Analytical/Technological Method

To avoid the dependency of statistical analysis on gross judgement, efforts are made to determine the rate of change of traffic, by the key variables such as price of air travel, or the travel time. This is termed as elasticity of traffic, represented by per cent changes in key variables :

$$\text{Elasticity of traffic} = \frac{\text{per cent change in traffic}}{\text{per cent change in key variable}}$$

If the volume of traffic is insensitive to change in a particular factor the elasticity will be close to zero. If the traffic changes more rapidly than the factor, the elasticity will be larger than plus or minue one, depending upon the nature of change. These analytical methods differ from mechanical methods in that additional thought is given to explaining reasons for change. For example in analytical forecasting, one would try to determine the reason for trip and how over time, these reasons change. The introduction of technological improvements can change the demand for air travel. For example, introduction of jet services has accelerated the demand for air transportation.

One major group of these methods deals with study of economic relationship between causes and effects. Before air traffic forecasts are made a thorough study is made on the influence of economic parameter of air travel and then a model which takes account of economic parameters and air traffic penetration as a mode of transport is developed a typical model is :

$$Y_t = (R_t)^k - (T_t)^r - (m)^{t-1}$$

Y_t = number of passenger in year t

R_t = average income in the country

T_t = air traffic fares

m = logistic function representing the penetration

k = elasticity of traffic volume

r = elasticity of traffic fare

3.6.6 Intutive Methods (Delphi Method)

Intutive methods are based essentially on the individual's feeling for the situation or on the expert's opinion. These methods are often employed either in very complex situations or situations in which the available data are irrelevant, unreliable, inadequate or obsolete and that it is not possible to collect the required data in time. One of the intutive methods which was used in forecasting air traffic is called 'Delphi' technique. Instead of relying on a single expert the Delphi technique is based on getting a consensus forecast from a number of experts from different organisations which directly or indirectly deal with air transport, such as ICAO, ICAA, IATA etc.

One major drawback of intutive methods is that it is very difficult and in some cases it is impossible to evaluate the standard error for future observations given the forecast. They rely on subjectivity. It is impossible to eliminate subjectivity even in most statistical methods.

3.6.7 Econometric Method

The econometric method attempts to relate traffic variations to the movement of logically relevant economic variables such as income, demographic variables such as population and service variables such as fare and trip time. This method explores and analyses parameters that have affected the historical travel demand pattern and that may influence future travel demand. An econometric demand model shows, through one or more equations, an economic relationship between demand and a number of predictor variables.

Factors contributing to travel demand change can be grouped in two broad categories : socioeconomic and transport related. Socioeconomic variables relate to the general economic, geographic, social and political environment. Transport-related variables, on the other hand, are those inherent in the transport mode such as cost, travel time, comfort safety and convenience. Passenger traffic volume is influenced by a complex interaction of one or more of these variables. This method though complicated, if performed by experts can be most effective for traffic projections.

3.7 CARGO TRAFFIC FORECASTING

In the past air transportation industry has focussed primarily on the movement of people. Air cargo has been given a secondary role to fill the available space on passenger flights as belly cargo. Air cargo is now growing faster than any other element of air traffic and its rate is constantly increasing, requiring more and more attention to predict the potentialities of air cargo.

The growth of air cargo depends on a number of factors such as the rate structure, industrial marketing, distribution process, cooperative arrangements with truckers and freight forwarders, aircraft technology, automation of cargo terminals, environmental improvement costs, domestic as well as world economy and changes in airline management attitude towards providing air cargo transportation services.

Though social variables which have been found so important in passenger der. and models are not important in analysis of freight movement, greatly simplifying the procedure, however forecasting of freight demand is currently in its infancy reflecting the great scarcity of historical data.

Because of lack of data, aggregated projections at the national level are more easily made than disaggregated forecasts of freight movements between specific locations.

As more origin destination and city-pair data become available, it will be possible to develop models for predicting regional airfreight movements, through development of regression and input-output analysis techniques that shows promise for the future. There will be more and more need in future to predict freight demands and plan for it.

Mail component of cargo is relatively small and volumes of mail are carried in combination aircraft, rather simple projections are sufficient to determine space in terminal building. At big airports, where mail handling facilities are required, the postal authorities do the forecast and supply the space.

The method(s) of forecasting depend upon the available data, on the time and resource available to carry out the forecast, and on the purpose for which forecast is required.

A major difference between above forecasting methods can be drawn as "top down" and "bottom up". 'Top down' methods project an aggregate figure and then use fractional coefficients and other approaches to break it down into lower level estimates. 'Bottom up' methods consist of synthesizing forecasts of categorised units/cells.

Where feasible, both approaches may be used and results compared. Forecasting for existing airports is simple and is largely based on historical data for the airport, the air transport system and the region. The forecasts in such cases may be made by projection of past trends. For a new airport, particularly if the transport environment is unstable, the assessment of future traffic is more difficult. The approaches for new airport include category analysis method, market research and surveys etc.

Within the limit of available resources, it is better to use more than one method in producing a forecast. Forecasts should be presented in consistent term, for periodic updating. It may be revised, if necessary, when influencing factors deviate from basic assumptions. Review of forecasting methods as well as of the forecasts may be required in such cases.

The forecasts are reduced to meaningful units expressed in number of passengers and aircraft movements for application in airport planning. Secondly passenger figures must be further subdivided by class and nature of carrier. Finally airports annual passenger and aircraft movement figures must be converted to daily and hourly flows to determining average and peak values to establish planning criteria and facility sizing.

3.8 DEVELOPMENT OF ALTERNATIVE SYSTEMS

After aeronautical demand has been determined, the next step is technical planning process is the assessment of facility requirements. From the forecasts, required physical dimensions of the aircraft, requirements of number of runways, taxiways, aprons, are known. Guidelines for all facets of airport development and operational control should be obtained through concerned organisations such as ICAO.

An analysis of air traffic capacity shall help in determining of how much and when additional capacity at existing airport or a new airport is required. Capacity has to be analysed in term of airfield, terminal area, airspace and ground access.

The expansion of existing airports as well as need for new airport, is theoretical at this point of planning process. Several alternative schemes should also be developed for evaluation for the following :

(a) Airfields

For working out the airport capacity (rate of aircraft movement on runway and taxiway system) a reasonable level of delay is assumed depending upon the type of airport, type of aircraft and type of service. This will ascertain number of runways, taxiway and their configuration which shall be discussed in following chapters.

(b) Terminal area

The capacity of terminal area is its ability to accept the passengers, cargo, hanger and other facilities. After determining the airport capacity required capacity for terminal elements should be determined. Terminal elements include airlines and cargo aprons, gate positions, hangers, passenger building, aprons, cargo buildings, automobile parking and aircraft maintenance facilities etc..

(c) The airspace

The proximity of airports to one another, the relationship of runway alignments and the nature of operation (IFR and VFR) need considerations for overall capacity of the airspace. Inter airport airspace relationship is examined for safe and efficient use of airspace.

(d) Ground access

The ground access capacity includes number of movements by people and automobiles. In determining the volume of people, it is necessary to include passengers, visitors and airport employees, and ascertain the proportion of those who are expected to travel by private cars and public transit system. Airport origin and destination study may be required to provide adequate access to airports.

(e) Air traffic control aids

The demand forecasts for existing and new airports will provide insight into the type of air traffic control facilities and air navigation aids that will be required. The airspace available, the volume of traffic projected and the nature of the expected operations shall determine the needs. Airports should be planned for the most sophisticated instrumentation available.

Having collected the data and thoughts about the required facilities as above, alternative system plans should be developed to meet the demands. A comparison of relative merits and demerits of developing new airports versus expansion of existing airports will be an important exercise. Both positive and negative impacts on the environment by development of airports must be considered. The principal positive impacts are economic growth, employment opportunities, against the negative impacts of noise, air pollution, and ecological compromises.

In developing alternatives considerations for ground transportation facilities are significant. The potential constraining factor should also be considered in developing alternative plans. They may be airspace limitations, land limitations (high cost of land, long and costly ground travel), environmental impacts, financial and political considerations.

Each alternative plan developed should provide complete information on technical aspects, alongwith their justification inview of the following factors :

- Consistency with national and regional planning.
- Costs of development including land acquisition costs and displacements, airport access costs.
- Rating about utilisation of airspace and efficiency of air traffic handling.
- Quantitative evaluation of constraints.
- Qualitative assessment of political and citizen acceptability.

3.9 EVALUATION OF ALTERNATIVES

Having developed several alternative system plans, it is now necessary to select the **"best"** alternative. Firm position on various alternatives should be established after consulting airlines and other concerned agencies, after selecting the best alternate and prior to its adoption.

The evaluation criteria to select the best are developed with an emphasis on safety, efficiency, economic justification, environmental impacts, cost and an equitable distribution of facilities. Models based on these criteria rank the plans with respect to their effects on airport system and in relation to the community service they provide.

(a) Safety criteria

Safety criteria are critical elements of the evaluation process. Safety criteria are divided into two categories : Airport safety and aviation safety. Relative impact of each different alternative would have on safety of an airport should be assessed. The safety rating emphasizes the magnitude of the benefits derived from the plan. An airport without safety problems may have a rating 1, otherwise it may have a factor one minus the effective safety factor. Establishing quantitative criteria for deciding whether an emergency landing strip is justified is difficult.

(b) Efficiency criteria

Efficiency criteria measures the extent to which a proposed plan will increase its transportation efficiency. For airports, two criteria relate to efficiency. The first is operational efficiency indicating the extent to which a plan will improve the capacity and capability of the airport and the second access it provides to remote locations.

(c) Environmental criteria

This rating reflects impacts on the community as opposed to impacts on the user. The best plan is one where no or minimal adverse environmental effects, no conflict with current or projected landuses, and no potentially serious construction, clearance or airport approach or departure problems are created.

(d) Economic development criteria

Economic development is a tool for planner to shape the growth. If accelerated growth is desired in a specific area, providing complete and modern transportation facilities will tend to further the goal, likewise if a reduction in the rate of growth in another area is considered beneficial reduced emphasis should be possible. The economic development rating allows the planner to increase or decrease the growth of development as per needs.

(e) Cost criteria

Cost criteria provide the means for reflecting the magnitude of annual capital cost of development and maintenance, for each of the alternative plan.

Since this is the final stage of selection among alternatives, views of participating and non-participating agencies, should be invited to make sure that proposed plan is within the agreed policies.

The format of the selected plan should include the following important features :
- A summary document well illustrated for general public.
- One or more technical drawings and documents to provide basic plan information on location, configurations, present and future land use and urban ground transportation systems, expansions proposed now and in future, navigation aids and control facilities along with a summary of costs and expected revenues.

3.10 IMPLEMENTATION OF PLAN AND CONTINUING PLANNING

After adoption of a plan, the recommended development should take place in the right way, proper financing in a timely fashion should be available. The programme for land acquisition of the site should be finalised, for long term needs.

In the changing environment, it is important that a continuing planning approach is used to determine if the assumptions made previously are holding over time. The main objective of continuing airport system planning process is the development of the capability to provide needed planning information and assistance to those responsible for plan implementation.

To conclude, in the master planning first process involves assembly of existing data and analysis of capacity problems. At the same time new data on passengers and cargo etc. is collected for forecasting purposes. In the second, the detailed capacities of runways, terminal areas and other facilities are determined on the basis of forecasts. The need and timing are determined specially. A concurrent environmental study specially on noise is done. Finally after considering all the alternatives, airport layout, terminal area, and airport access plans are made for the 5, 10 and 20 years development for each airport. Development stages are shown for a typical airport in Fig. 3.2, 3.3 and 3.4.

For realisation of maximum benefits from the heavy capital investments that are made for provision of new airports or development of an existing one, to ensure safety of aircraft operations and to avoid hazards or discomfort to the surrounding community, sites should be selected which offers best land areas for long term development at least financial and social costs.

3.11 AEROPLANE AND ITS PARTS

The chief parts of an aeroplane are :
1. Wings
2. Fuselage

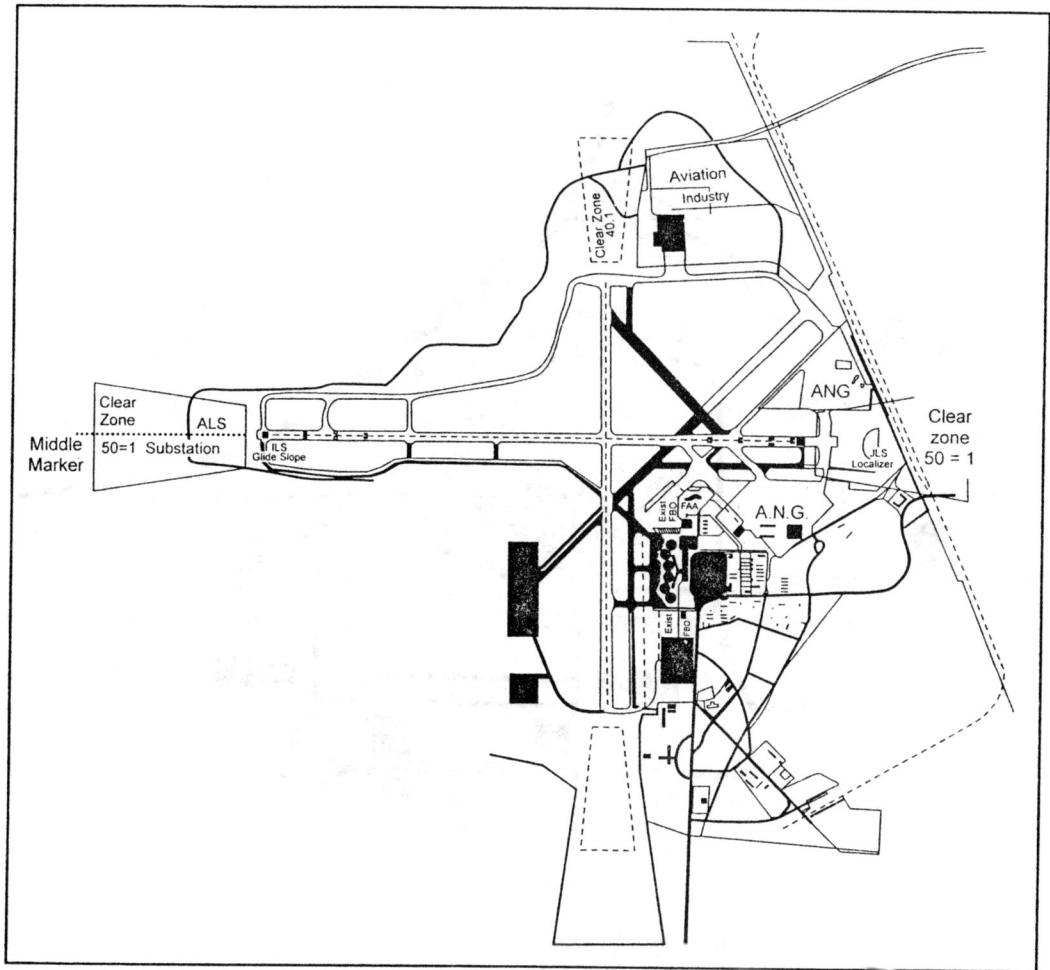

Fig. 3.2. First stage development of an airport.

3. Engine (one or more)
4. Air screw
5. Controls
    (a) Elevator
    (b) Rudder
    (c) Aileron
    (d) Flaps
6. Tri-cycle under carriage

1. The wings

An aeroplane may have single pair, double pair, triple or quadruple pair of wings. The standard plane is monoplane having one wing on either side of fuselage. With more wings planes are termed biplane, triplane, quadriplane.

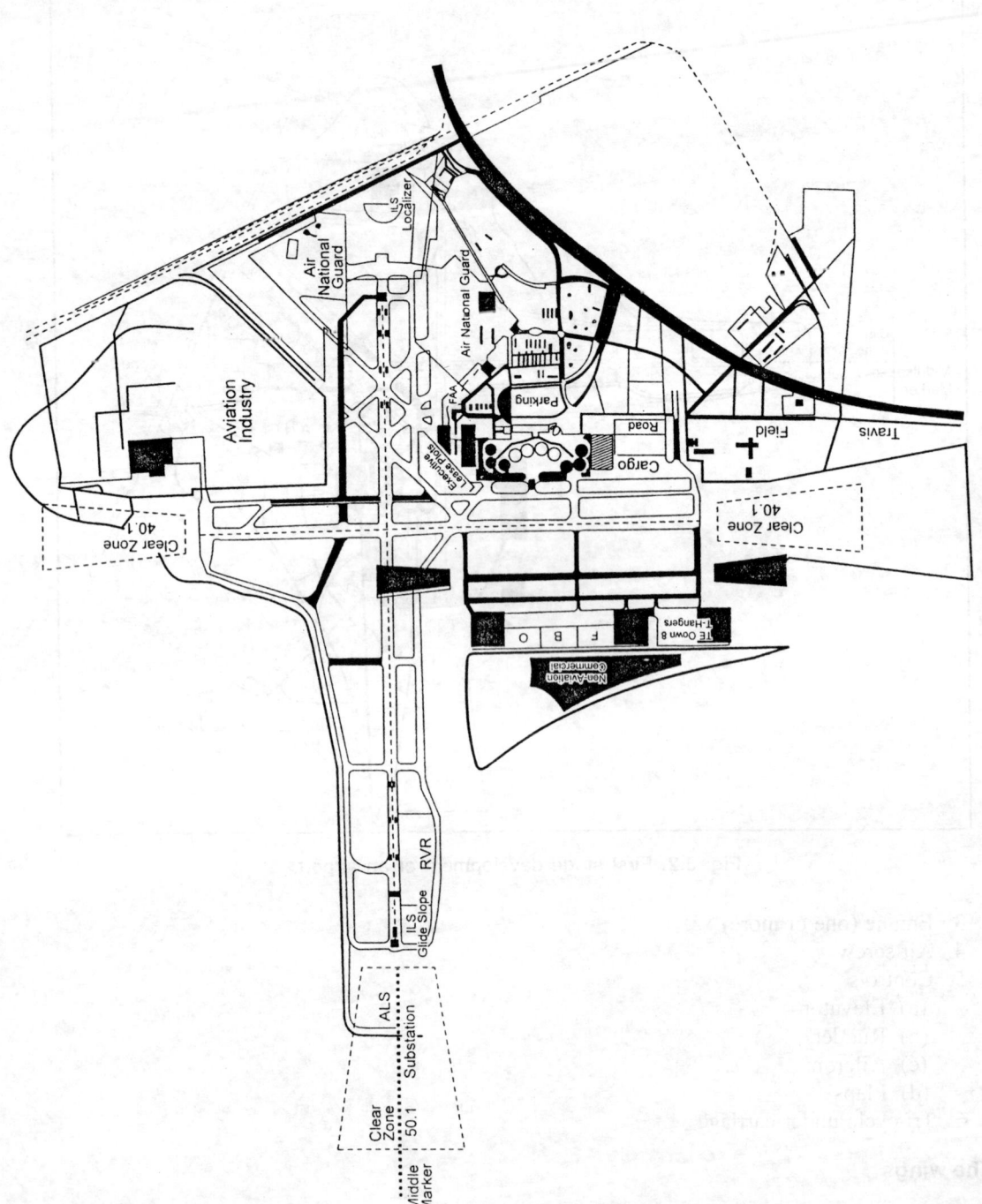

Fig. 3.3. Second stage development of an airport.

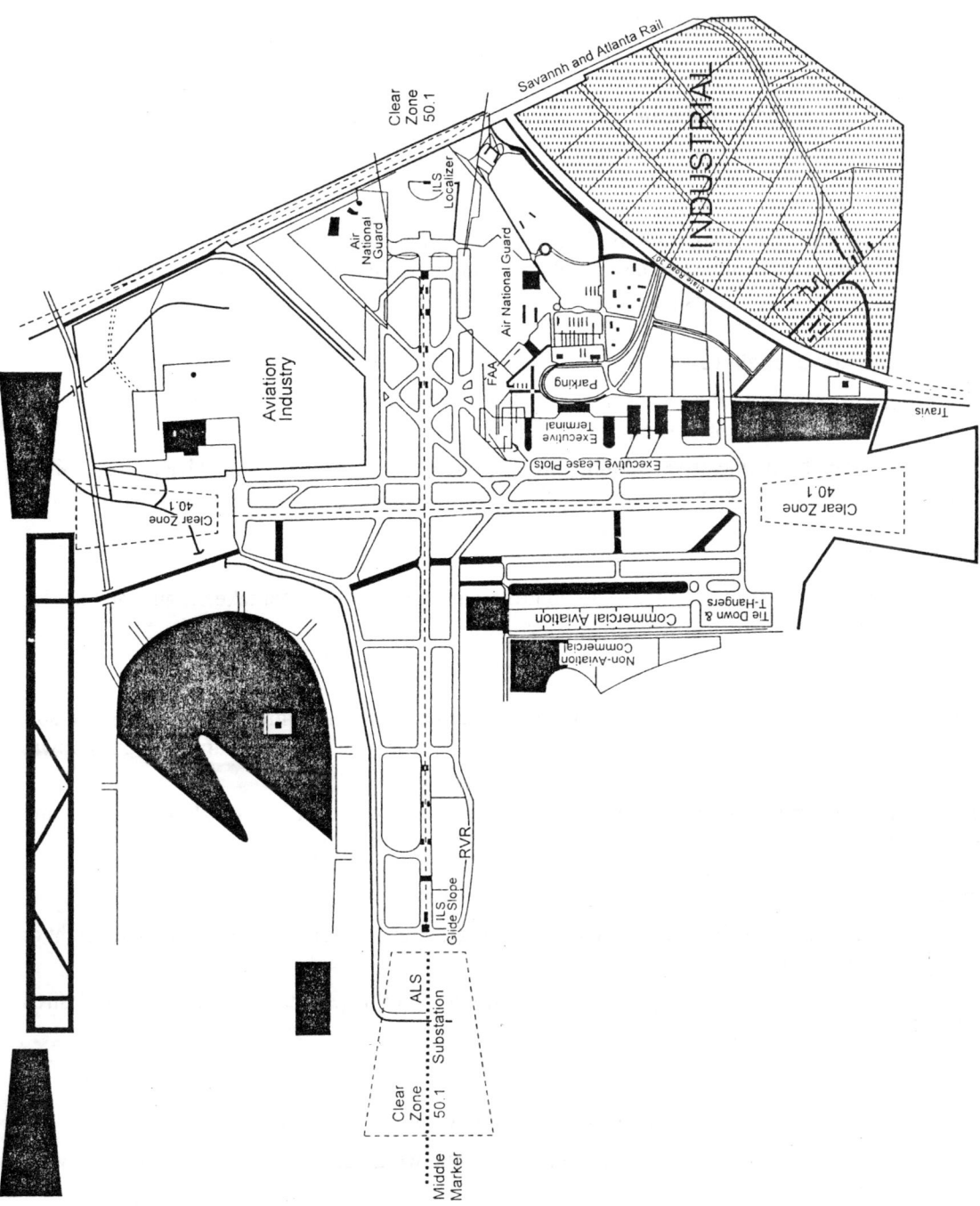

Fig. 3.4. Third stage development of an airport.

Monoplanes are lighter, less air resistance and are, therefore, speedier than bi-planes and others. Biplanes have wings one above another have greater stability.

2. Fuselage

The fuselage is the main body of the air craft. It consists of pilots cockpit, space for passengers, cargo, mail and tail.

3. The engine

An aeroplane may have one, two, three or four engines. If single engine is used, it is provided in the nose of the air-craft. Two or four engines are placed symmetrically about the nose on wings.

In case of three engines, one is placed in nose, one each of the two wings.

4. The air screw or propellers

Air screw helps in converting the power of the engine into thrust to drive the aeroplane forward. The screw is provided in the front of the engine. It has two or more blades.

In jet propulsion, the engine drives a kind of turbine impeller from which a tunnel leads to the tail of the aeroplane or rear end of the wing. The tunnel emits a stream of air that derives the plane forward.

5. The controls

The aircraft is provided with four principal air controls. Controls are meant for take off and landing, for losing and gaining the speed.

(a) Elevator

It is hinged flap to the rear frame member on tail. It can be moved upwards through an arc of 50° to 60°. It is operated by pilot from his cabin. With flap raised, air pressure is increased and tails goes down, making nose point up.

Elevator is most important control during take off and manoeuvre. It helps in loosing speed and turning of aeroplane.

(b) Rudder

Rudder is like a rudder of a ship or boat, usually placed at the rear end of fuselage. It can move right or left along vertical axis by 30°.

It is operated from pilots cabin and enables the pilot to point nose of the aeroplane in the desired direction.

(c) Ailerons

They are hinged flaps mounted near the rear ends of the wing on either side.

It is operated by the pilot such that when aileron on one wing rises, the aileron on the other wing goes down and vice versa.

This produces rolling motion and enables the pilot to restore it to level flight. They are stabilising agents for plane.

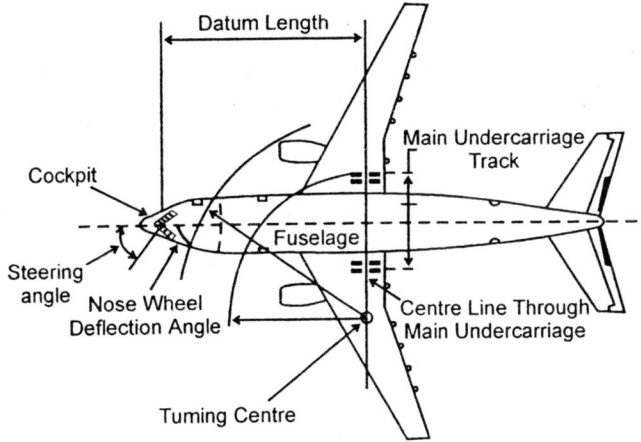

A. Term—Aircraft with two main undercarriage legs

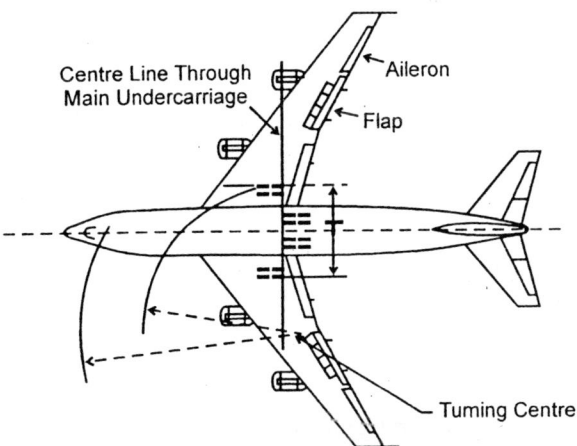

B. Terms—Aircraft with four main undercarriage legs

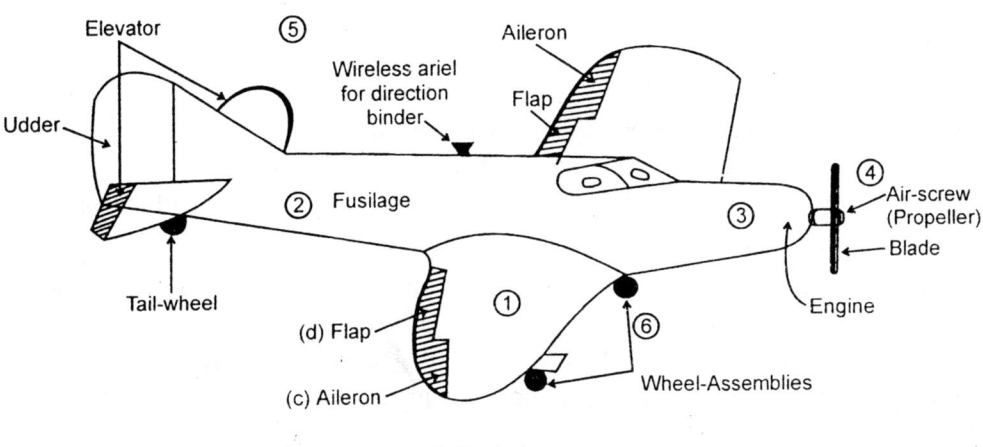

C. Symbols

Fig. 3.5. Component (parts) of an aeroplane.

(d) Flaps

Flaps at the rear end of wings are used as air brakes when turned down under the wings to provide powerful resistance causing quick drop in the flying speed.

(e) Tri-cycle under carriage

A pair of wheels is provided in the fuselage or in the wings, near the junction of wings and fuselage.

A third wheel, if provided at the TAIL, is to keep nose position up, and if provided at the NOSE, keeps the nose position down.

3.12 AIRCRAFT CHARACTERISTICS RELATED TO AIRPORT DESIGN

Planner and designer of airports must know the following principal characteristics of the air-crafts, which can be accommodated in design :

1. Type of propulsion
2. Size of aircraft
3. Weight of the aircraft
4. Capacity of aircraft
5. Range of aircraft
6. Speed
7. Turning radius
8. Tyre-pressure and contact area
9. Fuel spillage
10. Heat and noise
11. Aircraft circling radius
12. Weight on gear system and gear arrangement
13. Speed of jet blast

1. Type of propulsion

Based on the type of propulsion and thrust generating medium aircrafts could be classified as :

1. *Piston engine* : have reciprocating engine which is fitted with propellers.
2. *Turbo jet* : have turbo-engines which are not driven by propellers.
3. *Turbo fan or turbo prop* : Turbo engines have a fan either at front or rear of the turbo engine.

Examples of different aircrafts based on propulsion are :

Piston engine : DC-3, DC-4, DC-6, DC-7, Convair 340, Lockheed 1649, 1049, G. Super-constellation.

Turbo-engine (prop) : Fair Child F-27, Lockheed's Electra, Vicker's Viscount 802.

Turbo-jet : De Havilland's Comet 4B, Convair's 880 and 990, DC-8, Boeing 720, 707-120 and 707-320.

Turbo-fan : DC-8-52, Boeing 720-B, 707-120B and 70-732 OB.

2. Size

Size of an aircraft can be defined by the following :
  (a) wing-span
  (b) length
  (c) the maximum height
  (d) distance between main gear

Wing span decides the apron size, taxiway clearance, turning clearance. Length decides the width of exit taxiway, apron size, length of hanger. Height effects the height of hanger and its gate.

Distance between main gear decides the geometrics of exit runway.

3. Capacity

Capacity of an aircraft includes :
  (a) fuel space
  (b) passenger space and
  (c) cargo space

Bigger the capacity, larger are dimension and greater is the weight.

4. Weight

Structural design of the airport is based on the total load of the aircraft. The weight of the aircraft may be classified into :

(a) Operating weight

Operating weight is the weight of empty aircrafts, its crew and all equipment ready for flight, excluding the passengers and fuel-load.

(b) Payload

It is revenue producing load which consists of passengers, mail and cargo.

(c) Fuel weight

It consists of weight of the fuel taken by the aircraft required for the trip and certain reserve. It may vary from 9 to 40% of the total gross weight.

(d) Maximum gross take-off weight

The maximum overall weight of the aircraft is that weight which is permitted for take-off manoeuvre. It consists of operating empty weight + payload + fuel reserve + fuel required for trip. Airport pavement is designed for this load.

(e) Maximum landing weight

At landing aircraft loses weight of fuel consumption in flight.

5. Range

The length of normal haul is called 'range'. The range has important influence on the frequency of operation, affecting peak hour traffic and runway capacity.

6. Speed

Although speed has nothing to do with direct planning of the airport, it gives an idea of the arrival of the aircrafts.

Aircraft speed may be defined as :

(a) ground speed, also called cruising speed

Ground or cruising speed is the speed of the aircraft relative to ground. It is important to passenger.

(b) Air speed

Air speed is the speed of the aircraft relative to the medium in which it is travelling. It is important for pilot.

If craft is moving at 1000 k.p.h. and air moving in opposite direction at 100 k.p.h. the air speed is 900 k p.h.

7. Turning radius

Turning radius would decide the radius of curves at the end of taxiway. It helps in location of aprons and other installations.

Turning radius is recommended by the manufacturers.

8. Tyre-pressure and contact area

Maximum tire pressure and contact areas help in arriving at the total thickness of the airport pavement and suitability of the type of pavement. Small contact area has the punching effect on pavement.

9. Fuel-spillage

Jet aircrafts issue hot blasts at high temperatures, which may melt the binder of the flexible pavement. Under repetitive loads binder comes out. Maximum temperatures may be of the order of 385°F.

Fuel spillage affects taxiways, aprons and ends of runway, as it occurs when engine is shut down or speeded down.

With a view to cope with these high temperatures asphaltic concrete, rubberised tar concrete or plain concrete may be used. "Epox" is another material used in USA, with asphalt concrete and is known as "Epoxy Asphalat Concrete".

10. Heat and noise

In supersonic planes, the speed and fuel economy favours the use of turbo fan engine which gives greater noise during ground stay. Noise is of importance in planning and site-selection of the airport. It is considered advantageous to locate the airports away from the residential areas.

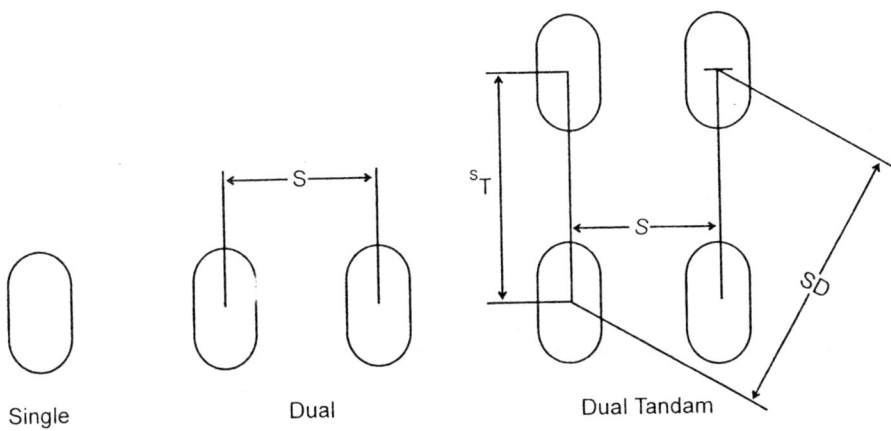

Fig. 3.6. Different wheel assemblies for aircrafts.

11. Circling radius

In landing operations, bigger aircrafts will have to take longer and bigger radius circles, before landing. Two aircrafts should be so spaced that manoeuvering path for the aircraft landing simultaneously one at each port do not interfere.

For jet planes it may be 50 miles (80 kms)

For other planes it may be 5 to 10 miles (8 kms to 15 kms)

12. Gear system and gear arrangements

Aircrafts are supported on nose or tail wheels (gears) and two main gears located in the wing area on each side of the fuselage.

When in stationary position, the load distribution of the 3 wheel-gear system vary with the type of plane.

For designing the pavements, it is assumed that :

10% of the weight is on the nose gear

90% of the weight is on the main gears

Gear combinations for aircrafts is shown in Fig. 3.6.

13. Speed of jet blast

The shoulders are badly affected due to tremendous speed of the jet blast. Maximum blast occurs at all turnings, taxiways, take-off ends of the runway.

Dense turf or a thin bituminous pavement may save the cohesionless soils from erosion.

CHAPTER 4

Site Selection for Airports

4.0 INTRODUCTION

The starting point in selection of an airport site is the assessment of the suitability of an existing site, for the purpose for which airport is required, its type and operational system. The following information, obtained in the planning and data collection phase should be available for selection of a suitable site.

- The quantity and type of traffic to be handled now and in future (peak hour volumes for passengers and cargo) through demand/capacity analysis.
- The type of aircraft which will use the airport and their characteristics.
- The type of airport, its size and shape, and general requirements of terminal and other facilities such as number and length of runways, number of gates, size of aprons and overall land area required.
- The type of operation controls used.

4.1 Steps in site selection

The goal of site selection is to find a suitable location to accommodate all functions of the airport through evaluation of feasibilities of possible locations from environmental, geographic, economic and engineering standpoints. The various steps which are involved in selecting the suitable site are :

- Requirement of land area.
- Evaluation of factors affecting airport location.
- Preliminary office study of possible sites.
- Site inspection.
- Environmental study.
- Review of potential sites.
- Preparation of outline plans and estimates of costs and revenues.
- Final evaluation and selection.
- Report and recommendations.

Airports should be sited at a place from where aircraft operations can be carried out efficiently and safely, cost of development is at optimum level, and it is an integral part of the national network of airports, as per regional planning requirements.

4.2 FACTORS AFFECTING AIRPORT SITE SELECTION

For evaluation of the different available sites following factors require considerations :

1. Presence of other airports
2. Land use and land values
3. Topography of the area
4. Obtructions
5. Wind considerations
6. Atmospheric factors
7. Geological factors
8. Environmental factors
9. Aircraft noise
10. Availability of construction material
11. Availability of utilities
12. Avoiding hazards
13. Ground access
14. Social considerations

1. Presence of other airport

Interference between the two airports in close proximity should be avoided. The distance between the airports will depend upon circling radius of largest aircrafts, and operational controls used. Locations of existing airports and their associated airspaces should be noted, and also their future plans (if any) to change them when two airports have to share same airspace, their combined aircraft movements are restricted. New airports should be so located to minimise such restrictions. A site close to demand with some restriction of an airspace may be preferable to a site with no airspace restrictions but so remote from demand centre. A best balance is to be achieved.

2. Land use and land values

Broad assessment of the land area required should be done for site selections. The site should provide adequate space to permit development and expansion on long term basis. Study should be made about general land values and its usage (residential, agricultural, etc.). The advantage and disadvantage of different sites will be influenced by surrounding form of land use. Existing land use should not be affected by aircraft operations. Costly land acquisitions may have to be avoided. Land values generally increase significantly as areas change from rural to urban use. Early reservation of suitable sites will often enable airports to be better located and at lower costs.

3. Topography of the area

Topography is important because the slope of the terrains, the location and variation of natural features can affect the requirements for clearing, filling, grading etc. Natural slopes and drainage of land are important for design and construction. Topographical maps should be studied to ascertain the suitability of site for drainage and grading. Terrain which conforms closely to the desirable levels and which is well drained may produce significant cost savings.

4. Obstruction

Obstructions in the approach area for landing or taking off of an aircraft should be removed. Long clearance areas (15 kms) are required on either side of runways. It is difficult to obtain sites which provide all the clearance desired. Features such as high terrain, trees and structures which constitute obstacles need to be avoided. Details of obstruction clearance requirements are described in other chapter.

5. Wind considerations

Runways should be oriented in the direction of the prevailing wind when it blows consistently from one direction. Runways should be oriented so that aeroplane may be landed atleast 95% time with cross wind components not exceeding prescribed limits.

The wind data at site, collected for several years in the past should be studied for arriving at suitable direction of runway and number of runways required. It is important to know the velocity and direction of wind along with their direction, throughout the year for several years. Wind distribution in association with visibility are of prime importance in deciding on runway orientation and the need to make provisions for operation under all weather or only visual conditions.

6. Atmospheric factors

The information on fog, haze, smoke etc. which may consequently reduce visibility should be collected. Any special weather factor such as rapid variations, low clouds, turbulence, snowfall and rainfall etc. affect the use, efficiency and capacity of an airport operation. Weather conditions can vary significantly between sites in the same general area. Non-visual electronic aids for guidance, specially under low cloud ceiling and restricted visibility conditions are significant for site of an airport.

7. Geological factors

The geological maps should be studied to know the distribution of soil and rock types. Soil sampling and testing may be required to establish characteristics of the soil. Classification of natural soils at potential sites is important from design point of view. Soil improvement techniques are quite costly, and may be avoided if possible.

8. Environmental factors

Locations of wild life reserves and migration of wildlife should be studied. Noise sensitivity and other environmental and ecological aspects of the site should be taken into account. Studies of the impact of the construction and operation of airport upon acceptable levels of air and water quality, noise levels, ecological processes and demographic development of the area must be conducted to determine how the airport requirements can be best accommodated.

Other environmental factors include air and water pollution, industrial wastes, domestic sewage, originating at the airport and the disturbance of natural environmental values. The environmental study must consider suitable method to overcome these problems.

9. Aircraft noise

The noise in the vicinity of airports is a serious problem to be considered in the development of airport facilities. It is not always feasible to site an airport sufficiently far away from population centres to prevent an adverse social reaction. Remotely located airports are costly, unrealistic and waste lot of time and energy. Proper site selection and adjacent land use planning can serve to greatly reduce the noise problem associated with airports. For an existing airport this may be difficult as the land may have already been built up. Every effort should be made to orient air traffic away from builtup area.

10. Availability of construction material

The materials of constructions, stone quarries if they are available nearly it will save transportation costs of such materials from distant places. The location of water supplies is also relevant because their availability and the distance over which they have to be carried will affect the cost of construction.

11. Availability of utilities

The utilities such as main power line, water supplies, sewage, telephone services, fuel etc. if available at or near the site, it will be a big advantage. Potential airport site, if possible should be close to utilities. Availability of these services may eliminate the need to provide them specifically for the airport and thus reduce cost.

12. Avoiding hazards

Local factors of site should be considered. For example, industry can produce smoke in certain direction of wind creating low visibility, restricting VFR operations. Sites close to wildlife reserves, lakes, rivers and coastal areas, refuse dumps and sewage outfalls etc. may not be desirable because of the danger of aircrafts collision with birds.

Location of sites close to migratory patterns and routes of birds such as swans and geese may be a potential bird hazard.

13. Ground access

Fast and efficient access facilities for passengers and freight are essential for an airport to provide efficient service. Sites offering convenient road network may be better than those with inefficient or inadequate transport system. Heavy expenditures may be required to overcome those deficiencies.

When ground travel time are approximately equal between several potential sites, the journey cost and convenience are the major factors, to be considered.

14. Social considerations

Airports need to be sited such that flight paths do not pass over concentrations of population while aircrafts are below certain heights. However, airports also need to be located not very far away from towns and commercial areas they serve. A compromise between the above two opposing requirements will be required, for a site with best overall merit.

4.3 REVIEW AND FINAL EVALUATION OF SITE

After study of the above factors, the next step is to analyse the available sites, in view of these factors. This process shall eliminate some undesirable sites, and help in arriving potential sites considered worthy of further investigation. This evaluation should rate remaining sites by preference and state reasons by the rating.

As the next step, thorough field and aerial inspection is required to assess the merits and demerits of each site. Considerations for the relative merits of the remaining sites require :
- detailed site surveys.
- preparation of outline airport layouts for each site.

• working out the cost estimates covering total capital and operating expenditures required, including the anticipated phasing of expenditures.

The total cost involved in each alternative site plays an important role in final selection. If all the potential sites were of equal merit, least cost will be the basis for selection.

The need for cost effectiveness, suggests that attention be given to measuring and weighing of benefits and costs through the technique known as benefit cost analysis. By analysing the estimated stream of benefits and costs over the anticipated useful life of the airport, it is possible to determine benefit-cost ratios which serve as a guide to the value of the project and the choice of the best site.

After finalising the site a comprehensive report supported by various drawings should be prepared, along with an airport layout of plan. The actual design and construction of the airport involves three major areas :

  (i) the runways, taxiways and aprons,

  (ii) the terminal, and

  (iii) the circulation road and parking.

The first consideration must be given to airport 'configuration' because that determines the number and orientation of runways and their location relative to the terminal area. Basically configuration is a function of the type and volume of aircraft to be accommodated in relation to the prevailing wind conditions. Wind direction determines which of the several runways will be active at any given time.

Taxiways and aprons provide the means of getting aircraft to and from runway and storing them preparatory to takeoff. Exit taxiways or turnoffs should be spaced to expedite the maneuvering of aircrafts from an active runway.

The terminal is the interface between 'landside' and 'airside'. Adequate facilities for handling passengers and cargo and for housing various airport administrative functions must be provided.

The design of roadways intersections and parking areas, as well as installation and operation of necessary signs, markings and other control devices, should then be done following standard and recommended practices.

The total design of an airport is a complex undertaking including geometric and structural designs, lighting, marking, signing, drainage etc. which shall be covered in the following chapters of this book.

4.4 AIRPORT LAYOUT PLAN

The airport layout plan is a graphic presentation to scale of existing and proposed airport facilities, their location on the airport, and the pertinent clearance and dimensional information. The development of the airport layout plan should establish the configuration of runways, taxiways and aprons and should set aside area for terminal facilities. Runway approach zones should be shown on the airport layout plan, with proper dimensions to a suitable scale.

It shows the airport location, clear zones, approach areas and other environmental features that may influence airport usage and expansion capabilities.

The airport layout plan consists of several components depending upon airport size and usage.

4.5 DRAWINGS FOR LAYOUT PLAN FOR AN AIRPORT

The following drawings are prepared :

4.5.1 Airport layout

This is a main drawing depicting existing and ultimate airport development and landuses drawn to a scale. It shows :
- Runways, taxiways, aprons, blast pads, shoulders, buildings, navaids, parking areas, roads, lighting runway marking, pipelines, fences, drainage facilities, wind indicators and beacons.
- Prominent natural and man made features such as trees, streams, ponds, rocks, outcrops, ditches, railroads, powerlines and towers.
- Areas reserved for existing and future aviation development and services such as heliport, cargo facilities, service and maintenance areas.
- Existing ground contours shown lightly.
- Fueling facilities and tiedown areas.
- Facilities that are to be phased out.
- Airport boundaries, township corners and bench marks.
- Approach and clear zone outlines, indicating height and location of controlling objects. This information may be given in a separate drawing as well.
- Airport reference point with latitude and longitude given.
- Elevation of runway ends, high and low points and runway intersections. For ILS runway changes in elevation within 1000 m of the threshold.
- North point and true azimuth of runways measured from true north.
- Pertinent dimensional data-runway and taxiway widths and lengths, taxiway-runway-apron clearances, apron dimensions, building clearance lines, clear zones and parallel runway separations.

4.5.2 Location map

This is the map drawn to a scale sufficient to show the airport, cities, railways, major highways and roads within 40 to 80 kms of the airport.

4.5.3 Vicinity map

This drawing shows the relationship of the airport to the city or cities, nearby airports, builtup areas, railways and roads etc.

4.5.4 Basic data required

Provision of the following information is required :
- Airport elevation - highest point of the landing areas.
- Runway identification e.g. 6/10.
- Percent effective runway gradient for each proposed and existing runway.
- Percent of wind coverage by principal runway, secondary runway and combined coverage.
- Instrumental landing system runway and other runways as designated.
- Normal or mean maximum daily temperatures of the hottest month.
- Pavement strength of each runway in gross weight and type of mean gear.
- Plan for obstruction removal, relocation of facilities etc.

4.5.5 Wind information

A wind rose diagram showing runway orientation superimposed, crosswind coverage and the source and period of data should be given. Wind information should be in terms of the all-weather conditions supplemented by IFR weather conditions when IFR operations are expected.

More than one drawing can be used to depict the various items as included above.

4.5.6 Approach and clear zone layout drawing

This drawing shows the following information :
- Areas under the imaginary surfaces and objects affecting navigational air space.
- Existing and ultimate approach slopes.
- A plan and profile of clear zones and approach areas showing the controlling structures and trees etc. Also railways, roads, pipelines that cross clear zones and approach areas, should be shown on the profile. Any plan concerning the alternation or removal of obstruction should be noted.
- For airport serving jet aircrafts and within the boundaries defined by the imaginary surfaces an outline of all areas with present or potential concentrations of people should be shown on the drawing, alongwith other landuses in these areas such as industrial, residential, parks, schools, hospitals etc. For other airports this information should be shown under approach surface atleast 300 m to either side of each runway.
- In the approach area, television and radio transmission towers, garbage dumps and other areas attracting large number of birds and other potential hazards to aircraft flight should be shown.

The above drawings, representing a complete master plan for the airport should also be accompanied by a airport layout plan report which explains reasons behind design features, computation of runway lengths and designs, basis for runway orientation and maximum wind coverage etc.

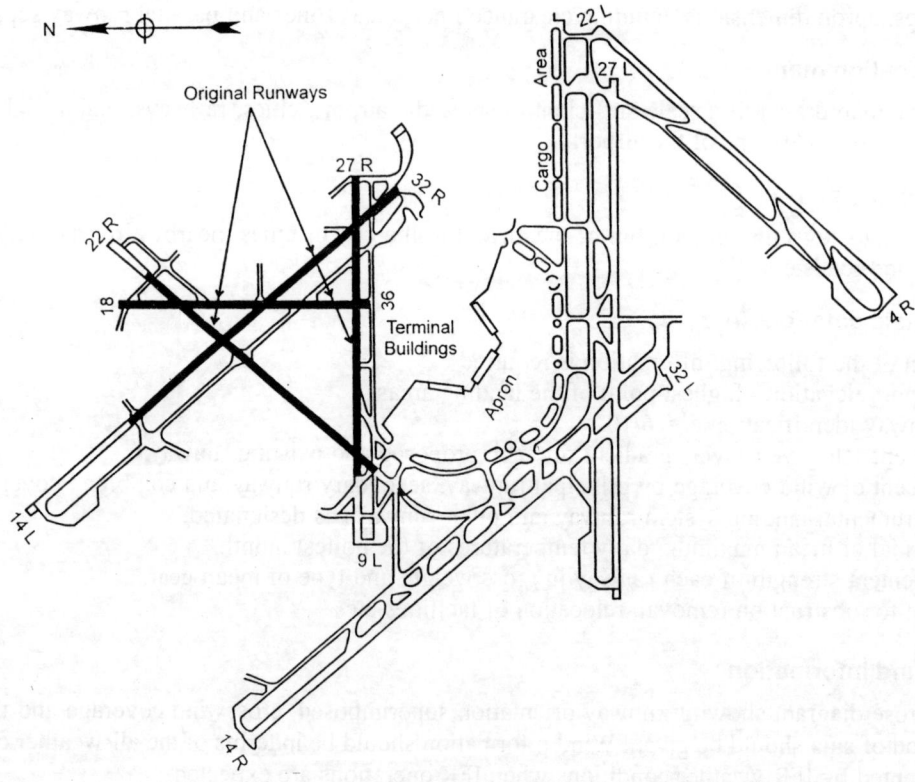

Fig. 4.1. Typical airport layout plan.

CHAPTER 5

Orientation of Runway

5.0 INTRODUCTION

Runway location and orientation are of the utmost importance to aviation safety, comfort and convenience of operation, environmental impacts, and the overall efficiency and economics of the airport. In establishing a new runway layout and/or evaluating existing layouts for improvements where runways are added and/or existing runways are extended, the factor influencing runway location and orientation should be considered. The weight and degree of concern to be given to each factor are in part dependent on the airplane types expected to utilise each runway, the meteorological conditions to be accommodated, the surrounding environment and the volume of air traffic expected to be generated on each runway. Following factors should be considered in locations and orienting new runways and/or establishing which end of existing runway should be extended.

1. Location of neighbouring airports.
2. Obstructions and topography.
3. Built up areas and noise.
4. Air traffic control technique.
5. Wind direction and visibility conditions.
6. Capacity (type and amount of traffic).

5.1 Location of neighbouring airports

The existence of other airports in the area and their operating procedures should be investigated to see if they are compatible with a proposed new facility.

5.2 Obstructions and topography

Obstructions and topography have severe impact on operating conditions and costs. Detailed surveys should be done for studying the topography and obstructions and considerations should be given to the topographical features of the airport and its surroundings. All runways should be oriented so that approach and departure areas are free of obstacles. Following aspects should be reviewed :

(a) Compliance with the obstacle limitation surfaces.

(b) The orientation and layout should be selected so as to protect as far as possible the particularly sensitive areas like school, hospitals etc.

(c) Current and future runway lengths to be provided.

(d) Construction costs.

5.3 Built up areas and noise

The surrounding landuse patterns should be investigated and outlined to ensure that the new facility will be environmentally compatible. In areas where no conflict exist, Zoning should be introduced to ensure future compatible usage. Runways should be oriented so that aircrafts are not directed over populated areas. Account should be taken of the effect of a particular runway alignment on wild life and the general ecology of the area.

The noise noise levels produced by aircrafts at and around the airport should be considered. Most noise exposure lies within the land area immediately below and adjacent to the aircraft approach and departure paths.

5.4 Air traffic control techniques

The air traffic control system whether instrumental or precision approach should be considered to ensure that obstacles will not restrict the operation of aeroplanes. Possibilities of installing suitable non-visual and visual aids for approach to land should be considered.

5.5 Wind directions and visibility conditions

An airport runway should be oriented so that an aircraft is not subjected to excessive cross winds. A wind analysis should be made considering wind by time of the day, velocity and direction. Runways should be oriented in the direction of prevailing winds when it blows consistently from one direction. Sufficient number of runways should be provided to meet air traffic demand to be accommodated in one hour during busiest period. Visibility, clouds and other environmental factors should be considered. Attention should be paid whether the airport is to be used in all meteorological conditions or only in visual meteorological conditions and whether it is intended for use by day and night or only by day.

5.5.1 Cross wind component and usability factor

If the direction in which an aircraft takes off or lands, is opposite to in which wind blows (provides greater lift on the wings when taking off and provides breaking effect while landing), it saves about ten per cent of the length of runway needed if the landing and take-off were to be made in the direction of wind. Runways should, therefore, be oriented in such a way that their direction concides with that of the prevailing winds. This may be possible for major part of the year, but certainly not always.

On certain days and hours wind may be blowing at a certain angle with the centre line of the runway as shown in Fig. 5.1.

If V is the velocity of wind and θ is the angle wind direction makes with centre line of the runway. The component V sin θ, perpendicular to the direction of runway is usually responsible for preventing safe usage of the runway. It is called the cross wind component and should be kept minimum. The maximum allowable cross wind component depends not only on the size of the aircraft but also on the wind configuration and the condition of the pavement surface. ICAO has specified that runways should be oriented so that aeroplanes may be landed at least 95 per cent of time with cross wind components as follows :

Aeroplane reference field length	Cross wind component
1500 m or over	37 km/hr (20 Kt)
1200 m to 1500 m	24 km/hr (13 Kt)
Less than 1200 m	19 km/hr (10 Kt)

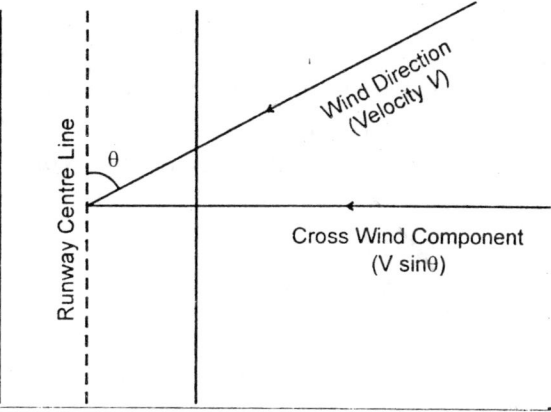

Fig. 5.1. Cross wind component.

An exception to the above for runways 1500 m or over is that when poor runway braking action owing to an insufficient longitudinal co-efficient of friction is experienced with some frequency, a cross wind component not exceeding 24 km/hr should be assumed.

The percent of time in a year during which cross wind component is not exceeded is called the usability factor (wind coverage). The number and orientation of runways at an airport should be such that the usability factor of the airport is not less than 95 per cent for the aeroplanes, the airport is intended to serve.

The maximum mean cross-wind components given above refer to normal circumstances. There are some factors which may require that a reduction of those maximum values be taken into account at a particular airport. These include :

- The wide variations which may exist, in present and future types of aircraft.
- Prevalence and nature of gusts.
- Prevalence and nature of turbulence.
- The availability of a secondary runway.
- The width of runways.
- The runway surface conditions - water, snow, slush and ice on the runway materially reduce the allowable cross-wind component.
- The strength of the wind associated with the limiting cross wind component.

The 95 per cent visibility factor is applicable to all conditions of weather, and it is useful to examine wind speed and direction for different visibility conditions.

5.5.2 Wind data and wind rose diagram

After the maximum permissible cross-wind component is selected, the most desirable direction of runway for wind coverage can be determined by examination of the wind data and its characteristics. The selection of data to be used for the calculation of the usability factor should be based on reliable wind distribution statistics that extend over as long period as possible, preferably of not less than five years. The observations used should be made at least eight times daily and spaced at equal intervals of time, and taken as close to the site of the airport as possible, specially in hilly areas.

The wind data : Direction, duration and velocity are graphically represented by a diagram called 'wind rose diagram'. The velocities are generally grouped into 22.5 degree increments. The weather records contain the percentage of time certain combinations of ceiling and visibility occur (for example ceiling 600 to 260 m, visibility 4.6 to 9 km) and the percentage of time wind of a specific velocity occur from different directions. The directions are relative to true north.

Weather and wind records are generally kept by the government departments, however, if suitable and enough data is not readily available, it is collected through consultation with local residents, installation of wind gauges and the keeping of wind records.

The wind characteristics is examined for following conditions :

(a) The entire wind coverage regardless of visibility or cloud ceiling.

(b) Wind conditions when ceiling is between 60 m and 300 m and/or the visibility is between 0.8 km and 4.8 km.

The first condition represents the entire range of visibility from excellent to very poor. The next condition represents the various degrees of poor visibility requiring the use of instruments for landing. Normally when the visibility approaches 0.8 km and the ceiling is 60 m, there is very little wind, the visibility being reduced by fog, haze, smoke or smog.

5.5.3 Wind rose diagram and orientation of runway

The Table 5.1 shows the typical wind data for all conditions of visibility. From these data wind rose diagram can be plotted as shown in Fig. 5.2. The percentage of wind which correspond to a given direction and velocity range is marked in the proper section of the wind rose.

Table 5.1. Wind data

| Wind direction | Percentage of winds | | | |
	7~24 km/h (4~13 kt)	26~17 km/h (14~20 kt)	39~76 km/h (21~41 kt)	Total
N	4.8	1.3	0.1	6.2
NNE	3.7	0.8	–	4.5
NE	1.5	0.1	–	1.6
ENE	2.3	0.3	–	2.6
E	2.4	0.4	–	2.8
ESE	5.0	1.1	–	6.1
SE	6.4	3.2	0.1	9.7
SSE	7.3	7.7	0.3	15.3
S	4.4	2.2	0.1	6.7
SSW	2.6	0.9	–	3.5
SW	1.6	0.1	–	1.7
WSW	3.1	0.4	–	3.5
W	1.9	0.3	–	2.2
WNW	5.8	2.6	0.2	8.6
NW	4.8	2.4	0.2	7.4
NNW	7.8	4.9	0.3	13.0
Calms – (0~6 km/hr (0~3 kt))				4.6
Total				100.0

The wind velocity less than 6.4 KPh is considered as calm. The total period of time when wind velocity is from 0 to 6.4 KPh is called calm period. The concentric circles on the wind rose diagram are drawn to scale and represent breaks in wind velocity. Optimum runway directions can be determined by the following steps :

(i) Plot the wind rose diagram as shown.

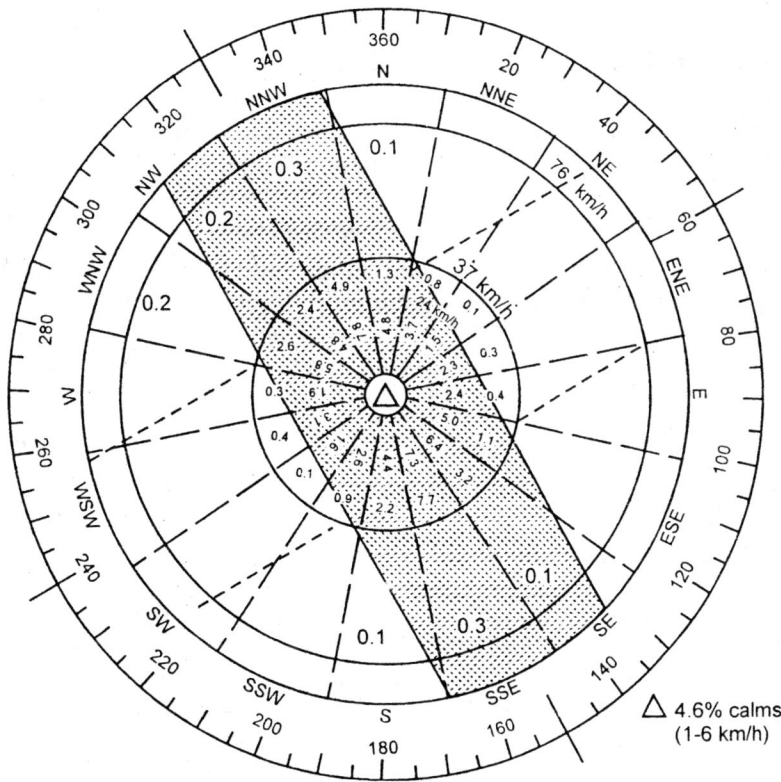

Fig. 5.2. Typical wind rose diagram (ICAO).

(ii) Take a transparent (tracing) strip and draw on it three parallel lines equally spaced, the spacing being equal to allowable cross wind component (say 24 km/hr) to some velocity scale of wind rose diagram. The middle line represents the runway centre line.

(iii) The transparent strip is placed over the wind rose diagram in such a manner that the centre line on the strip passes through the centre of the wind rose diagram.

(iv) Select some point near the middle of centre line and using the centre of wind rose as a pivot point, the transparent strip is rotated until the sum of the percentages included between the outside lines is a maximum.

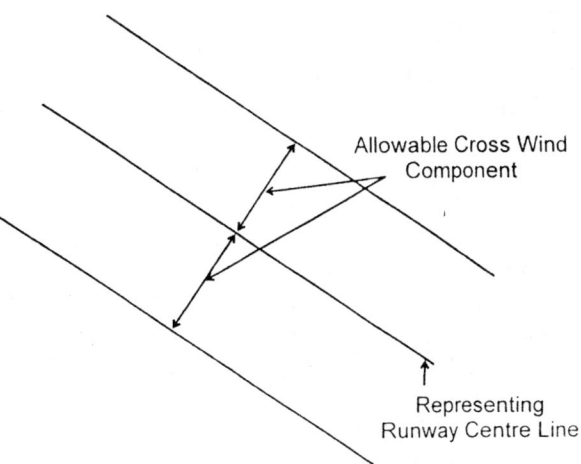

Fig. 5.3. Equally spaced lines on transparent paper.

(v) When one of the outside lines on the transparent strip divides a segment of wind direction, the fractional part is estimated visually to the nearest 0.1 per cent.

(vi) The next step is to read the bearing of the runway on the outer scale of the wind rose where centre line on the transparent strip crosses the directional scale.

(vii) The centre line of the tracing paper shall indicate the desired direction of the runway. As seen in the figure, the direction of runway is 150° to 330° (S 30° E).

(viii) Add visually the areas falling outside the outerlines of the transparent strip, say it is x.

(ix) The wind coverage i.e., the per cent of time during which winds shall be favourable is given by (100 – x). In this case runway orientation S 30° E will permit operations 95 per cent of time with the cross wind component not exceeding 24 km/h.

(x) Rotating the transparent paper, other suitable directions and coverages can be determined and the one most suitable is selected.

It may not be possible in all situations to orient the runway exactly along the direction determined by wind rose, due to other factors such as topography, obstructions or noise etc. suitable adjustments to the directions are then necessary.

5.5.4 Visibility conditions

Often wind characteristics under poor visibility conditions are quite different from those experienced under good visibility conditions. Therefore, a study should be conducted, on the wind conditions occurring with poor visibility and/or low clouds at the airport. Account should be taken of the frequency of occurrence as well as the accompanying wind directions and velocities. A wind rose can be plotted for this condition. From this analysis it can be ascertained whether the runways are capable of accepting aircraft at least 95 per cent of time when restricted visibility conditions prevail. This analysis will also yield information on the percentage of the total time each of the conditions prevail. Fig. 5.4 represents observations of winds in one compass direction only i.e. northeast in this case. To complete the analysis charts of this type would have to be plotted for other compass directions. The circle number 7 means that there were seven observations made when wind was from northeast with velocities varying from 8-15 km/h, ceiling between 0-30 m and visibility between 0-400 m.

5.6 Capacity (type and amount of traffic)

The orientation and location of runways at an airport has a major impact on the airport capacity. Fig. 5.4 shows various runway configuration and their associated capacity. A sufficient number of runways should be provided to meet air traffic i.e. the number of aircrafts, mixture of aircraft types and the mixture of arrivals and departures, to be accommodated in one hour during the busiest periods. The decision as to the total number of runways to be provided should also take into account the airport usability factor and economic considerations.

95 per cent usability with regard to surface cross wind velocity is a minimum. To meet for the remaining 5 percent time (18 days per year) in addition to primary runways, one or more secondary runways may be planned to accommodate aircraft traffic under strong wind conditions. However, as cross-wind runways would require to be used only under high head wind conditions, their length can be considerably shorter than the main runways. The number and orientation of runways at an airport should be such that the usability factor of the aerodrome is not less than 95 percent for the aeroplanes that the aerodrome is intended to serve.

Ceiling Groups in Metres	Velocity Groups in km	NE Wind							Total Observations: 24 081
		Visibility — Metres							Total obs.
		0~400	400~800	800~1200	1200~1600	1600~2400	2400~4800	4800+	
300	1~7	4		1	2	4	14	202	227
	8~15	1	5	1	3	6	17	383	416
	16~23	2			1		5	277	285
	24~47							114	114
	48+								
	Total	7	5	1	6	10	36	976	1042
180 thru 270	1~7		1			1		1	3
	8~15			1	1	1	1	8	12
	16~23				1		3	4	8
	24~47								
	48+								
	Total		1	1	2	2	4	13	23
150	1~7			1				1	2
	8~15						2		2
	16~23								
	24~47								
	48+								
	Total			1			2	1	4
120	1~7			1					1
	8~15				1	1	2		4
	16~23						1		1
	24~47								
	48+								
	Total			1	1	1	3		6
90	1~7	1	1		1	1	1		5
	8~15	1						1	2
	16~23						1	1	2
	24~47								
	48+								
	Total	2	1		1	1	2	2	9
60	1~7					1			4
	8~15	1	1	1			1	1	5
	16~23						1		1
	24~47				1				1
	48+								
	Total	1	1	1	1	1	2	1	8
30	1~7	3							3
	8~15	⑦	1						8
	16~23		3						3
	24~47								
	48+								
	Total	10	4						14
	% by Velocity Groups		1.6~7 km 10	8~15 km 19	16~25 km 12	24~47 km 5	48 km		

Observations to be considered because of ceiling conditions

Observations to be considered because of visibility conditions

Observations to be considered because of ceiling and visibility conditions

Fig. 5.4. Sample of data for analysing wind coverage in a specific direction during periods of restricted visibility (ICAO).

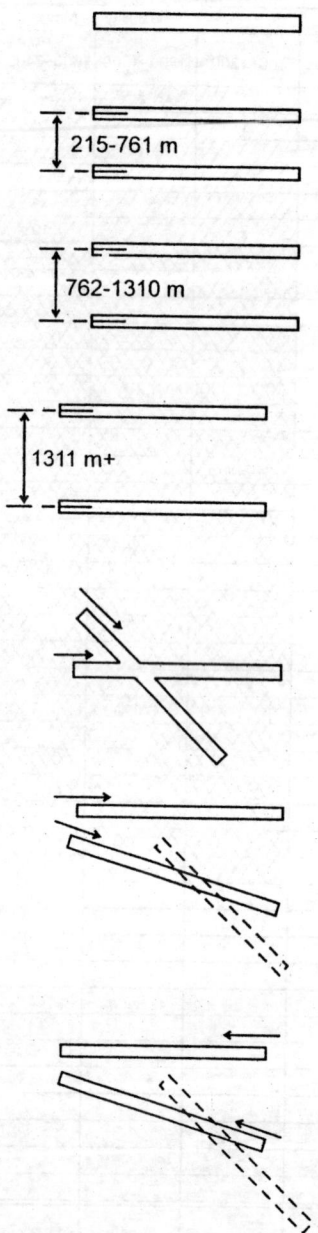

Fig. 5.5. Hourly capacity and annual volume for long-range planning (ICAO).

CHAPTER 6

Airport Obstruction Clearance and Restrictions

6.0 INTRODUCTION

The airspace around airports should be maintained free from obstacles so as to permit the intended aeroplane operations at the airport to be conducted safely and to prevent the airport becoming unusable by the growth of obstacles around the airport. Zoning laws, therefore, established to prevent the possibility of future development of obstacles, close to an airport. The permissible heights and type of structures depends upon the size and type of airport, the aeroplane types which use the airport and the control system adopted. Zoning regulations should be consistant with the requirements, without being unreasonably restrictive.

6.1 Imaginary surfaces (obstacle limitation surfaces)

A series of obstacle limitation surfaces are established to define the limits to which objects may project into the airspace.

As shown in Fig. 6.1 following imaginary surfaces of different types are used. Their characteristics and description is covered in following paragraphs, while table 6.1 shows the various dimensions and slopes based on airports and runways.

1. Take-off climb surface.
2. Approach surface and inner approach surface.
3. Inner horizontal surface.
4. Conical surface.
5. Transitional surface and inner transitional surface.
6. Outer horizontal surface.
7. Balked landing surface.

6.2 Take-off climb surface

Take off climb surface is an inclined plane or other specified surface beyond the end of a runway or clearway, for take-off operations of aeroplanes.

The limits of take-off climb surface shall comprise of :

(a) An inner edge horizontal and perpendicular to the centre line of runway and located either at a

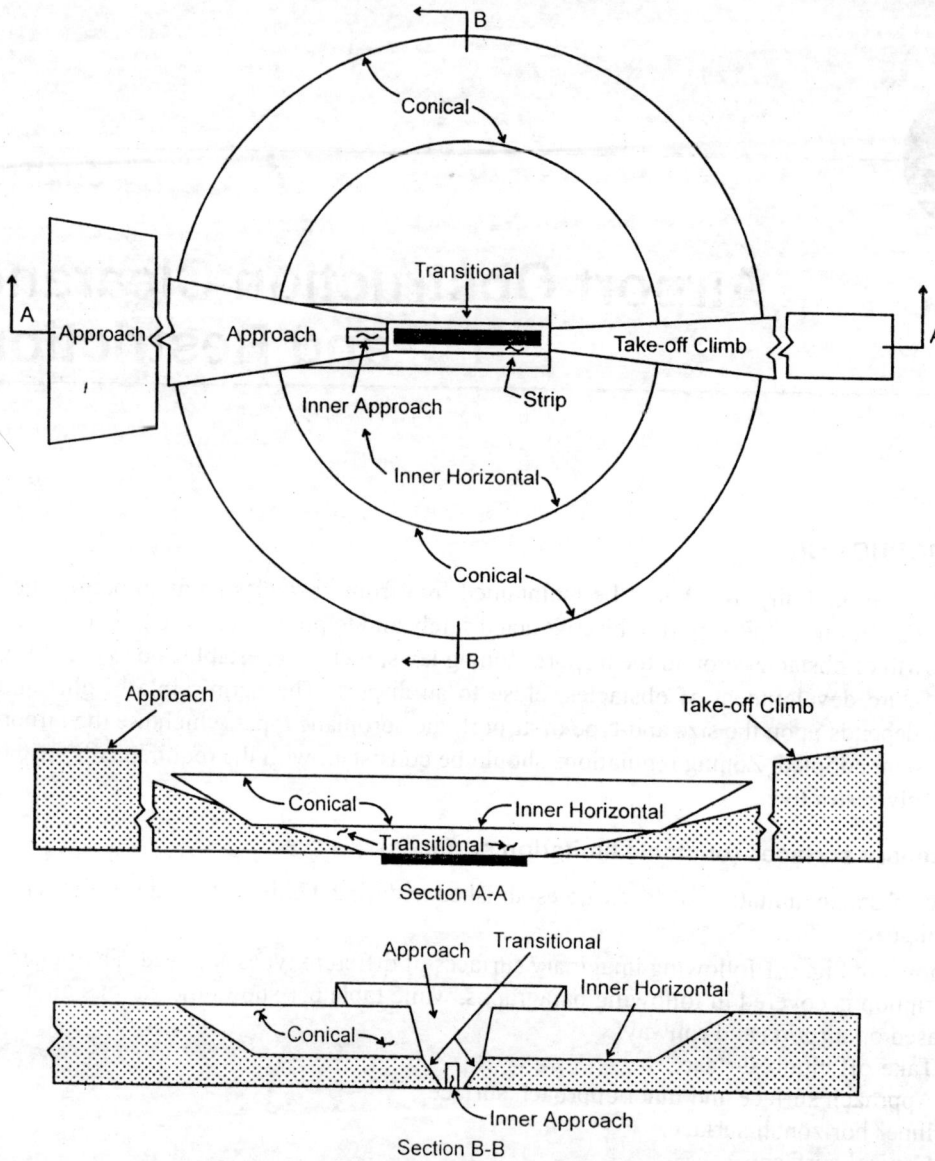

Fig. 6.1. Obstruction limitation surfaces (ICAO).

specified distance beyond the end of the runway or at the end of the clearway when such is provided. Its length equals the specified distance depending upon the category of the airport.

(b) The two sides originating at the ends of the inner edge, diverging uniformly at a specified rate from the take-off track to a specified final width and continuing thereafter at that width for the remainder of the length of the take-off climb surface. The surface extends at a specified slope upwards and outwards.

(c) An outer edge horizontal and perpendicular to the specified take-off track.

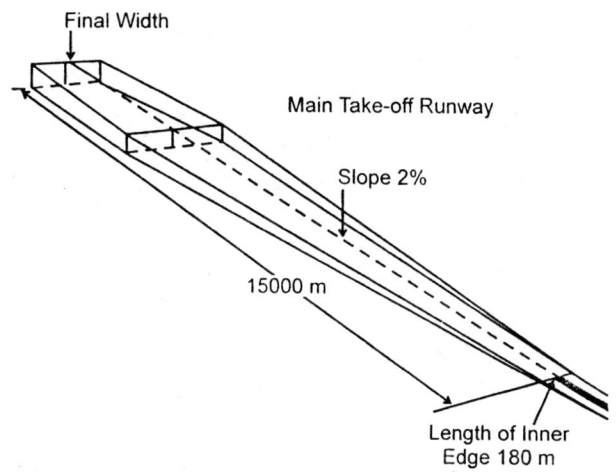

Fig. 6.2. Take-off climb surface (ICAO).

The elevation of the inner edge shall be equal to the highest point on the extended runway centre line between the end of the runway and the inner edge, except that when a clearway is provided the elevation shall be equal to the highest point on the ground on the centre line of the clearway.

In case of a straight take-off flight path, the slope of the take-off climb surface shall be measured in the vertical plane containing the centre line of the runway.

In the case of a take-off flight path involving a turn, the take-off climb surface shall be a complex surface containing the horizontal normals to its centre line, and the slopes of the centre line shall be the same as that for a straight take-off flight path.

6.3 Approach surface

Approach surface is an inclined plane or combination of planes preceeding the threshold. It is longitudinally centered on the extended runway centre line and extending outwards and upwards.

The limits of the approach surface shall comprise of :

(a) An inner edge of specified length, horizontal and perpendicular to the extended centre line of the runway and located at a specified distance before the threshold.

(b) Two sides originating at the ends of the inner edge and diverging uniformly at a specified rate from the extended centre line of the runway,

(c) An outer edge parallel to the inner edge. An isometric view of approach surface is shown in Fig. 6.3.

The elevation of the inner edge shall be equal to the elevation of the mid-point of the threshold. The slope(s) of the approach surface shall be measured in the vertical plane containing the centre line of the runway.

6.3.1 Inner approach surface

The inner approach surface is a rectangular portion of the approach surface immediately preceeding the threshold.

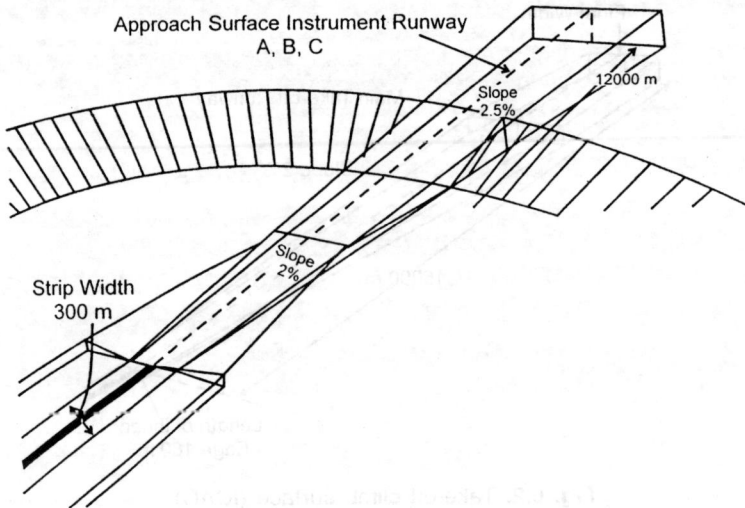

Fig. 6.3. Approach surface.

The limits of the inner approach surface shall comprise of :

(a) An inner edge concident with the location of the inner edge of the approach surface but of its own specified length.

(b) Two sides originating at the ends of the inner edge and extending parallel to the vertical plane containing the centre line of the runway.

(c) An outer edge parallel to inner edge.

6.4 Inner horizontal surface

Inner horizontal surface is a surface located in a horizontal plane above the aerodrome and its environ. The radius or outer limits (the shape of inner horizontal surface need not necessarily be circular) of the inner horizontal surface shall be measured from a reference point or points established for such purposes.

The height of the inner horizontal surface shall be measured above an elevation datum established for such purpose.

6.5 Conical surface

The conical surface is a surface sloping upwards and outwards from the periphery of the inner horizontal surface. The limits of the conical surface shall comprise of :

(a) A lower edge concident with the periphery of the inner horizontal surface.

(b) An upper edge located at a specified height above the inner horizontal surface.

The slope of the conical surface shall be measured in a vertical plane perpendicular to the periphery of the horizontal surface.

6.6 Transitional surface

Transitional surface is a complex surface along the side of the strip and part of the side of approach surface that slopes upwards and outwards to the inner horizontal surface. It controls obstacles like buildings etc.

The limits of a transitional surface shall comprise of :

(a) A lower edge beginning at the intersection of the side of the approach surface with inner horizontal surface and extending down the side of the approach surface to the inner edge of the approach surface and from there along the length of the strip parallel to the runway centreline.

(b) An upper edge located in the plane of the inner horizontal surface.

The elevation of a point on the lower edge shall be :

(a) Along the side of the approach surface, equal to the elevation of the approach surface at that point, and

(b) Along the strip, equal to the elevation of the nearest point on the centre line of the runway or its extension. As a result, the transitional surface along the strip will be curved if the runway profile is curved or a plane if the runway profile is a straight line. The intersection of the transitional surface with the inner horizontal surface will also be a curved or a straight line depending on the runway profile.

The slope of transitional surface shall be measured in a vertical plane at right angles to the centre line of the runway.

6.6.1 Inner transitional surface

Inner transitional surface is a surface similar to the transitional surface, but closer to the runway. It is intended that the inner transitional surface be the controlling obstacle limitation surface for navigation aids, aircraft and other vehicles that must be near the runway and which is not penetrated except for frangibly mounted objects.

The limits of an inner transition surface shall comprise of :

(a) a lower edge beginning at the end of inner approach surface and extending down the side of the inner approach surface, from there along the strip parallel to the runway centre line to the inner edge of the balked landing surface and from there up the side of the balked landing surface to the point where the side intersects the inner horizontal surface, and

(b) an upper edge located in the plane of the inner horizontal surface.

The elevation of a point on the lower edge shall be :

(a) along the side of the inner approach surface and balked landing surface, equal to the elevation of the particular surface at that point, and

(b) along the strip equal to the elevation of the nearest point on the centre line of the runway or its extension. As a result of this the inner transitional surface along the strip will be curved if the runway profile is curved or a plane if the runway profile is a straight line. The intersection of the inner transitional surface with the inner horizontal surface will also be a curved or straight line depending on the runway profile.

The slope of the inner transitional surface shall be measured in a vertical plane at right angles to the centre line of the runway.

6.7 Outer horizontal surface

Outer horizontal surface is circular in plan and it is not necessary for airports with runways of length less than 900 m. The centre of outer horizontal surface is Airport Reference Point (ARP).

Where the longest runway is more than 900 m in length but less than 1500 m the outer horizontal surface shall extend to 9900 m from ARP. For airports where the length of the longest runway is 1500 or more, the outer horizontal surface shall extend 15000 m from ARP. The height of outer horizontal surface is 150 m above ARP elevation.

6.8 Balked landing surface

The balked landing surface is an inclined plane located at a specified distance after the threshold, extending between the inner transitional surface as shown in Fig. 6.4.

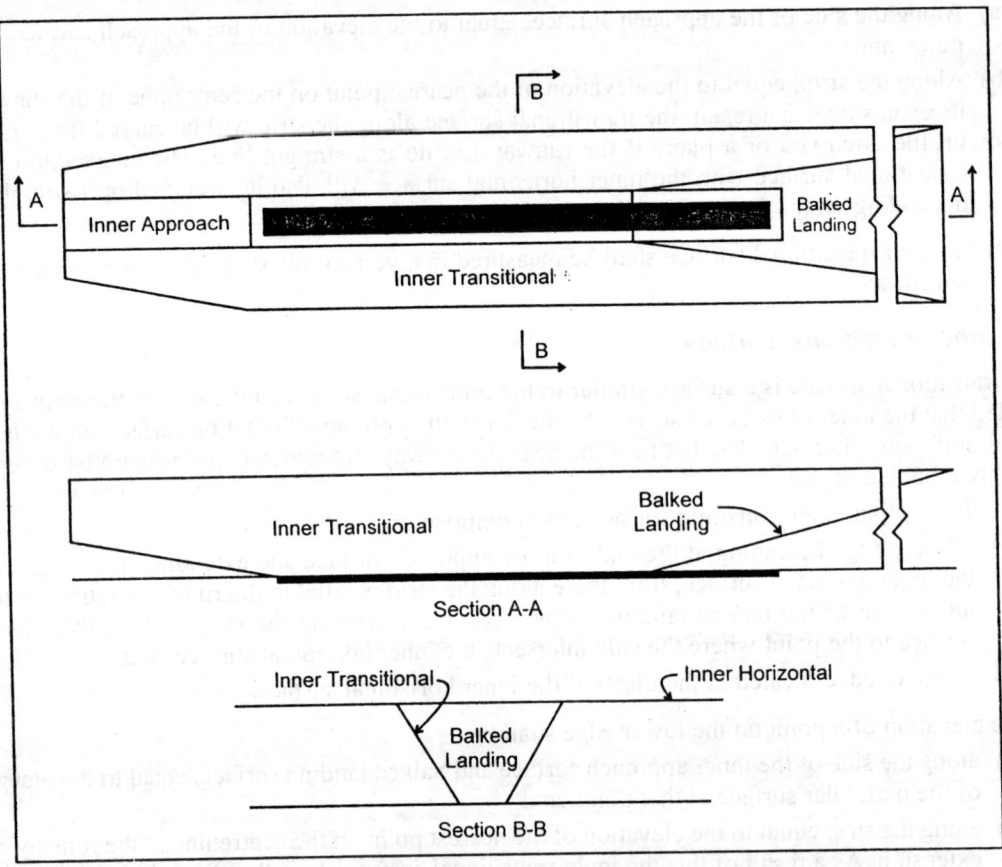

Fig. 6.4. Inner approach, inner transitional and balked landing surfaces (ICAO).

The limits of the balked landing surface shall comprise of :

(a) an inner edge horizontal and perpendicular to the centre line of the runway and located at a specified distance after the threshold.

(b) Two sides originating at the ends of the inner edge and diverging uniformly at a specified rate from the vertical plane containing the centre line of the runway, and

(c) An outer edge parallel to the inner edge and located in the plane of the inner horizontal surface.

The elevation of the inner edge shall be equal to the elevation of the runway centre line at the location of the inner edge.

The slope of the balked landing surface shall be measured in the vertical plane containing the centre line of the runway.

6.9 Requirements of obstruction limitation surfaces

The various types of imaginary surface were described in the previous paragraphs, however, they are specified on the basis of the intended use of a runway that is – take-off or landing, and type of approach. In cases where operations are conducted to or from both directions of a runway, then the functions of certain surfaces may be nullified because of more stringent requirements of another lower surface.

Various requirements of surfaces and obstruction limitations are considered for following types of runway operations :

1. Non-instrumental runways
2. Non-precision approach runways
3. Precision approach runways
4. Take-off runways

6.9.1 Non-instrument runways

The following obstacle limitation surfaces shall be established for a non-instrument runway :

- Conical surface
- Inner horizontal surface
- Approach surface
- Transitional surfaces

The heights and slopes of the surfaces are given in Table 6.1.

- New objects or extension of existing objects should not be permitted above the conical surface or inner horizontal surface, unless an aeronautical study determines that the object would not adversely affect the safety or the regularity of operation of aeroplanes.
- Existing objects above any of these surfaces should be removed, until or unless the object is shielded and aeronautical study determines that safety and regularity of operation of aeroplanes will not be affected.

In considering proposed construction, account should be taken of the possible future development of an instrument runway and consequent requirements for more stringent obstacle limitation surface.

6.9.2 Non-precision approach runway

The following obstacle limitation surfaces shall be established for a non-precision approach runway.

- Conical surface
- Inner horizontal surface
- Approach surface, and
- Transitional surfaces.

The heights and slopes of the surfaces shall not be greater than and their other dimensions not less than those specified in Table 6.1 except in following case of horizontal section of approach surface.

Table 6.1. Dimensions and slopes of obstacle limitation surfaces – approach runways

Approach runways

Surface and dimensions	Non-instrument Code number				Non-precision approach Code number			Precision approach category I Code number		II or III Code number
	1	2	3	4	1,2	3	4	1,2	3,4	3,4
Conical										
Slope	5%	5%	5%	5%	5%	5%	5%	5%	5%	5%
Height	35 m	55 m	75 m	100 m	60 m	75 m	100 m	60 m	100 m	100 m
Inner horizontal										
Height	45 m	45 m	45 m	45 m	45 m	45 m	45 m	45 m	45 m	45 m
Radius	2000 m	2500 m	4000 m	4000 m	3500 m	4000 m	4000 m	3500 m	4000 m	4000 m
Inner approach										
Width	–	–	–	–	–	–	–	90 m	120 m	120 m
Distance from threshold	–	–	–	–	–	–	–	60 m	60 m	60 m
Length	–	–	–	–	–	–	–	900 m	900 m	900 m
Slope								2.5%	2%	2%
Approach										
Length of inner edge	60 m	80 m	150 m	150 m	150 m	300 m	300 m	150 m	300 m	300 m
Distance from threshold	30 m	60 m	60 m	60 m	60 m	60 m	60 m	60 m	60 m	60 m
Divergence (each side)	10%	10%	10%	10%	15%	15%	15%	15%	15%	15%
First section										
Length	1600 m	2500 m	3000 m	3000 m	2500 m	3000 m	3000 m	3000 m	3000 m	3000 m
Slope	5%	4%	3.33%	2.5%	3.33%	2%	2%	2.5%	2%	2%
Second section										
Length	–	–	–	–	–	3600 m[b]	3600 m[b]	12000 m	3600 m[b]	3600 m[b]
Slope	–	–	–	–	–	2.5%	2.5%	3%	2.5%	2.5%
Horizontal section										
Length	–	–	–	–	–	8400 m[b]	8400 m[b]	–	8400 m[b]	8400 m[b]
Total length	–	–	–	–	–	15000 m	15000 m	15000 m	15000 m	15000 m
Transitional										
Slope	20%	20%	14.3%	14.3%	20%	14.3%	14.3%	14.3%	14.3%	14.3%
Inner transitional										
Slope	–	–	–	–	–	–	–	40%	33.3%	33.3%
Balked landing surface										
Length of inner edge	–	–	–	–	–	–	–	90 m	120 m	120 m
Distance from threshold	–	–	–	–	–	–	–	c	1800 m[d]	1800 m[d]
Divergence (each side)	–	–	–	–	–	–	–	10%	10%	10%
Slope	–	–	–	–	–	–	–	4%	3.33%	3.33%

a. All dimensions are measured horizontally unless specified otherwise.
b. Variable length
c. Distance to the end of strip
d. Or end of runway whichever is less.

The approach surface shall be horizontal beyond the point at which the 2.5 per cent slope intersects:

(a) a horizontal plane 150 m above the threshold elevation, or

(b) The horizontal plane passing through the top of any object that governs the obstacle clearance altitude or height.

Whichever between (a) and (b) is higher.

New objects or extensions of existing objects shall not be permitted above an approach surface within 3000 m of the inner edge or above a transitional surface.

Existing objects above any of these surfaces should be removed or shielded as per aeronautical study which determines safety and regularity of operation of aeroplanes.

6.9.3 Precision approach runway

The following obstacle limitation surfaces shall be established for a precision approach runway category I :

- Conical surface,
- Inner horizontal surface,
- Approach surface,
- Transitional surfaces,
- Inner approach surface,
- Inner transitional surfaces, and
- Balked landing surface.

The following obstacle limitation surfaces shall be established for a precision approach runway category II or III.

- Conical surface,
- Inner horizontal surface,
- Approach surface and inner approach surface,
- Transitional surfaces,
- Inner transitional surfaces, and
- Balked landing surface.

The heights and slopes of these surfaces are shown in Table 6.1.

The approach surface shall be horizontal beyond the point at which the 2.5 per cent slope intersects:

(a) a horizontal plane 140 m above the threshold elevation, or

(b) the horizontal passing through the top of any object that governs the obstacle clearance limit.

Fixed objects shall not be permitted above the inner approach surface, the inner transitional surface or the balked landing surface. New objects or extensions of existing objects shall not be permitted above an approach surface or a transitional surface, the conical surface and the inner horizontal surface.

6.9.4 Take-off runways

The following obstacle limitation surface shall be established for a take-off runway :

- Take-off climb surface

The dimensions of the surface shall be not less than the dimensions specified in Table 6.2 except that a lesser length may be adopted for the take-off climb surface where such lesser length would be consistent with procedural measures used to govern the outward flight of aeroplanes.

The operational characteristics of aeroplanes for which the runway is intended should be examined to see if it is desirable to reduce the slope specified in Table 6.2 when critical operating conditions are to be catered to. If the specified slope is reduced, corresponding adjustment in the length of take-off climb surface should be made so as to provide protection to a height of 300 m.

Table 6.2. Dimensions and slopes of obstacle limitation surface "Take-off runway"

Surface and dimensions[a]	Code numbers		
	1	2	3 or 4
Take-off climb			
Length of inner edge	60 m	80 m	180 m
Distance from runway end	30 m	60 m	60 m
Divergence (each side)	10 %	10 %	12.5 %
Final width	380 m	580 m	1200 m
			1800 m[c]
Length	1600 m	2500 m	15000 m
Slope	5 %	4 %	2 %

a All dimensions are measured horizontally unless specified otherwise.
b The take-off climb surface starts at the end of the clearway if the clearway length exceeds the specified distance
c 1800 m when the intended track includes changes of heading greater than 15° for operations in night

When local conditions differ widely from sea level standard atmospheric conditions, it may be advisable for the slopes specified in Table 6.2 to be reduced. The degree of this reduction depends upon the divergence between local conditions and on the performance characteristics and operational requirements of aeroplanes for which the runway is intended.

New objects or extensions of existing object shall not be permitted above a take-off climb surface, new objects should be limited to preserve the existing obstacle free surface or a surface down to a slope of 1.6 per cent (1 : 62.5).

Existing objects that extend above a take-off surface should as far as practicable be removed except when the object is shielded and satisfactory from the aeronautical study to determine that object would not adversely affect the safety or significantly affect the regularity of operation of aeroplanes.

In areas beyond the limit of the obstacle limitation surfaces, at least those objects which extend to a height of 150 m or more above ground elevation should be regarded as obstacles, unless a special aeronautical study indicates that they do not constitute a hazard to aeroplanes.

Other objects which do not project through the approach surface but which would nevertheless adversely affect the optimum siting or performance of visual or non-visual aids should as far as possible be removed.

Fig. 6.5 shows the obstacle limitation surfaces at an airport with two runways, an instrument runway and a non-instrument runway. Both are also take-off runways.

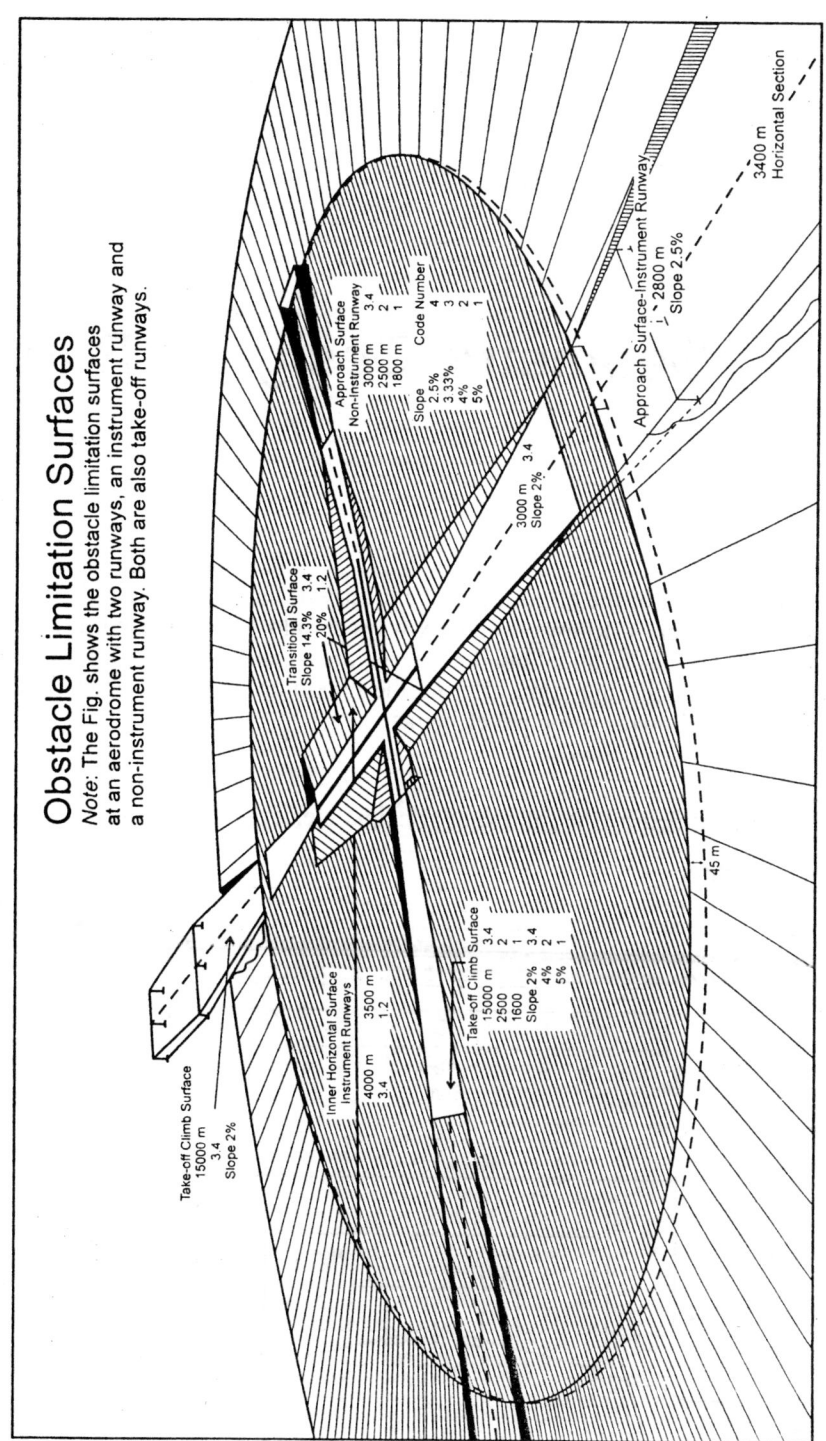

Obstacle Limitation Surfaces

Note: The Fig. shows the obstacle limitation surfaces at an aerodrome with two runways, an instrument runway and a non-instrument runway. Both are also take-off runways.

Fig. 6.5. Obstacle limitation surfaces (ICAO).

CHAPTER 7

Geometric Design of Runway

7.0 INTRODUCTION

The runway is a major element of the airport. It is the clearly defined area of an airport prepared for landing and/or take off of aircraft. Runways and taxiways should be so planned in relation to other major operating elements such as terminal building, cargo areas, aprons, air traffic services and parking etc. to provide an airport configuration offering the maximum overall efficiency. Runways are normally identified by the following principal elements.

7.1 Elements of a runway

Runways are normally identified by the following principal elements shown in Fig. 7.1.

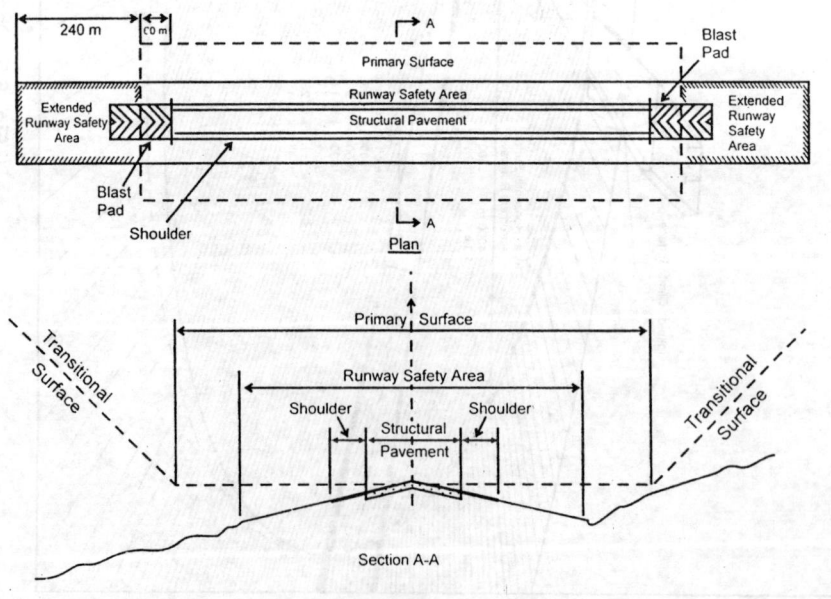

Fig. 7.1. Design elements of a runway (FAA).

(a) *The structural pavement* supports the aeroplane load. It is the paved area whose length and width is designed to ensure a safe operating surface.

(b) *The shoulders* are adjacent to the structural pavement, which are designed to resist erosion due to jet blast and to accommodate maintenance equipment and patrol.

(c) *The runway strip* includes the structural pavement, shoulders and an area that is cleared, drained and graded. This area should be capable of supporting fire, crash, rescue and snow removal equipment under normal conditions as well as providing support for aircraft in case they veer off the pavement.

(d) *The blast pad* is an area designed to prevent erosion of surfaces adjacent to the ends of runways which are subjected to sustained or repeated jet blasts and to minimise excess ground maintenance. This area is either paved or planted with turf.

(e) *The runway end safety area* is an area intended to reduce accidents of aircraft undershooting or over running the runway. It is rectangular symmetrical about the runway centre line, which includes the runway structural pavement, shoulders and stopway (if present). The portion abutting the edges of the shoulder is cleared, drained, graded and usually turfed. Under normal (dry) conditions, it should be capable of accommodating the occasional passage of aircraft without causing a major damage, as well as capable of supporting fire, crash and snow removal equipment.

(f) *A stop way* is an additional length of pavement which extends beyond the end of the runway. The stopway pavement must have adequate strength to support occasional aircraft loadings. It is used for decelerating the aircraft and bringing it to a stop during an aborted take-off (Fig. 7.2).

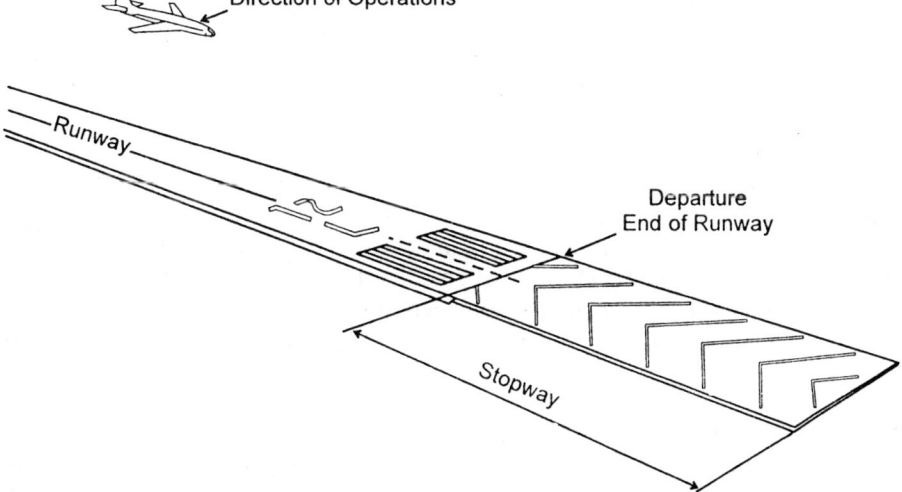

Fig. 7.2. Stopway location.

(g) *A clearway* is an unobstructed, unpaved area also beyond the end of the runway which is controlled by and maintained by the airport authority. No obstruction should exist in the clearway (Fig. 7.3). It is an area beyond the runway not less than 150 m wide, centrally located beyond the extended centreline of the runway. The clearway is expressed in terms of a clearway plane extending from the end of the runway with an upward slope not exceeding 1.25% about which no other object nor any terrain protrudes. However threshold lights may protrude above the plane if their height above the end of the runway is 0.66 m or less and if they are located to each side of the runway.

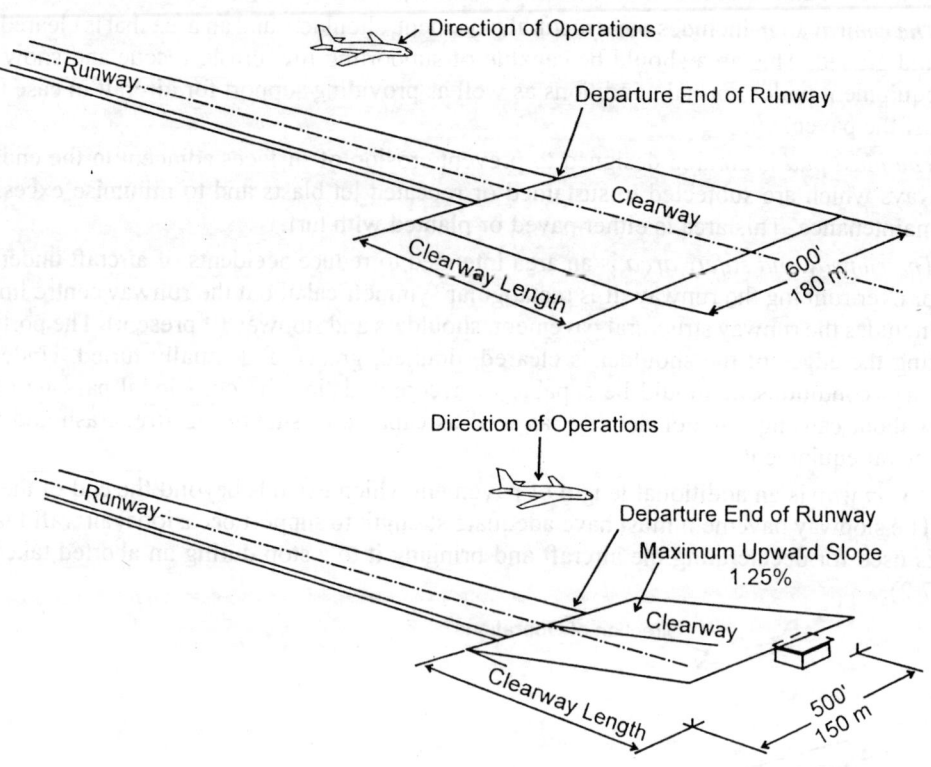

Fig. 7.3. Clearway location.

In view of the vital functions of runways and its above associated elements in providing for safe and efficient aircraft landings and take-offs, it is imperative that in designing these facilities account should be taken of the operational and physical characteristics of the aeroplanes expected to use the runway as well as engineering and economic considerations. Additional runways may be necessary to accommodate operational demands, minimise adverse wind conditions or overcome environmental impacts.

7.2 Airport reference code

For the purpose of identifying standards for various sizes of airports and the functions they serve, reference codes have been developed by I.C.A.O. The intent of the reference code is to provide a simple method for inter-relating the numerous specifications concerning design characteristics so as to provide a series of airport facilities that are suitable for the aeroplanes that are intended to operate at the airport.

The code is composed of two elements as shown in Table 7.1.

1. Element one is a number based on the runway basic length, and
2. Element two is a letter based on the aeroplane wing span and outer main gear wheel span.

Table 7.1. Airport reference code

Code element 1		Code element 2		
Code No.	*Aeroplane reference field length*	*Code letter*	*Wing span*	*Outer main gear wheel span[a]*
1	Less than 800 m	A	Up to but not including 15 m	Up to but not including 4.5 m
2	800 m up to but not including 1200 m	B	15 m up to but not including 24 m	4.5 m up to but not including 6 m
3	1200 m up to but not including 1800 m	C	24 m up to but not including 36 m	6 m up to but not including 9 m
4	1800 m and over	D	36 m up to but not including 52 m	9 m up to but not including 14 m
		E	52 m up to but not including 65 m	8 m up to but not including 14 m

[a] Distance between the outside edges of the main gear wheels.

The code letter or number within an element selected for design purposes is related to the critical aeroplane characteristics for which the facility is provided. While applying the relevant specifications of ICAO, the aeroplane which the airport is intended to serve are first identified and then the two elements of the code.

A classification of representative aeroplanes by the code number and code letter are included in Table 7.2.

The aeroplane reference field length (basic runway length) given by element-1 Code numbers in the table, is defined as the "minimum field length required for take-off at maximum certificated take-off mass, sea-level, standard atmospheric conditions, still air and zero runway slope, as shown in appropriate aeroplane manufacturer manuals.

The code letter for element-2, is determined from above table column 3, by selecting the code letter which corresponds to the greatest wing span or the greatest outer main gear wheel span, whichever gives the more demanding code letter of the aeroplane for which facility is required. For example, if code letter 'C' corresponds to the aeroplane with greatest wing span and code letter 'D' corresponds to the aeroplane with greatest outer main gear wheel span, the code letter selected would be 'D'.

7.3 Location of threshold

The threshold is normally located at the extremity of a runway, if there are no obstacles penetrating above the approach surface. In some cases however, due to local conditions, it may be desirable to displace the threshold permanently.

When studying the location of a threshold, considerations should also be given to the height of the ILS reference datum and the determination of the obstacle clearance limit.

In determining that no obstacle penetrate above the approach surface, account should be taken of mobile objects such as vehicles on road, train etc., at least within that portion of the approach area within 1200 m longitudinally from the threshold and of an overall width of not less than 150 m.

7.3.1 Displaced threshold

If an object extends above the approach surface which cannot be removed considerations should be given to displacing the threshold permanently.

The threshold should ideally be displaced down the runway for the distance necessary to provide that the approach surface is cleared of obstacles.

Table 7.2. Typical aeroplane classification by code number and letter

Aircraft model	Code	Aeroplane reference field length (m)	Wing span (m)	Outer main gear wheel span (m)
1	2	3	4	5
Beechcraft A36	1A	670	10.2	2.9
Neechcraft B55	1A	457	11.5	2.9
Cessna 152	1A	408	10.0	–
Cessna 310	1A	518	11.3	–
Lear Jet 28/29	2A	912	13.4	2.5
Fokker F28-1000	3B	1646	23.6	5.8
F28-2000	3B	1646	23.6	5.8
Convair 580	3C	1341	32.1	8.6
DC-3	3C	1204	28.8	5.8
Fokker F27-600	3C	1670	29.0	7.9
Fokker 100	3C	1840	28.1	6.0
Airbus A300 B2	3D	1676	44.8	10.9
B-727-200	4C	3176	32.9	6.9
B-737-200	4C	2295	28.4	6.4
B-737-300	4C	2749	28.9	6.4
B-737-400	4C	2499	28.9	6.4
DC-9-10	4C	1975	27.2	5.9
DC-9-80	4C	2195	32.9	6.2
Viscount 800	4C	1859	28.6	7.9
Airbus A310	4D	1845	43.9	10.9
Airbus A320-200	4D	2480	33.9	8.7
B-707-100	4D	2454	39.9	7.9
B-707-200	4D	2697	39.9	7.9
B-707-400	4D	3277	44.4	7.9
B-767-200	4D	1981	47.6	10.8
DC-8-61	4D	3048	43.4	7.5
Lockheed L-1011-1	4D	2426	47.3	12.8
B-747-100	4E	3060	59.6	12.4
B-747-400	4E	3383	64.9	12.4

The displacement of the threshold from runway end will cause reduction of the landing distance available and this may be of greater operational significance than penetration of the approach surface by marked and lighted obstacles. A decision to displace the threshold and the extent of such displacement, should be considered to obtain an optimum balance between the considerations of clear approach surfaces and adequate landing distance. In deciding this, factors to be considered are the type of aeroplanes which runway is intended to serve, the limiting visibility and cloud base, position of obstacles in relation to the threshold, and in the case of a precision approach runway, the significance of the obstacles to the determination of the obstacle clearance limit.

The selected position for the threshold should not be such that the obstacle-free surface to the threshold is steeper than 3.3 per cent where code number is 4 or steeper than 5 per cent where the code number is 3.

Where the displacement of threshold is due to an unserviceable runway condition, a cleared and graded area of at least 60 m in length should be available between the unserviceable area and the displaced threshold. Additional distance should also be provided to meet the requirements of the runway end safety area as appropriate.

7.4 Declared distances

The term declared distances indicates the various physical distances available and suitable for the landing and take-off of aeroplanes, using stopway, clearway and displaced threshold. The following four distances associated with a particular runway are :

(a) *Take-off run available (TORA)* : The length of runway declared available and suitable for the ground run of an aeroplane take-off.

(b) *Take-off distance available (TODA)* : The length of take of run available plus the length of clearway if provide.

TODA = TORA + CWY

(c) *Accelerate stop distance available (ASDA)* : The length of the take-off run available plus the length of the stopway, if provided.

ASDA = TORA + SWY

(d) *Landing distance available (LDA)* : The length of runway which is declared available and suitable for the ground run of an aeroplane landing.

The decision to provide a stopway and/or a clearway as an alternative to an increased length of runway depends upon physical characteristics of the area beyond the runway, and the operating requirements of the prospective aeroplane. The runway, stopway and clearway lengths to be provided are determined by aeroplane takeoff performance, but a check should be made of the landing distance required by the aeroplanes, for landing. The length of a clearway, however, cannot exceed half the length of take-off run available. Fig. 7.4 further explains the declared distances.

(a) Where a runway is not provided with a stop way or clearway and the threshold is located at the extremity of the runway, the four declared distances should normally be equal to the length of runway.

(b) Where a runway is provided with a clearway (CWY) then TODA will include the length of clearway.

(c) Where a runway is provided with stopway (SWY) then ASDA will include the length of stopway.

(d) Where a runway has a displaced threshold, then LDA will be reduced by the distance the threshold is displaced. A displaced threshold affects only the LDA for approaches made to that threshold.

(e & f) Illustrates the declared distance when stopway, cleanwayand displace threshold, all the three features exist using the same principles illustrated above.

7.5 Separation of parallel runways

Where parallel runways are provided for simultaneous use under visual meteorological conditions only, the minimum distance between centrelines should be :

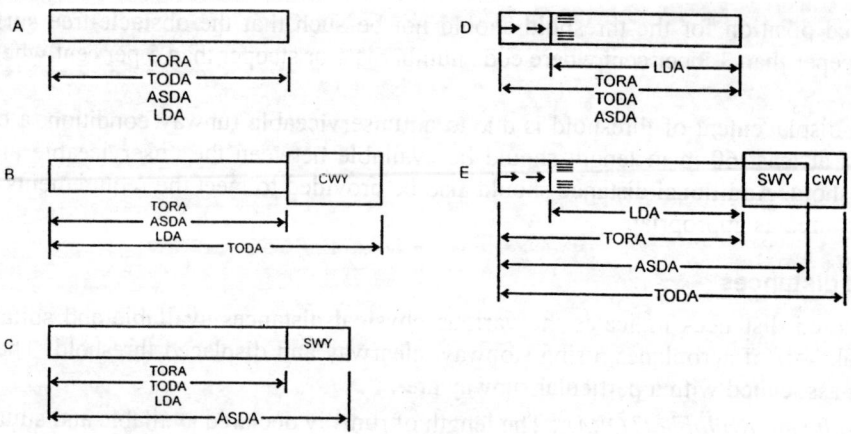

Note: All decleared distances are illustrated for operations from left to right.

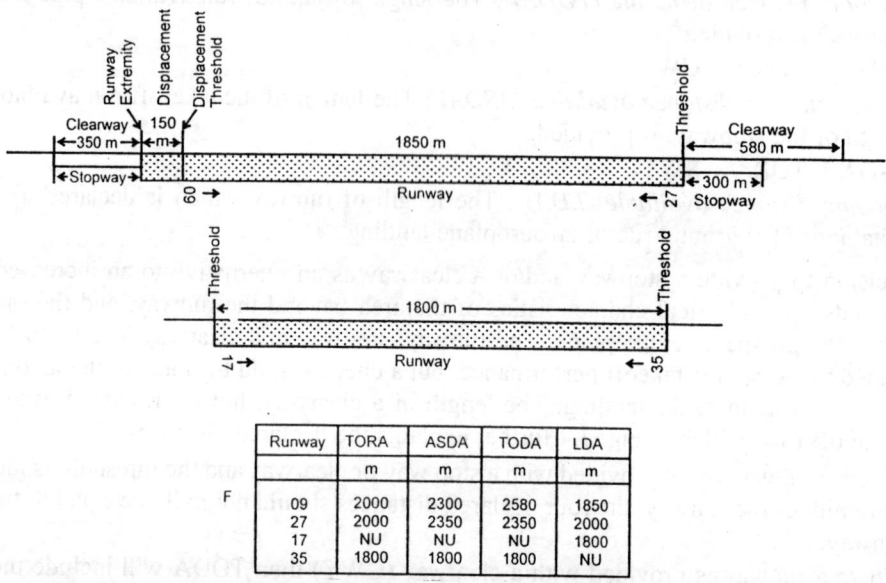

Runway	TORA	ASDA	TODA	LDA
	m	m	m	m
09	2000	2300	2580	1850
27	2000	2350	2350	2000
17	NU	NU	NU	1800
35	1800	1800	1800	NU

Fig. 7.4. Illustration of declared distances (ICAO).

210 m where the code number is 3 or 4,

150 m where the code number is 2,

120 m where the code number is 1.

Where parallel runways are provided for simultaneous operations under instrumental meteorological conditions, the minimum separation distance between their centre lines should be :

1525 m for independent parallel approaches

915 m for dependent parallel approaches

760 m for independent parallel departures

760 m for segregated parallel operations

Except that :

(a) For segregated parallel operations the specified separation distance :

    (i) may be decreased by 30 m for each 150 m that the arrival runway is staggered towards the arriving aeroplane, to a minimum of 300 m, and

    (ii) should be increased by 30 m for each 150 m that the arrival runway is staggered away from the arriving aeroplane.

(b) Lower separation distance than those specified above may be applied if, after aeronautical study it is determined that such lower separation distances would not affect the safety of operations of aircraft.

7.6 Runway length requirement and aeroplane performance

The aeroplane performance operating limitation require a length which is enough to ensure that the aeroplane can, after starting a take-off either :

(a) be brought safely to a stop (engine failure case), or

(b) completes the take-off safely.

The lengths of runway, stopway and clearway should provide for these situations.

For each take-off there is a speed called the decision speed (V_1), below this speed the take-off must be abandoned if an engine fails, while above it the take-off must be completed.

The decision speed is the speed at which the pilot having recognised the failure of the critical engine, decides whether to continue the flight or initiate the application of the retarding device. If the engine failure occurs before the decision speed is reached, the pilot should stop, but if failure occurs later, the pilot should not stop but continue the take-off. The decision speed is either lower or equal to the 'take-off safety speed' (V_2).

'Take-off safety speed' (V_2) is the minimum speed at which the pilot allowed to climb after attaining a height of 10.7 m, to maintain at least the required climb gradient, during a take-off with one engine inoperative.

The lift-off speed (V_{LoF}) is the speed at which the aeroplane first becomes airborne.

The decision speed is not a fixed speed for an aeroplane, but can be selected by the pilot within limits to suit the accelerate-stop and take-off distance available, take off weight of the aeroplane, runway characteristics and ambient atmospheric conditions at the airport.

The most common case is where the decision speed is such that take-off distance required is equal to the accelerate-stop distance required. This length is called the 'balance field length" where 'stopway' and 'clearway' are not provided these distances are both equal to the runway length.

In view of economic considerations, stopway and clearway are to be provided. The minimum runway length and the maximum stopway or clearway length to be provided may be determined as follows:

(a) if a stopway is economically possible, the length to be provided are those for the balanced field length. The runway length is the take-off run required or the landing distance required, whichever is the greater. If the accelerate-stop distance required is greater than the runway length so determined, the excess may be provided as stopway, usually at each end of the runway. In addition, a clearway of the same length as the stopway must also be provided;

(b) if a stopway is not to be provided, the runway length is the landing distance required, or if it is greater, the accelerator stop distance required which corresponds to the lowest practical value of the decision speed. The excess of the take-off distance required over the runway length may be provided as clearway, usually at each end of the runway.

Concept of clearway in certain circumstances can be applied to a situation where take-off distance required for all engines operating exceeds that required for the engine failure case.

The economy of a stopway can be entirely lost if after each usage, it must be regraded and compacted. It, therefore, should be designed to withstand at least a certain number of loadings of the aeroplane without inducing structural damage to the aeroplane.

The following figures 7.5 and 7.6 illustrate the above discussion :

Taking as a schematic illustration Fig. 7.5, the case of an aeroplane standing at the entrance end A of a runway, the pilot starts his take-off, the aeroplane accelerates and approaches the decision speed (V_1) point B. A sudden and complete failure of an engine is assumed to occur and is recognised by the pilot as the decision speed (V_1) is attained. The pilot can either :

- brake until the aeroplane comes to a standstill at point Y (the accelerate-stop distance); or
- continue accelerating until he reaches the rotation speed (V_R), point C, at which time he rotates and becomes airborne at the lift-off speed (V_{LoF}), point D, after which the aeroplane reaches the end of the take-off run, point X, and continues to the 10.7 m (35 ft) height at the end of the take-off distance, point Z.

Fig. 7.6 illustrates a normal all-engines operating case where d_1' and d_3' are similar to d_1 and d_3, respectively, in Fig. 7.5.

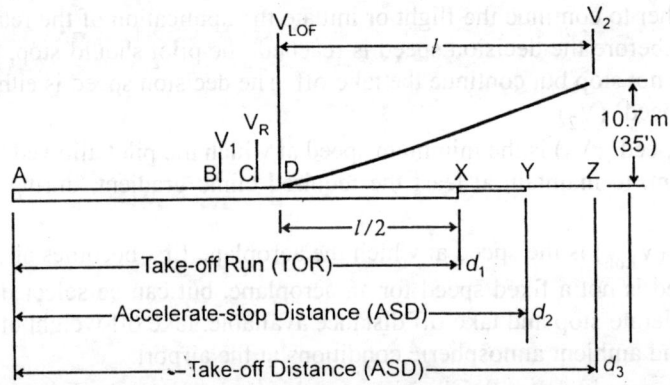

Fig. 7.5. Critical engine inoperative (ICAO).

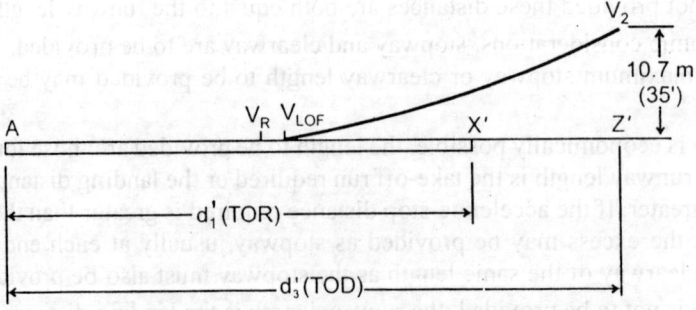

Fig. 7.6. All engines operating (ICAO).

The engine-inoperative take-off and accelerate-stop distance will vary according to the selection of the decision speed (V_1). If the decision speed is reduced, the distance to point B (Fig. 7.5) is reduced, as is the accelerate-stop distance; but the take-off run and take-off distance are increased as a larger part of the take-off manoeuvre is carried out with an engine inoperative.

In certain circumstances, the construction of runways with stopways and clearways may prove to be more advantageous than construction of conventional runways, depending upon local physical and economical conditions, size and clearances of the site, soil characteristics, possibilities of acquiring land, plans for future development, nature and cost of available materials, maintenance costs etc. In particular, the construction of stopway at each end of the runway (since there are normally two directions for take-off) may be an economical first stage in the extension of an existing runway.

7.7 Landing length requirements

Landing length requirements are normally not critical, however, aeroplane landing performance charts should be consulted to check that runway length requirements for take-off provide adequate runway length for landing.

Where the length of runway required for landing is greater than that required for take-off run, this will determine the minimum length of runway required.

7.8 Elements of runway geometric design

The ICAO recommendations for the following elements of the runway geometrics are summarised in the following paragraphs :
   (i) Actual length of runway
   (ii) Runway width
  (iii) Longitudinal slope (gradient)
  (iv) Longitudinal slope changes
   (v) Sight distance
  (vi) Distance between slope changes
 (vii) Transvers slope (gradient)
(viii) Runway surface
  (ix) Runway surface texture
   (x) Runway shoulders
  (xi) Runway strips
 (xii) Runway end safety areas
(xiii) Clearway
(xiv) Stopways

7.8.1. Actual length of runway

The actual runway length to be provided for a primary runway should be adequate to meet operational requirements of the aeroplanes for which the runway is intended, and should not be less than the longest length determined by applying the corrections for the local conditions. Both take off and landing requirements of the selected aeroplane should be considered. The following local conditions need considerations :
   (i) Elevation (above mean sea level)

(ii) Temperature

(iii) Runway slope (gradient)

(iv) Humidity

(v) Runway surface characteristics

The length of the secondary and other runways is determined similar to primary runway except that it need only to be adequate for those aeroplanes which require to use that secondary runway in addition to the other runway or runways to obtain a usability factor of at least 95 per cent.

Correction to runway length

1. Correction for elevation

As the height above sea level increases, the air pressure and density become less. The consequence of these factors upon aeroplane operation is a reduction of lift for a given true airspeed, a reduction of power and a reduction of propellar efficiency. The combined result of these reductions is that it takes longer to attain the forward speed to produce lift. Thus the length required to take off for a given aeroplane becomes progressively longer as the altitude of the airport increases.

Similarly at higher altitudes, true landing speeds are greater and less dense air reduces the drag available to assist in decelerating during landing, thus requiring a longer runway.

"ICAO has recommended that the basic length selected for the runway should be increased at the rate of 7 per cent per 300 m rise in elevation above mean sea level".

Elevation of the airport and runways is measured and expressed in nearest meters.

2. Correction for temperature

The performance of an aeroplane depends upon the temperature. At a given pressure, higher temperature results in lower air density and so has an adverse effect on both piston engined and jet aeroplanes. This effect is greater when taking off. The efficiency of turbo-jet engine depends on the difference between the outside air temperature and the maximum temperature attainable in the combustion chamber. As the outside temperature increases above certain value, depending upon the altitude, the engine efficiency is decreased and, therefore, aeroplane's performance is reduced. The effect of temperature is considerably greater on the take-off distance than landing distance. Take off distances are, therefore, determined taking into account the influence of the ambient air temperature. The higher the temperature, the longer will be the runway required because higher temperatures create lower air densities resulting in lower output of thrust and reduced lift.

"ICAO recommends that the length corrected for elevation, should further be increased at the rate of one per cent for every one degree centigrade by which the 'Airport Reference Temperature' exceeds the temperature in the 'Standard Atmosphere' for the airport elevation."

Airport reference temperature

Airport reference temperature can be calculated by the following formula :

$$\text{Airport reference temperature } T = \left(T_1 + \frac{T_2 - T_1}{3} \right)$$

where T_1 = Monthly mean of the mean daily temperatures for the hottest month of the year i.e., if maximum temperature of a day is t_2 and minimum temperature of same day is t_1

then $T_1 = \dfrac{1}{30} \left(\text{Thirty values of } \dfrac{t_2 + t_1}{2} \text{ for each day of the hottest month added together} \right)$

and $T_2 =$ The monthly mean of the maximum daily temperature for the same hottest month of the year

$= \dfrac{1}{30}$ [Thirty values of t_2 for each day of the hottest month added together]

Standard temperature

Standard temperature at the site can be determined by reducing the standard sea level temperature of 15°C at the rate of 6.5°C per 1000 metre rise in elevation.

Check for elevation and temperature corrections

If the total correction for elevation and temperature exceeds 35 per cent, the required correction should be obtained by means of a specific study, prescribed by ICAO.

3. Correction for gradient

Where the basic length determined by take-off requirement is 900 m or more, that length should be further increased at the rate of 10 per cent for each one per cent of the runway effective gradient.

Effective runway gradient is computed by dividing the difference between the maximum and minimum elevation along the runway centre line, by the length of runway.

$$\text{Effective runway gradient} = \frac{\text{Maximum difference in runway centreline elevation}}{\text{Length of runway}}$$

Example 1

Find out the actual length of the runway required if the length required for landing and take-off under standard atmospheric conditions at sea level are 2000 m and 1700 m respectively. The elevation of the airport site is 200 m above sea level and the airport reference temperature is 22°C. Effective runway gradient may be taken as 0.5%.

Solution

Apply correction for elevation, temperature and gradient to get the actual length of the runway required for both landing as well as take-off.

(a) Correction to runway take-off length

(i) Correction for elevation (C_e)

7% increase for 300 m elevation above M.S.L.

$$C_e = 1700 \times 0.07 \times \frac{200}{300} = 80 \ m$$

Length corrected for elevation = 1700 + 80 m = 1780 m

(ii) Correction for temperature (C_t)

Airport reference temperature = 22°C

Standard temperature = $15^0 - \dfrac{6.5}{1000} \times 200 = 13.7$

Correction for temperature is one per cent increase for one degree difference between Airport reference temperature and standard temperature.

$C_t = 1780 \times (22 - 13.7) \times 0.01 = 148$ m

Take off length corrected for temperature and elevation = 1780 + 148 = 1928 m

Check : Total increase in length = $\dfrac{1928 - 1700}{1700} = 13.5\%$ OK

(iii) Correction of gradient (Cg)

10% increase of the length corrected for elevation and temperature for each 1% of the effective runway gradient.

Cg = 1928 × 0.5 × 0.10 = 96 m

Runway take off length corrected for elevation, temperature and gradient = 1928 + 96 = 2024 m

(b) Correction to runway landing length

(i) Correction for elevation (C_e)

$$C_e = 2000 \times 0.07 \times \dfrac{200}{300} = 93 \text{ m}$$

Corrected length for elevation = 2000 + 93 = 2093 m.

Actual runway length == 2093 m

Example 2

(a) *Data*

1. Runway length required for landing at sea level in standard atmospheric conditions = 2100 m.
2. Runway length required for take-off at a level site at sea level in standard atmospheric conditions = 2500 m.
3. Aerodrome elevation = 150 m.
4. Aerodrome reference temperature = 24°C.
5. Temperature in the standard atmosphere for 150 m = 14.025°C.
6. Runway slope = 0.5%

(b) *Correction to runway take-off length*

1. Runway take-off length corrected for elevation = $\left[2500 \times 0.07 \times \dfrac{150}{300} \right] + 2500 = 2587$ m

2. Runway take-off length corrected for elevation and temperature
$$= [2587 \times (24 - 14.025) \times 0.01] + 2587 = 2845 \text{ m}$$

3. Runway take-off length corrected for elevation, temperature and slope

$$= [2845 \times 0.05 \times 0.10] + 2845 = 2985 \text{ m}$$

(c) *Correction to runway landing length*

Runway landing length corrected for elevation $= \left[2100 \times 0.07 \times \dfrac{150}{300} \right] + 2100 = 2175 \text{ m}$

(d) Actual runway length = 2985 m.

7.8.2 Runway width

As per ICAO, the width of a runway should not be less than the dimensions specified in Table 7.3.

Code number	Code letter				
	A	B	C	D	E
1[a]	18 m	18 m	23 m	–	–
2[a]	23 m	23 m	30 m	–	–
3	30 m	30 m	30 m	45 m	–
4	–	–	45 m	45 m	45 m

a. The width of a precision approach runway should be not less than 30 m where the code number is 1 or 2.

7.8.3 Longitudinal slopes

The slope computed by dividing the difference between the maximum and minimum elevation along the runway centre line by the runway length should not exceed :
- 1 per cent where the code number is 3 or 4; and
- 2 per cent where the code number is 1 or 2.

Along no portion of a runway should the longitudinal slope exceed :
- 1.25 per cent where the code number is 4 except that for the first and last quarter of the length of the runway the longitudinal slope should not exceed 0.8 per cent;
- 1.5 per cent where the code number is 3 except that for the first and last quarter of the length of a precision approach runway category II or III the longitudinal slope should not exceed 0.8 per cent; and
- 2 per cent where the code umber is 1 or 2.

7.8.4 Longitudinal slope changes

Where slope changes cannot be avoided, a slope change between two consecutive slopes should not exceed :
- 1.5 per cent where the code number is 3 or 4; and
- 2 per cent where the code number is 1 or 2.

The transition from one slope to another should be accomplished by a curved surface with a rate of change not exceeding :
- 0.1 per cent per 30 m (minimum radius of curvature of 30000 m) where the code number is 4.

- 0.2 per cent per 30 m (minimum radius of curvature of 15000 m) where the code number is 3; and
- 0.4 per cent per 30 m (minimum radius of curvature of 7500 m) where the code number is 1 or 2.

7.8.5 Sight-distance

Where slope changes cannot be avoided, they should be such that there will be an unobstructed line of sight from :

- Any point 3 m above a runway to all other points 3 m above the runway within a distance of at least half the length of the runway where the code letter is C, D or E;
- Any point 2 m above a runway to all other points 2 m above the runway within a distance of at least half the length of the runway where the code letter is B; and
- Any pooint 1.5 m above a runway to all other points 1.5 m above the runway within a distance of at least half the length of the runway where the code letter is A.

7.8.6 Distance between slope changes

Undulations or appreciable changes in slopes located close together along a runway should be avoided. The distance between the points of intersection of two successive curves should not be less than :

(a) The sum of the absolute numerical values of the corresponding slope changes multiplied by the appropriate value as follows :

30000 m where the code number is 4;

15000 m where the code number is 3; and

5000 m where the code number is 1 or 2; or

(b) 45 m;

whichever is greater.

7.8.7 Transverse slope

To promote the most rapid drainage of water, the runway surface should, practicable, be cambered except where a single crossfall from high to low in the direction of the wind most frequently associated with rain would ensure rapid drain. The transverse slope should ideally be :

1.5 per cent where the code letter is C, D or E; and

2 per cent where the code letter is A or B;

but in any event should not exceed 1.5 per cent or 2 per cent, as applicable, nor be less than 1 per cent except at runway or taxiway intersections where flatter slopes may be necessary. For a cambered surface the transverse slope on each side of the centre line should be symmetrical. On wet runways with cross-wind conditions the problem of aquaplanning from poor drainage is apt to be accentuated.

The transverse slope should be substantially the same throughout the length of a runway except at an intersection with another runway or a taxiway where an even transition should be provided taking account of the need for adequate drainage.

Combined slopes

When a runway is planned that will combine the extreme values for the longitudinal slopes and changes in slope combined with extreme transverse slopes, a study should be made to ensure that the resulting surface profile will not hamper the operation of aeroplanes.

7.8.8 Runway surface

The surface of runway should be without irregularities that could result in loss in braking action or adversely effect the take-off or landing of an aeroplane, by causing excessive bouncing, pitching, vibration or other difficulties in the control of an aeroplane. Adequate smoothness of the surface should be maintained.

7.8.9 Runway surface texture

The surface of a paved runway should be so constructed as to provide good friction characteristic when runway is wet. Proper engineered and maintained Bituminous or Portland cement concrete surface meet these criteria. Runway surface texture is improved by groovings or scorings, the grooves and scorings should be either perpendicular to the runway centre line or parallel to non-perpendicular transverse joints, where applicable.

The average surface texture depth of a new surface should not be less than 1.0 mm. This normally will require some form of special treatment.

7.9.10 Runway shoulders

Runway shoulders should be provided for a runway where code letter is D or E and the runway width is less than 60 m. The shoulder of a runway or stopway should be prepared or constructed so as to minimize any hazard to an aeroplane running off the runway or stopway.

Runway shoulders depending upon the local soil, and the weight of the aeroplane the runway is intended to serve may be stabilised, paved or surfaced.

When designing the shoulders, attention should be paid to prevent ingestion of stones or other objects by turbine engines. Good visual contrast in the surfacing of the runway and shoulder should be provided. Runway side stripe markings help in separating the two paved surfaces.

The runway shoulders should extend symmetrically on each side of the runway so that overall width of the runway and its shoulder is not less than 60 m. It should be able to support the aeroplane without any structural damage to the aeroplanes and vehicles.

The surface of the shoulder that abuts the runway should be flush with the surface of the runway and its transverse slope should not exceed 2.5 per cent.

7.8.11 Runway strips

Runway strip consists of a runway and any associated stopway.

(a) Length of runway strip

A strip should extend, before the threshold and beyond the end of the runway or stopway, for a distance of at least :

    60 m where the code number is 2, 3 or 4;

    60 m where the code number is 1 and the runway is an instrument one; and

    30 m where the code number is 1 and the runway is a non-instrument one.

(b) Width of runway strip

A strip including a precision approach runway shall, wherever practicable, extend laterally for a distance of at least :

150 m where the code number is 3 or 4; and

75 m where the code number is 1 or 2;

on each side of the centre line of the runway and its extended centre line throughout the length of the strip.

A strip including a non-precision approach runway should extend laterally to a distance of at least:

150 m where the code number is 3 or 4; and

75 m where the code number is 1 or 2;

on each side of the centre line of the runway and its extended centre line throughout the length of the strip.

A strip including a non-instrument runway should extend, on each side of the centre line of the runway and its extended centre line throughout the length of the strip, for a distance of at least :

75 m where the code number is 3 or 4;

40 m where the code number is 2; and

30 m where the code number is 1.

(c) Objects on runway strip

Any equipment or object or installation required for air navigation which have to be located on the runway strip should be of minimum weight and height, frangibly designed and mounted, and located in such a manner as to reduce the hazard to aircraft to a minimum. No fixed objects, other than visual aids required for air navigation should be permitted on a runway strip.

    (a) Within 60 m of runway centre line of a precision approach runway where code number is 3 or 4.

    (b) Within 45 m of runway centreline of a precision approach runway when code number is 1 or 2.

(d) Grading of runway strip

The portion of a strip of an instrument runway within a distance of at least :

75 m where the code number is 3 or 4; and

40 m where the code number is 1 or 2;

from the centre line of the runway and its extended centre line should provide a graded area for aeroplanes which the runway is intended to serve in the event of an aeroplane running off the runway.

For a precision approach runway it may be desirable to adopt a greater width where the code number is 3 or 4. Fig. 7.7 shows the shape and dimensions of a wider strip that may be considered for such a runway. This strip has been designed using information on aircraft running off runways. The portion to be graded extends to a distance of 105 m from the centre line except that the distance is gradually reduced to 75 m from the centre line at both ends of the strip, for a length of 150 m from the runway end.

That portion of a strip of a non-instrument runway within a distance of at least :

75 m where the code number is 3 or 4;

40 m where the code number is 2; and

30 m where the code number is 1;

from the centre line of the runway and its extended centre line should provide a graded area for aeroplanes which the runway is intended to serve in the event of an aeroplane running off the runway.

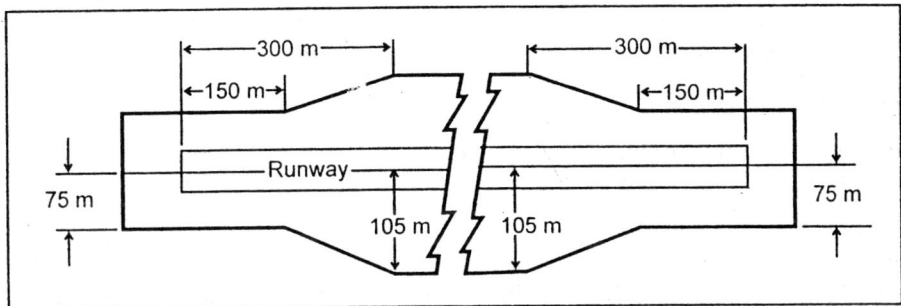

Fig. 7.7. Graded portion of a strip including a precision approach runway where the code number is 3 or 4 (ICAO).

The surface of that portion of a strip that abuts a runway, shoulder or stopway shall be flush with the surface of the runway, shoulder or stopway.

That portion of a strip to at least 30 m before a threshold should be prepared against blast erosion in order to protect a landing aeroplane from the danger of an exposed edge.

(e) Longitudinal slopes on runway strip

A longitudinal slope along that portion of a strip to be graded should not exceed :

1.5 per cent where the code number is 4;

1.75 per cent where the code number is 3; and

2 per cent where the code number is 1 or 2.

(f) Longitudinal slope changes on runway strip

Slope changes on that portion of a strip to be graded should be as gradual as practicable and abrupt changes or sudden reversals of slopes avoided.

In order to accommodate aeroplanes making auto-coupled approaches and automatic landings (irrespective of weather conditions) it is desirable that slope changes before the threshold of a precision approach runway should be avoided or kept to a minimum on that portion of the strip within a distance of at least 30 m on each side of the extended centre line of the runway. This is desirable because these aeroplanes are equipped with a radio altimeter for final height and flare guidance and when the aeroplane is above the terrain immediately prior to the threshold the radio altimeter will begin to provide information to the automatic pilot for auto-flare. Where slope changes cannot be avoided on this portion, the rate of change between two consecutive slopes should not exceed 2 per cent per 30 m.

(g) Transverse slopes on runway strips

Transverse slopes on that portion of a strip to be graded should be adequate to prevent the accumulation of water on the surface but should not exceed :

2.5 per cent where the code number is 3 or 4; and

3 per cent where the code number is 1 or 2;

except that to facilitate drainage the slope for the first 3 m outward from the runway, shoulder or stopway edge should be negative as measured in the direction away from the runway and may be as great as 5 per cent.

The transverse slopes of any portion of a strip beyond that to be graded should not exceed an upward slope of 5 per cent as measured in the direction away from the runway.

(h) Strength of runway strips

That portion of a strip of an instrument runway within a distance of at least :

    75 m where the code number is 3 or 4; and

    40 m where the code number is 1 or 2;

from the centre line of the runway and its extended centre line should be so prepared or constructed as to minimize hazards arising from differences in load bearing capacity to aeroplanes which the runway is inteded to serve in the event of an aeroplane running off the runway.

    That portion of strip containing a non-instrument runway within a distance of at least :

    75 m where the code number is 3 or 4;

    40 m where the code number is 2; and

    30 m where the code number is 1;

from the centre line of the runway and its extended centre line should be so prepared or constructed as to minimize hazards arising from differences in load bearing capacity to aeroplanes which the runway is intended to serve in the event of an aeroplane running off the runway.

7.8.12 Runway end safety areas

A runway end safety area should be provided at each end of a runway strip where the :

    Code number is 3 or 4; and

    Code number is 1 or 2 and the ruunway is an instrument one.

(a) Length of runway and safety area

A runway end safety area should extend from the end of a runway strip for as great a distance as practicable, but at least 90 m.

    When deciding the length to be provided, consideration should be given to providing an area long enough to contain overruns and undershoots resulting from a reasonably probable continuation of adverse operational factors on a precision approach runway, the ILS localiser is normally the first upstanding obstacle and the runway end safety area should extend up to this facility. In other circumstances and on a non-precision approach or non-instrument runway, the first upstanding obstacle may be a road, a railroad or other manmade or natural feature. In such circumstances, the runway end safety area should extend as far as the obstacle.

(b) Width of runway end safety area

The width of a runway end safety area should be at least twice that of the associated runway.

(c) Objects on runway end safety area

An object other than equipment or installation required for air navigation purposes, situated on a runway end safety area which may endanger aeroplanes should be regarded as an obstacle and should, as far as practicable, be removed. Any equipment or installation required for air navigation purposes which must be located on the runway end safety area should be of minimum mass and height, frangibly designed and mounted, and sited in such a manner as to reduce the hazard to aircraft to a minimum.

(d) Clearing and grading of runway end safety area

A runway safety area should provide a cleared and graded area for aeroplanes which the runway is inteded to serve in the event of an aeroplane undershooting or overrunning the runway. The surface of the ground in the runway end safety area does not need to be prepared to the same quality as the runway strip.

(e) Combined slopes of runway end safety area

The slopes of a runway end safety area should be such that no part of the runway end safety area penetrates the approach or take-off climb surface.

(f) Longitudinal slopes of runway end safety area

The longitudinal slopes of a runway end safety area should not exceed a downward slope of 5 per cent. Longitudinal slope changes should be as gradual as practicable and abrupt changes or sudden reversible of slopes avoided.

In order to accommodate aeroplanes making auto-coupled approaches and automatic landing (irrespective of weather conditions) it is desirable that slope changes be avoided or kept to a minimum on an area symmetrical about the extended runway centre line approximately 60 m wide, and 300 m long before the threshold of a precision approach runway. This is desirable because these aeroplanes are equipped with a radio altimeter for final height and flare guidance and when the aeroplane is above the terrain immediately prior to the threshold the radio altimeter wil begin to provide information to the automatic pilot for auto-flare. Where slope changes cannot be avoided, the rate of change between two consecutive slopes should not exceed 2 per cent per 30 m.

(g) Transverse slopes of runway end safety area

The transverse slopes of a runway end safety area should not exceed an upward or downward slope of 5 per cent. Transitions between differing slopes should be as gradual as practicable.

(h) Strength of runway end safety areas

A runway end safety area should be so prepared or constructed as to reduce the risk of damage to an aeroplane undershooting or overrunning the runway and facilitate the movement of rescue and fire fighting vehicles.

7.8.13 Clearways

The origin of a clearway should be at the end of take-off run available.

(a) Length of clearways

The length of a clearway should not exceed half the length of the take-off run available.

(b) Width of clearways

A clearway should extend laterally to a distance of at least 75 m on each side of the extended centre line of the runway.

(c) Slopes of clearways

The ground in a clearway should not project above a plane having an upward slope of 1.25 per cent, the lower limit of this plane being a horizontal line which :

(a) is perpendicular to the vertical plane containing the runway centre line; and

(b) passes through a point located on the runway centre line at the end of the take-off run available.

Because of transverse or longitudinal slopes on a runway, shoulder or strip, in certain cases the lower limit of the clearway plane specified above may be below the corresponding elevation of the runway, shoulder or strip. It is not intended that these surfaces be graded to conform with the lower limit of the clearway plane nor is it intended that terrain or objects which are above the clearway plane beyond the end of the strip but below the level of the strip be removed unless it is considered they may endanger aeroplanes.

Abrupt upward changes in slope should be avoided when the slope on the ground in a clearway is relatively small or when the mean slope is upward. In such situations, in that portion of the clearway within a distance of 22.5 m on each side of the extended centre line, the slopes, slope changes and the transition from runway to clearway should generally conform with those of the runway with which the clearway is associated except that isolated depressions such as ditches running across the clearway may be permitted.

(d) Objects on clearways

An object other than equipment or installation required for air navigation purposes situated on a clearway which may endanger aeroplanes in the air should be regarded as an obstacle and should be removed. Any equipment or installation required for air navigation purposes which must be located on the clearway should be of minimum mass and height, frangibly desiged and mounted, and sited in such a manner as to reduce the hazard to aircraft to a minimum.

7.8.14 Stopways

A stopway should have the same width as the runway with which it is associated.

(a) Slopes on stopways

Slopes and changes in slope on a stopway and the transition from a runway to a stopway should comply with the specifications for the runway (as covered earlier) with which the stopway is associated except that :

(a) The limitation of a 0.8 per cent slope for the first and last quarter length of a runway need not be applied to the stopway; and

(b) At the junction of the stopway and runway and along the stopway the maximum rate of slope change may be 0.3 per cent per 30 m (minimum radius of curvature of 10,000 m) for a runway where code number is 3 or 4.

(b) Strength of stopways

A stopway should be prepared or constructed so as to be capable, in the event of an abandoned take-off of supporting the aeroplane which the stopway is intended to serve without inducing structural damage to the aeroplane.

(c) Surface of stopways

The surface of a paved stopway should be so constructed to provide a good coefficient of friction when the stopway is wet. The friction characteristics of an unpaved stopway should not be substantially less than that of the runway with which the stopway is associated.

CHAPTER 8

Geometric Design of Taxiway

8.0 INTRODUCTION

Taxiways form the transition from the runway to the terminal and other elements of the airport. Runways, passenger and cargo terminals, aircraft storage and servicing areas are all linked by the taxiway system. The components of the taxiway system serve as the transitional media between the aerodrome functions, necessary for optimum utilisation, maximum capacity and efficiency of an airport.

The taxiway system should be designed to minimize restriction to aeroplane movement to and from the runway and apron areas. A properly designed taxiway should be capable of maintaining a smooth, continuous flow of aeroplaes ground traffic at the maximum practical speed with minimum acceleration or decelerations. This requirement will ensure that the taxiway system will operate at the highest level of safety and efficiency.

For a given airport, the taxiway system should be able to accommodate, the demand for aircraft arrivals and departures on the runway system, without significant delay. The taxiway system capacity should be enough to avoid becoming the limiting aerodrome capacity factor. On very busy airports the taxiway system should allow aeroplanes to exist the runway as soon as possible after landing and to enter the runway just before take-off. This will enable aeroplane movement on the runway to be maintained at the minimum separation distance.

8.1 PLANNING OF TAXIWAYS

Runways and taxiways are least flexible elements of an airport and therefore, need considerations at the beginning of airport planning and development, in view of future demands of traffic, nature of traffic, type of aeroplanes, and other factors affecting the layout and dimensions of the runway and taxiway systems.

In planning the layout of the taxiway system the following factors should be considered :

- Taxiway routes should connect the various elements of an airport, by the shortest distances thus minimising taxiing time and cost.
- Taxiway routes should be as simple as possible to avoid the need for complicated instructions and confusion to pilots.

- Taxiways should be as straight as possible where changes in direction are necessary. Curves of sufficient radii, as well as fillets or extra taxiway width should be provided to permit taxiing at the maximum practical speed.
- Taxiway crossing the runways and other taxiways should be avoided if possible, in the interest of safety and to reduce potential for delays in taxiing.
- Taxiway routings should have one-way operations as far as possible to minimise aeroplane conflicts, delays and interference.
- Movement on taxiway system should be analysed for each configuration under which runway(s) will be used.
- Taxiway system should be planned for the phased development so that future additions are easily possible.
- Taxiway route should avoid areas where public could have easy access. This is for security of the taxiing aeroplane from sabotage.
- Taxiway layout should be planned to avoid interference with navigational aids by taxiing aeroplane or ground vehicles using the taxiway.
- All sections of the taxiway system should be visible from the aircraft control tower.
- The effects of jet blast on areas adjacent to the taxiways should be mitigated by stabilising loose soil and erecting blast fences.
- Entrance and exit taxiways serving a specific runway should be of sufficient number to accommodate the take-offs and landings. Additional exits and entrances may be planned for future as per expected growth of runway utilisation.
- Exit taxiways should be designed for hight turn-off speeds without restrictions to reduce the runway occupancy time of aeroplanes, and allowing another operation to take place on the runway as soon as possible.
- An exit taxiway should be at right angle to runway or at an acute angle. This will allow aeroplanes to decelerate little, thus reducing the time acquired on the runway and increasing the runway capacity.
- A single runway entrance at each end of the runway is generally sufficient to accommodate the demand for take-offs. However, it traffic volumes demand, use of by passes, holding bays or multiple runway entrances should be considered.

An elevation of alternative taxiway systems must be made to judge their operating efficiency in combination with runway and apron layouts, it is intended to serve, operating costs can be minimized through a comparison of alternative taxiway systems.

8.2 TAXIWAYS ON APRONS

Taxiways located on aprons are of two types :

(i) Apron taxiway is a taxiway located on an apron and intended either to provide a thorough taxi route across the apron or to gain access to an aircraft stand taxi lane; and

(ii) 'Aeroplane stand taxi lane' is a portion of an apron designated as a taxiway and intended to provide access to aircraft stands only (Fig. 8.1).

The requirements for apron taxiways are same as for any other type of taxiway. For 'aircraft stand taxi lanes' the transverse slope of taxi lane is governed by the apron slope requirements.

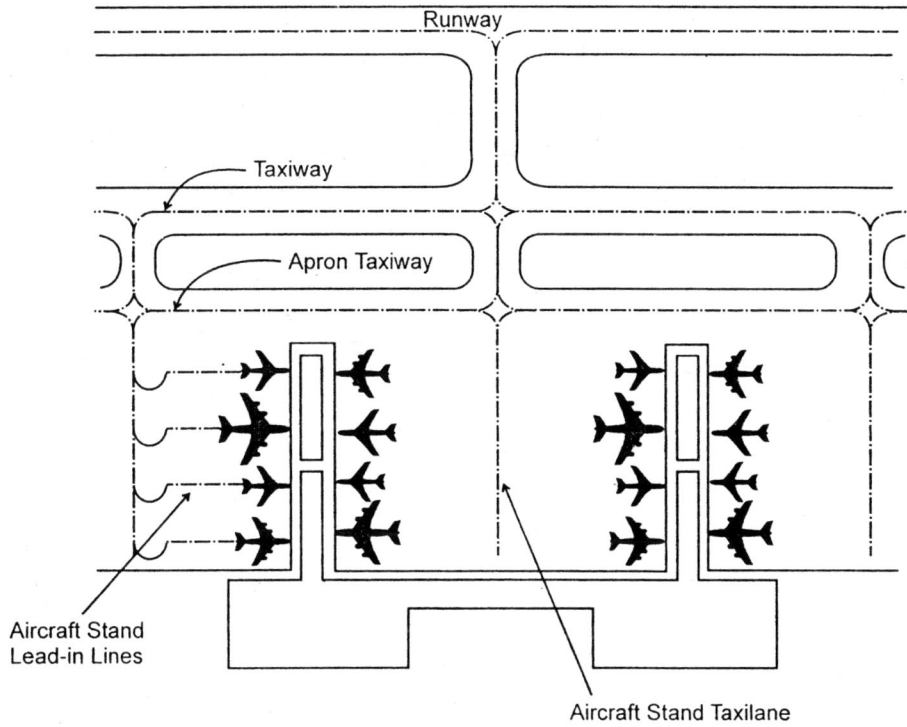

Fig. 8.1. Taxiways on aprons (ICAO).

8.3 GEOMETRIC DESIGN OF TAXIWAYS

Design criteria for taxiways are less stringent than those for runways since aeroplane speeds on taxiways are much lower than those on runways. Fig. 8.2 shows typical cross-section of a taxiway.

The ICAO recommendations for the following elements of taxiway geometrics are summarised in the following paragraphs :

- (i) Taxiway width
- (ii) Taxiway curves
- (iii) Taxiway junctions and intersections
- (iv) Taxiway minimum separation distance
    - – Separation distance between taxiways, and taxiways or objects
    - – Separation distance between taxiways and runways
    - – Separating distance between parallel taxiways
- (v) Slopes on taxiways
    - – Longitudinal slopes
    - – Longitudinal slope changes
    - – Transverse slopes
- (vi) Sight distance
- (vii) Surface of taxiways
- (viii) Strength of taxiways

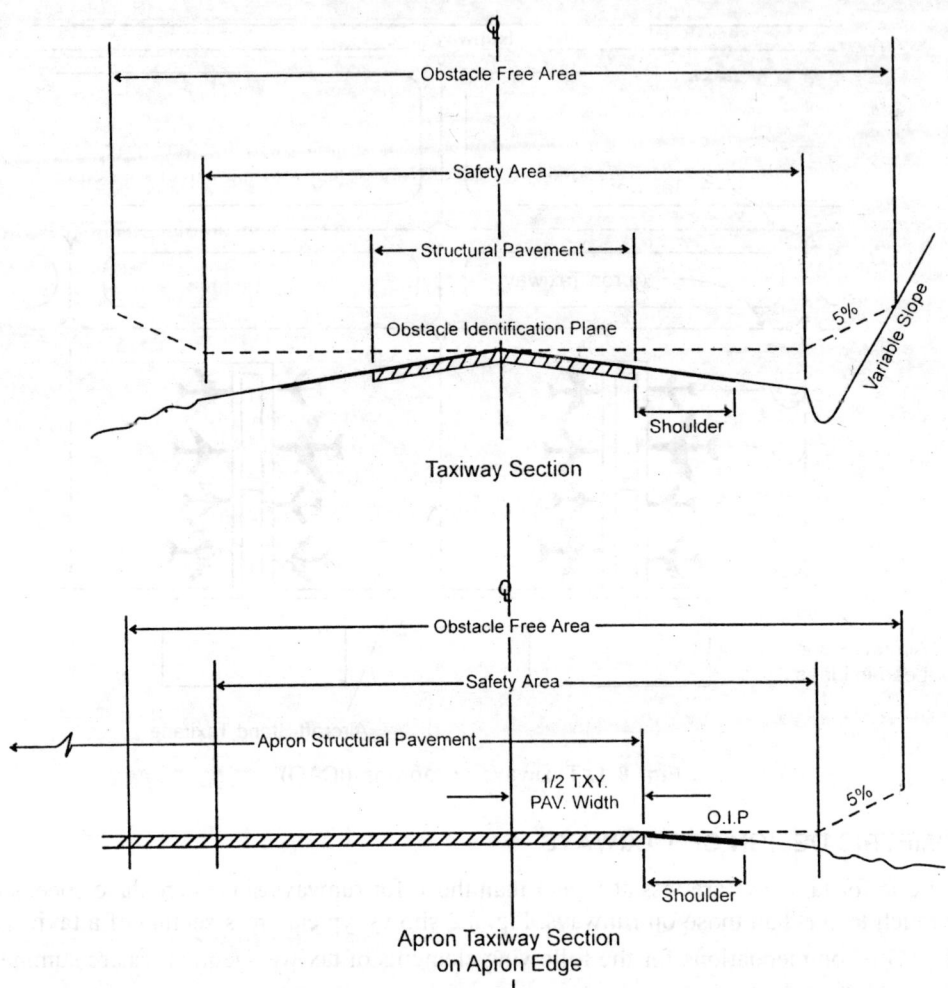

Fig. 8.2. Typical cross-section of taxiways (ICAO).

8.3.1 Taxiway width

The width of taxiways are lower than runways, because of low speed and ease of manoeverability. The values of minimum taxiway width recommended by ICAO are based on adding clearance distance from wheel to pavement edge to the main aircraft track. A straight portion of taxiway should have a width of not less than that given below.

Table 8.1 shows the varied widths of taxiways as recommended by ICAO, based on the relationship:

$$W_T = T_m + 2C$$

where W_T = the taxiway width

T_m = outer main gear wheel span

and C = clearance between the outer main gear wheel and the taxiway edge

Code letter	Taxiway width
A	7.5 m
B	10.5 m
C	15 m if the taxiway is intended to be used by aeroplanes with a wheel base less than 18 m 18 m if the taxiway is intended to be used by aeroplanes with a wheel base equal to or greater than 18 m
D	18 m if the taxiway is intended to be used by aeroplanes with an outer main gear wheel span less than 9 m 23 m if the taxiway is intended to be used by aeroplanes with an outer main gear wheel span equal to or greater than 9 m.
E	23 m

This geometry is shown in Fig. 8.3. If for future new aircrafts the outer main gear wheel span of 20 m and a wheel to edge clearance of 5 m, the taxiway width required shall be 30 m.

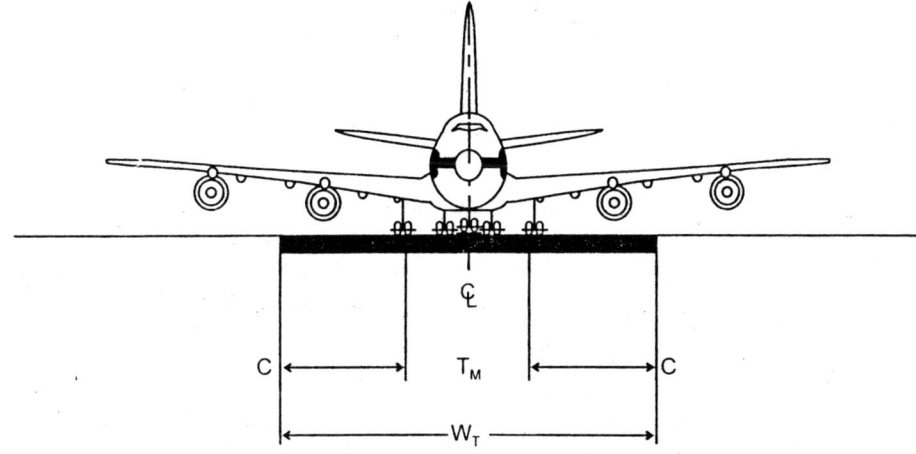

Fig. 8.3. Taxiway width geometry (ICAO).

8.3.2 Taxiway curves

Changes in direction of taxiways should be as few and small as possible. The radii of the curve should be compatible with manoeuvring capability and normal taxiing speeds of the aeroplanes for which the taxiway is intended. The design of the curve should be such that, when the cockpit of the aeroplane remains over the taxiway centreline markings, the clearance distance between the outer main wheels of the aeroplane and the edge of the taxiway should not be less than those specified below.

Code letter	Taxiway width
A	1.5 m
B	2.25 m
C	3 m if the taxiway is intended to be used by aeroplanes with a wheel base less than 18 m 4.5 m if the taxiway is intended to be used by aeroplanes with a wheel base equal to or greater than 18 m
D	4.5 m
E	4.5 m

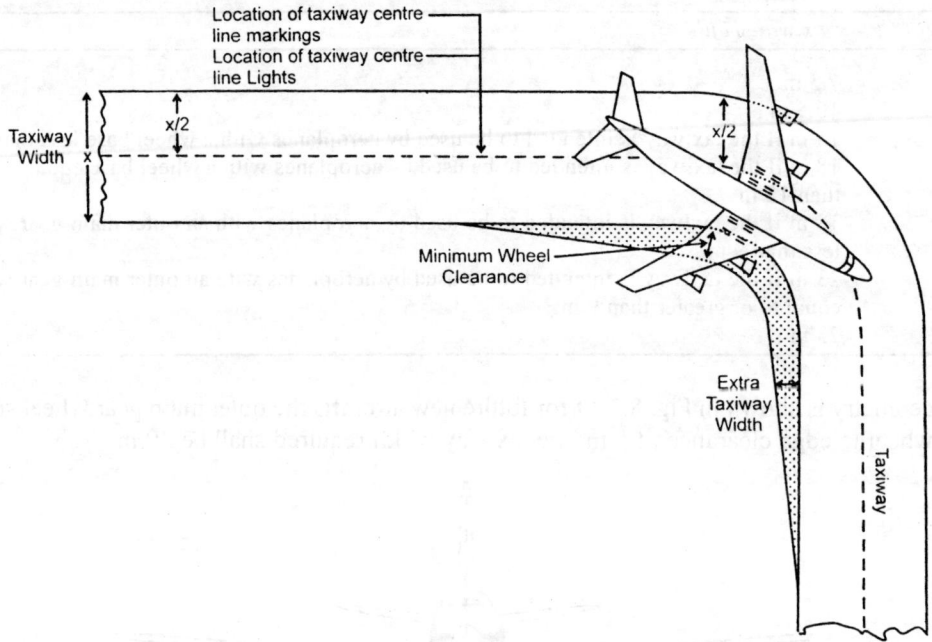

Fig. 8.4. Taxiway curve (ICAO).

An example of widening taxiways to achieve the wheel clearance specified is shown in Fig. 8.4.

8.3.3 Taxiway junctions and intersections

To facilitate the movement of aeroplanes, fillets should be provided at junctions and intersections of taxiways with runways, aprons and other taxiways. The design of fillets should ensure that minimum wheel clearances specified are maintained when aeroplanes are manoeuvring through the junctions or intersections.

The extra paved surface to be provided through fillets should have the same strength as that of the taxiway.

8.3.4 Taxiway minimum separation distances

The separation distances between the centreline of a taxiway and the centre line of a parallel taxiway or an object, and the centreline of a runway (instrumental and non-instrumental are given in Table 8.1 and 8.2.

(a) Separation distances between taxiways and taxiways or objects

These formula are based on wing span (Y), the maximum lateral deviation (X), (lesser speed and more attention of pilot may have lesser deviation off the centre line) and an increment (Z), (as a safety buffer intended to avoid accidents when aircraft go beyond the taxiway to facilitate taxiing and to account for other factors influencing taxiing speed) as show in Fig. 8.5 and Fig. 8.6.

(b) Separation distances between taxiway and runways

The separation distances as shown in Table 8.3 are based on the concept of the wing of an aircraft centred on a parallel taxiway remaining clear of the strip. This distance S is represented by the relationship :

Table 8.1. Design criteria for a taxiway

Physical characteristics		Code letter				
		A	B	C	D	E
Minimum width of	Taxiway pavement	7.5 m	10.5 m	18 m[a] 15 m[b]	23 m[c] 18 m[d]	23 m
	Taxiway pavement and shoulder	–	–	25 m	38 m	44 m
	Taxiway strip	27 m	39 m	57 m	85 m	93 m
	Graded portion of taxiway strip	22 m	25 m	25 m	38 m	44 m
Minimum clearance distance of outer main wheel to taxiway edge		1.5 m	2.25 m	4.5 m[a] 3 m[b]	45 m	4.5 m
Minimum separation distance between taxiway centre end:	Centre line of instrument runway code number					
	1	82.5 m	87 m	–	–	–
	2	82.5 m	87 m	–	–	–
	3	–	–	118 m	176 m	–
	4	–	–	–	176 m	180 m
	Centre line of non-instrument runway					
	Code number 1	37.5 m	42 m	–	–	–
	2	47.5 m	52 m	–	–	–
	3	–	–	93 m	101 m	–
	4	–	–	–	101 m	105 m
	Taxiway centre line Object	21 m	31.5 m	46.5 m	68.5 m	76.5 m
	Taxiway[e]	13.5 m	19.5 m	28.5 m	42.5 m	46.5 m
	Aircraft stand taxilane	2 m	16.5 m	24.5 m	36 m	40 m
Maximum longitudinal Slope of taxiway	Pavement	3 %	3 %	1.5 %	1.5 %	1.5 %
	Change in slope	1% per 25 m	1% per 25 m	1% per 30 m	1% per 30 m	1% per 30 m
Maximum transverse Slope of :	Taxiway pavement	2 %	2 %	1.5 %	1.5 %	1.5 %
	Graded portion of Taxiway strip-upwards	3 %	3 %	5 %	2.5 %	2.5 %
	Graded portion of taxiway strip-downwards	5 %	5 %	5 %	5 %	5 %
	Ungraded portion of strip-upwards	5 %	5 %	5 %	5 %	5 %
Minimum radius of longitudinal vertical curve		2500 m	2500 m	3000 m	3000 m	3000 m
Minimum taxiway sight distance		150 m from 1.5 m above	200 m from 2 m above	300 m from 3 m above	300 m from 3 m above	300 m from 3 m above

a. Taxiway intended to be used by aeroplanes with a wheel base equal to or greater than 18 m.
b. Taxiway intended to be used by aeroplane with a wheel base less than 18 m.
c. Taxiway intended to be used by aeroplanes with an outer main gear wheel span equal to or greater than 9 m.
d. Taxiway intended to be used by aeroplanes with an outer main gear wheel span less than 9 m.
e. Taxiway other than an aircraft stand taxilane.

Table 8.2. Minimum separation distance between taxiways and taxiways or objects (dimensions in m)

Between	Formula		A	B	C	D	E
Taxiway centre line and taxiway centre line (apron taxiway centre line and taxiway centre line)	Wing span (Y)		15	24	36	52	60
	+ 2x maximum lateral deviation (X)		3	4.5	6	9	9
	+ increment (Z)		3	3	4.5	7.5	7.5
	= V	Total	21	31.5	46.5	68.5	76.5
Taxiway centre line and object	½ wing span Y + maximum		7.5	12	18	26	30
	lateral deviation (X)		1.5	2.25	3	4.5	4.5
	+ increment (Z)		4.5	5.25	7.5	12	12
	= V	Total	13.5	19.5	28.5	42.5	46.5
Apron taxiway centre line and object	½ wing span (Y)		7.5	12	18	26	30
	+ maximum lateral deviation (X)		1.5	2.25	3	4.5	4.5
	+ increment (Z)		4.5	5.25	7.5	12	12
	= V	Total	13.5	19.5	28.5	42.5	46.5
Aircraft stand taxilane centre line and object	½ wing span (Y)		7.5	12	18	26	30
	+ gear deviation		1.5	1.5	2	2.5	2.5
	+ increment (Z)		3	3	4.5	7.5	7.5
	= V	Total	12	16.5	24.5	36	40

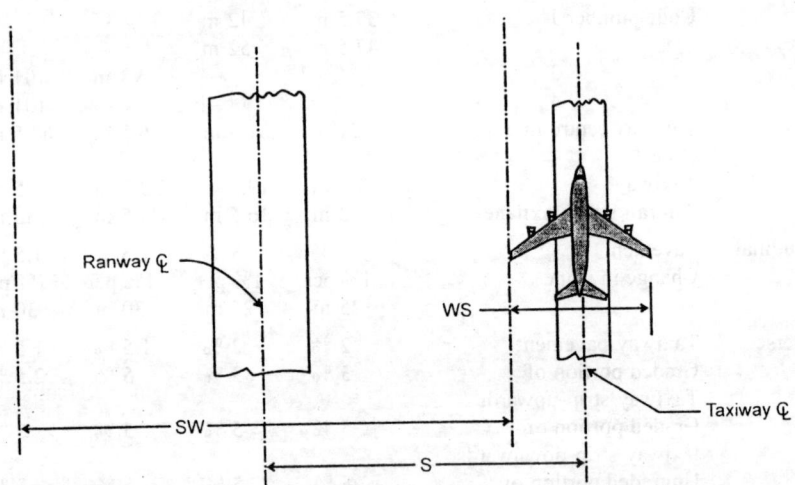

Fig. 8.5. Parallel runway-taxiway separation geometry (ICAO).

$$S = \frac{1}{2}(SW + WS)$$

where SW = Strip width

WS = Wing span as illustrated in Fig. 8.7.

The separation distance for planning of the largest aircraft predicted by the future trend data is 192 m. This value is based on the assumption that this aircraft having a wing span of 8.4 m can safely operate in the current 300 m runway strip width required for a precision approach.

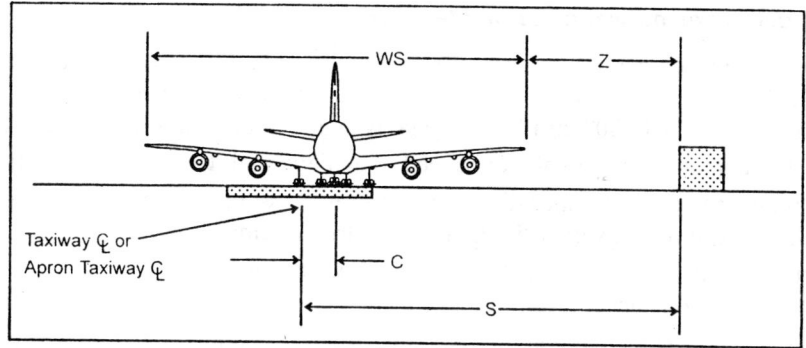

Fig. 8.6. Taxiway/apron taxiway-to-object geometry (ICAO).

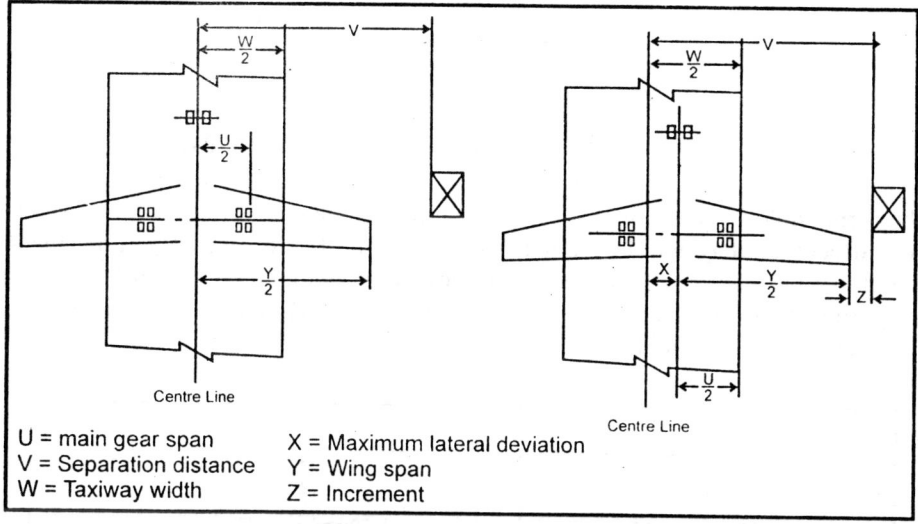

U = main gear span   X = Maximum lateral deviation
V = Separation distance   Y = Wing span
W = Taxiway width   Z = Increment

Fig. 8.7. Separation distance to an object (ICAO).

Table 8.3. Minimum separation distance between taxiways and runways (dimensions in m)

Between	Code number	1		2		3				4		
	Code letter	A	B	A	B	A	B	C	D	C	D	E
Taxiway centre line and runway centre line (apron taxiway centre line and runway centre line)	Formula											
	½ wing span (Y)	7.5	12	7.5	12	7.5	12	18	26	18	26	30
	(a) + ½ strip width (non-instr. RWY)	30	30	40	40	75	75	75	75	75	75	75
	Total	37.5	42	47.5	52	82.5	87	93	101	93	101	105
	or											
	(b) + ½ strip with (instr. approach RWY)	75	75	75	75	150	150	150	150	150	150	150
	Total	82.5	87	82.5	87	157.5	162	168	176	168	176	180

(c) Separation distances between parallel taxiways

Separation distance between parallel taxiways as shown in Table 8.2 are on the basis of desired wing tip clearances. There are other factors which should also be taken into account when evaluating the capability of making a normal 180° turn from one taxiway to another parallel taxiway. These include :

 (i) Ability to maintain a reasonable taxi speed to achieve high taxiway system utilisation.

 (ii) Maintaining specified clearance distances between the outer main wheel and the taxiway edge when the cockpit remains over the taxiway centre marking.

 (iii) Manoeuvering at a steering angle that is within the capability of the aircraft and which will not subject the tires to unacceptable wear.

To evaluate the taxi speed when making the 180° turn it is assumed that the radii of curvature are equal to one-half of the separation distance as below :

Code letter	Radius (m)	Allowable speed in a 180° turn KPh
A	10.5	13.32
B	15.75	16.32
C	23.25	19.82
D	34.25	24.06
E	38.25	25.41

It may be permissible to operate with lower separation distances at an existing airport, if they would not adversely affect the safety and the regularity of operation of aeroplanes. Short distances may also be possible in those parts where a taxiway joins a runway or another taxiway. The separation distance for planning purposes, that result from the future aircraft span of 84 m, a lateral deviation C of 5 m and a wing tip clearance of 11 m would be 105 m, as shown in Fig. 8.8.

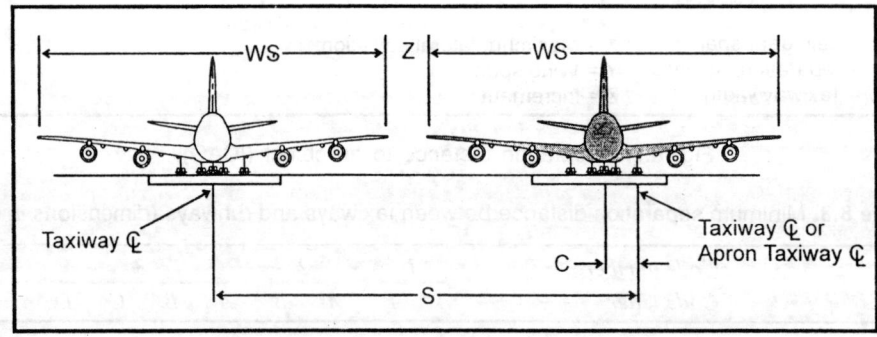

Fig. 8.8. Parallel taxiway separation geometry (ICAO).

In the Fig. 8.8 :

$$S = WS + 2C + Z$$

where S = Separation distance

  WS = Wing span

  C = Clearance between the outer main gear wheel and the taxiway edge (maximum allowable lateral deviation)

  Z = Wing tip clearance

(d) Aircraft stand taxilane-to-object

The lower taxiing speed of an aircraft in a stand taxilane permits a smaller lateral deviation to be considered than with other taxiways. The geometry of this relationship is shown in Fig. 8.9 below and given by formula :

$$S = \frac{WS}{2} d + Z$$

where   S = Separation distance
           WS = Wing span
           d = Lateral deviation
           Z = Wing tip clearance of an object

Application of the above results in an object separation distance for planning purposes, for the future large aircraft in a stand taxilane of 56.5 m. This value is based on a wing span of 84 m, a deviation of 3.5 m and wing tip clearance of 11 m.

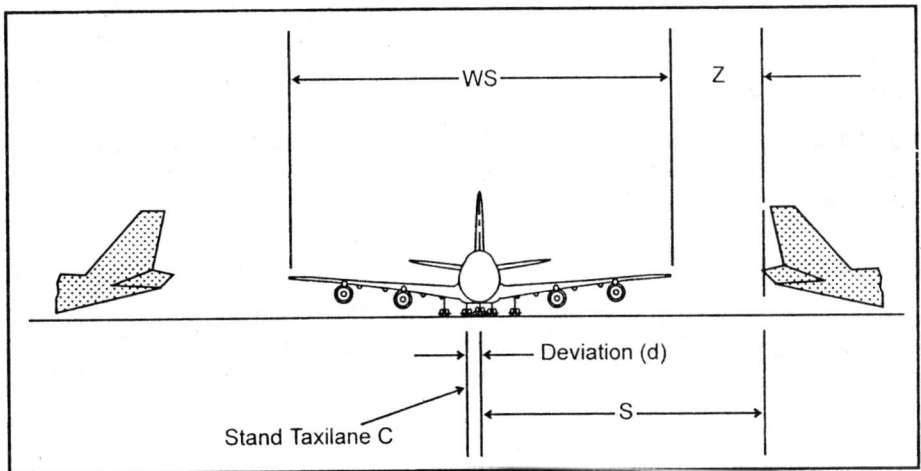

Fig. 8.9. Aircraft stand taxilane-to-object geometry (ICAO).

8.3.5 Slopes on taxiways

(a) Longitudinal slopes

The longitudinal slope of a taxiway should not exceed :
- 1.5 per cent where the code letter is C, D or E, and
- 3 per cent where the code letter is A or B.

(b) Longitudinal slope changes

Where slope changes on a taxiway can not be avoided, the transition from one slope to another slope should be accomplished by a curved surface with a rate of change not exceeding :
- 1 per cent per 30 m (minimum radius of curvature of 3000 m) where the code letter is C, D or E, and
- 1 per cent per 25 m (minimum radius of curvature of 2500 m) where the code letter is A or B.

(c) Transverse slopes

The transverse slopes of a taxiway should be sufficient to prevent the accumulation of water on the surface of the taxiway, but should not exceed :

- 1.5 per cent where the code letter is C, D or E, and
- 2 per cent where the code letter is A or B.

8.3.6 Sight distance

Where a change in slope on a taxiway cannot be avoided, the change should be such that from any point :

- 3 m above the taxiway, it will be possible to see the whole surface of the taxiway, for a distance of at least 300 m from that point, where code letter is C, D or E.
- 2 m above the taxiway, it will be possible to see the whole surface of the taxiway, for a distance of at least 200 m from that point, where code letter is B.
- 1.5 m above the taxiway, it will be possible to see the whole surface of the taxiway for a distance of at least 150 m from that point where the code letter is A.

8.3.7 Surface of taxiway

The surface of a taxiway should not have irregularities that cause damage to aeroplane structures. The surface of a paved taxiway should be so constructed as to provide good friction characteristics when the taxiway is wet.

8.3.8 Strength of taxiways

The strength of a taxiway should be at least equal to that of the runway it serves, due considerations being given to the fact that a taxiway will be subjected to a greater density of traffic and as a result of slow moving and stationary aeroplanes, to higher stresses than the runway it serves.

8.4 RAPID EXIT TAXIWAYS

The capacity and efficiency of a runway is directly related to the capabilities of the taxiway system to facilitate the movement of traffic from the runway. Ideally, exit taxiways should be located, such as to minimise runway occupancy time.

A rapid exit taxiway is a taxiway connected to a runway at an acute angle and designed to allow landing aeroplanes to turn off at higher speeds than are achieved on other exit taxiways and thus minimising the runway occupancy time. The main purpose of these taxiways is to increase airport capacity by reducing the runway occupancy time. The right angle taxiway may suffice in cases where peak hour density is about 25 operations. The construction of right angle taxiway is cheaper and can efficiently handle this volume of traffic.

For pilots to become familiar with tthe configuration of rapid exit runways it is necessary to establish design standards. ICAO has prescribed design parameters for a grouping of exit taxiways associated with a runway where the code number is 1 or 2 and another grouping for code numbers 3 or 4.

For safety reasons 93 km/h has been taken as the reference for determining curve radii and adjacent straight portions for rapid exist taxiways where the code number is 3 or 4, regardless of the design speed which the planner may choose for computing the optimum exit locations along the runway. Benefits and convenience obtained by rapid exit taxiways will increase their use in future.

LOCATION AND NUMBER OF RAPID EXIT TAXIWAYS

The location of exit taxiways with respect to aircraft operational characteristics is determined by the deceleration rate of the aircraft, after crossing the threshold. To determine the distance from threshold, the following basic conditions should be taken into account :

   (i) threshold speed, and

  (ii) initial exit speed or turn off speed at the point of tangency of the central (exit) curve-point A in Fig. 8.10 and Fig. 8.11.

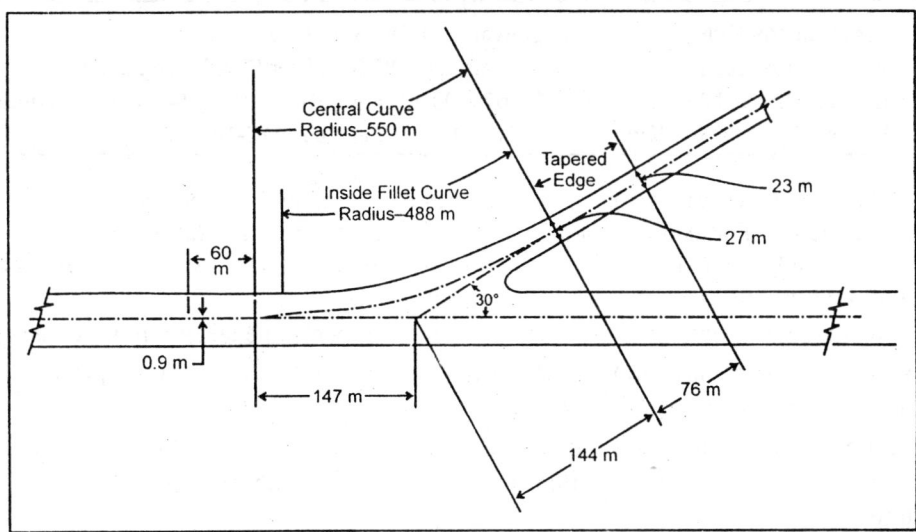

Fig. 8.10. Design for rapid exit taxiways I (ICAO).

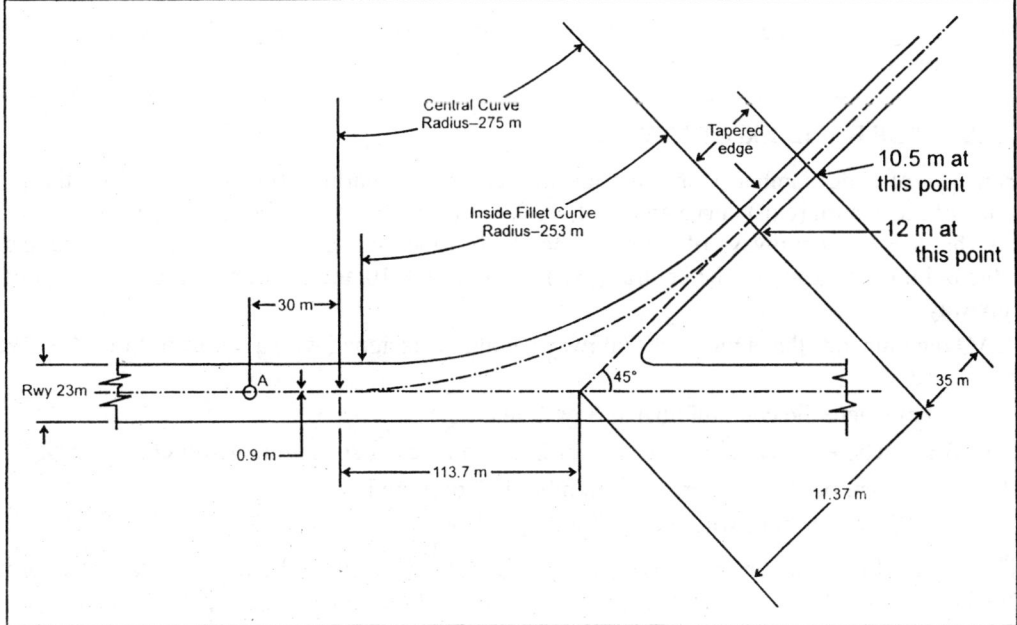

Fig. 8.11. Design for rapid exit taxiways II (ICAO).

For the purpose of exist taxiway design, the aircrafts are assumed to cross the threshold at an average of 1.3 times the stall speed in the landing configuration with an average gross landing mass of about 85 per cent of the maximum. Table 8.4 shows the four groups of aircrafts on the basis of threshold speed at sea level.

Table 8.4. Threshold speed and aircraft types

Group	Threshold speed	Aircraft types (examples)
A	Less than 169 KPh	Convair 240, DC-3, DHC-7
B	Between 169-222 KPh	Convair 600, DC-6, Fokker F27, Viscount 800
C	Between 224 and 259 KPh	B-707 (300 and 400), B-727, DC-8 (43 and 55), Trident (1 and 2)
D	Between 261 and 306 KPh	B-747, DC-8 (62 and 63), DC-10

The number of exit taxiways depends on the types of aircraft and number of each type that operate during the peak period. For example, at a very large aerodrome most aircraft will likely be in group C or D and for them only two exits may be required. On the other hand, an aerodrome having a balanced mixture of all four groups of aircraft may require four exists.

In locating the rapid exit taxiway following factors should also be taken innto account :

- weather conditions affectiing the distance required to slow down to exit speed.
- location of the terminal/apron area.
- location of other runways and their exists.
- optimisation of traffic flow within the taxiway system with respect to traffic control procedures.
- avoidance of unnecessary taxi detours.
- additional exit taxiways, for long runways depending upon local conditions and requirements.

Considering various factors, it is recommended that the exist taxiway be located 450 m to 600 m from the threshold.

8.5 DESIGN OF RAPID EXIT TAXIWAY

For runways of code number 3 or 4 the taxiway centreline marking begins 60 m from the point of tangency of the central (exit) curve and is offset 0.9 m to facilitate the pilot's recognition of the beginning of the curve. For runways of code number 1 or 2, the taxiway centre line marking begins 30 m from the point of tangency of the central (exit) curve. Fig. 8.10 and 8.11 show typical design of rapid exit taxiway.

ICAO recommends that a rapid exit taxiway should be designed with a radius of turn-off curve of at least :

- 550 m where the code number is 3 or 4, and
- 275 m where the code number is 1 or 2 to enable exist speed under wet condition of :
  - 93 km per hour where code number is 3 or 4, and
  - 65 km per hour where code number is 1 or 2.

The radius of the fillet on the inside curve at a rapid exit taxiway should be sufficient to provide a widened throat in order to facilitate recognition of the entrance and turn-off onto the taxiway.

A rapid exit taxiway should include a straight distance after the turn-off curve sufficient for an existing aircraft to come to a full stop clear of any intersecting taxiway (Fig. 8.12) and it should not be less than 35 m for code number 1 or 2 and 75 m for code number 3 or 4 when the intersection angle is 30°.

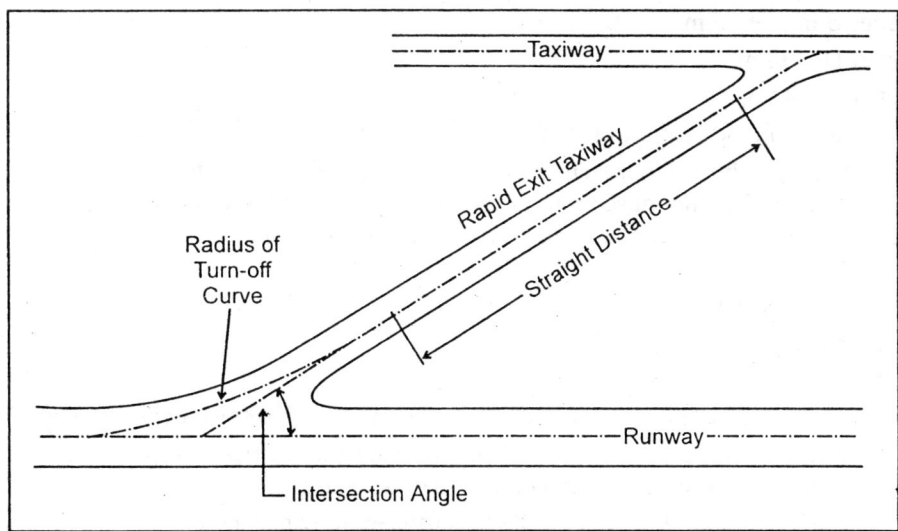

Fig. 8.12. Straight exit taxiway (ICAO).

The intersection angle of a rapid exit taxiway with the runway should not be greater than 45° nor less than 25° and preferably should be 30°. General requirements for taxiways also apply to rapid exit taxiways.

8.6 TAXIWAYS ON BRIDGES

In certain situations, the design and layout of an airport, may involve that taxiways are located over a bridge on roadways, railways, canals etc. or open waters (rivers, sea bays). The detail description of design of bridge and taxiway requirements are beyond the scope of this book, however, the following basic principles should be kept in mind.

- Taxiway bridge should be designed so that they do not impose any difficulties for taxiing aircraft.
- Strength, dimensions, grades and clearances should allow unconstrained aircraft operations day and night as well as under varying season conditions such as heavy rains, snow, low visibility or gusty winds.
- Requirements of taxiway maintenance, cleaning and snow removal should be taken into account when bridges are being designed.
- The bridge should be located on a straight portion of a taxiway with a straight portion provided on both ends of the bridge to facilitate the alignment of the aeroplanes approaching the bridge.
- Rapid exit taxiways should not be located on the bridge.
- Bridge locations that have adverse effect upon the instrument landing system or the approach lighting or runway/taxiway lighting system should be avoided.

- The bridge width measured perpendicularly to the taxiway centre line should not be less than the width of the graded portion of the strip provided for that taxiway, unless a proven method of lateral restraint is provided which shall not be hazardous for aeroplanes for which the taxiway is intended. The minimum width requirements will normally be :

  Code letter A = 22 m

  Code letter B or C = 25 m

  Code letter D = 38 m

  Code letter E = 44 m

  With the taxiway in the centre of the strip.

- Aircrafts should be able to approach to and depart from bridges on straight portions of the taxiway. The length of the straight section should be at least twice the wheel base (distance from the nose gear to the geometric centre of the main gear) of the most demanding aircraft and not less than :

  - 15 m for code letter A
  - 20 m for code letter B
  - 25 m for code letters C, D or E.

  Future aircraft in code E may have wheel base of 35 m, requiring a straight distance of 70 m.

- Normal taxiway transverse slopes should be provided for drainage purposes and longitudinal gradients should not exceed as prescribed for other taxiways.

- The strength of the bridge should normally be sufficient to withstand the traffic of the aeroplanes the taxiway is intended to serve.

- Lateral restraint devices may be provided as additional safety measures.

- Where taxiway passes over another transport mode some kind of protection against aircraft engine blast may need to be provided.

8.7 FILLETS

Fillets are used at the junction or intersection of a taxiway with a runway, another taxiways or apron, to provide the clearance distances. To meet these requirements when an aircraft is negotiating a turn it may be necessary to provide additional pavement on taxiway curves and on taxiway junctions and intersections. The extra taxiway area provided to meet the recommended clearance distance requirement is part of the taxiway, and therefore, the term 'extra taxiway width' is used, rather than fillet. The strength of the extra paved surface to be provided should be the same as that of the taxiway.

8.7.1 Design of fillets

For manoeuvering, it is assumed that cockpit of the aircraft remains over the taxiway centreline markings. Following techniques are used for designing of the fillets :

  (a) Simulation of the aircraft movement by use of a model.
  (b) Calculations of the fillets.
  (c) Utilisation of prepared graphs which provide satisfactory approximation of the path followed by the undercarriage centre.

Whatever technique is used to design the fillet, it is first necessary to establish the path of the main undercarriage centre. The clearance distances given (Table 8.1) are recommended minimum.

(a) Model simulation

The path of the outer main wheels of an aircraft during a turn may be obtained by the use of a model. It involves moving a scale model of an aircraft on a plan of the taxiways and runways. A reasonably large scale (e.g. 1/250) is required and the model must be well made to prevent excessive errors when the outlines are scaled up. Such a condition makes this method less practical.

(b) Calculation of the fillets

The fillet can be determined mathematically but the process is fairly complicated and degree of accuracy obtained is in excess of that required for the actual engineering construction of fillets. This method is practical when computer is available for which a programme of calculations can be prepared to provide a numerical solution of equations dealing with determination of the path.

(c) Use of graphs

A close approximation can simply be obtained, from the use of prepared graphs. Depending on their design, these graphs can either be used for all types of aircrafts or be adopted to one particular type.

For the detailed design applications of the above methods, readers may refer to ICAO's publication 'Aerodrome Design Manual Part II'.

8.8 TAXIWAY SHOULDERS AND STRIPS

A shoulder is an area adjacent to the edge of a paved surface so prepared as to provide a transition between the pavement and the adjacent surface. The main purposes of the provision of a taxiway shoulder are :

  (a) to prevent jet engines that overhang the edge of a taxiway from ingesting stones or other objects which might damage the engine.

  (b) to prevent erosion of the area adjacent to the taxiway.

A taxiway strip is an area including a taxiway intended to protect an aircraft operating on the taxiway and to reduce the risk of damage to an aircraft accidently running off the taxiway. The ICAO recommendations for widths to be provided for taxiway shoulders and strips are given in Table 8.1. The 10.5 m wide shoulder on both sides of the taxiway for code letter E, and 7.5 m wide shoulder on both sides for code letter D, is considered to be suitable, on the assumption that the distance between the outboard engines of the critical aircraft normally using that taxiway will not exceed 30 m.

The surface of the shoulder that abuts the taxiway should be flush with the surface of the taxiway while the surface of the strip should be flush with the edge of the taxiway or shoulder, if provided. For a code letter C, D or E, the graded portion of the taxiway strip should not rise more than 2.5 percent or slope down at a gradient exceeding 5 per cent. The respective slopes for a code letter A or B are 3 per cent and adjacent taxiway surface the downward slope is measured with reference to the horizontal. There should, furthermore, be no holes or ditches tolerated within the graded portion of the taxiway strip.

No obstacles should be allowed on either side of a taxiway within the distance shown in Table 8.1 for separation of fixed objects. However, signs and any other objects which because of their functions must be maintained within the taxiway strip in order to meet air navigation requirements may remain but they should be constructed and sited in such a manner as to reduce the hazard to an aircraft striking them to a minimum. Such objects should be sited so that cannot be struck by propellers, engine pods and wings of aircraft using the taxiway. As a guide they should be so sited that there is nothing higher than 0.30 m above taxiway edge level within the taxiway strip.

8.8.1 Treatment for shoulders

Taxiway shoulders and graded portions of strips provide an obstacle free area intended to minimize the probability of damage to an aircraft using these areas accidentally or in an emergency. These areas should thus be so prepared or constructed as to reduce the risk of damage to an aircraft running off the taxiway and be capable of supporting access by rescue and fire fighting vehicles and other ground vehicles, as appropriate, over its entire area. When a taxiway is intended to be used by turbine-engined aircraft, the jet engines may overhang the edge of the taxiway while the aircraft is taxiing and may not ingest stones or foreign objects from the shoulders. Further, blast from the engines may impinge on the surface adjacent to the taxiway and may dislodge material with consequent hazard to personnel, aircraft and facilities. Certain precautions must, therefore, be taken to reduce these possibilities. The type of surface of the taxiway shoulder will depend on local conditions and contemplated methods and cost of maintenance. While a natural surface (e.g., turf) may suffice in certain cases, in others, an artificial surface may be required. In any event, the type of surface selected should be such as to avoid the blowing up of debris as well as dust.

Under most taxiing conditions, blast velocities are not critical except at intersections where thrusts approach those on breakaway. With the present criteria of 23 m wide taxiways, the outboard engines of the larger jets extend beyond the edge of the pavement. For this reason, treatment of taxiway shoulders is recommended to prevent their erosion and to prevent the ingestion of foreign material into jet engines or the blowing of such material into the engines of following aircraft.

8.8.2 Thickness of shoulders

The thickness of taxiway shoulder should be able to support the critical aircraft axle loads occasionally or in emergency and maintenance vehicles should be able to pass over them. The following criteria should be used :

(a) The minimum design thickness required for shoulder can be taken as one half of the total thickness required for the adjacent paved area.

(b) The critical axle load of the heaviest emergency vehicle or maintenance vehicle, should determine the pavement thickness. If this thickness is more that based on (a) above, then this thickenss should be used for shoulders.

(c) For aircrafts such as Boeing 707, DC-8, DC-10, L-11011 or smaller the recommended minimum surface thickness, if bituminous concrete on an aggregate base is used should be 5 cm on shoulder. For service by aircraft such as Boeing 747 and L-500 etc. an increase of 2.5 cm is recommeded.

(d) A minimum of 5 cm of bituminous concrete surface is recommended on a bituminous stabilised base.

(e) If Portland cement concrete and a granular sub-base is used for shoulders, a 15 cm minimum thickness should be applied.

(f) The compaction and construction standard for the subgrade and pavement courses should be the same as for full strength pavement areas. It is recommended that approximately a 2.5 cm drop-off be used at the edge of the full strength pavement, and shoulder to provide a definite line of demarcation.

8.8.3 Taxiway strip

A taxiway strip should extend symmetrically on each side of the centre line of the taxiway throughout the length of the taxiway to at least the distance from centre line as given in Table 8.1. The taxiway strip should provide an area clear of objects which may endanger taxiing aeroplanes.

Grading of taxiway strips

The centre portion of a taxiway strip should provide a graded area to a distance from the centreline of the taxiway of at least :

- 11 m for code letter A
- 12.5 m for code letter B or C
- 19 m for code letter D, and
- 22 m for code letter E

as given in Table 8.1.

8.8.4 Slopes on taxiway strips

The surface of the strip should be flush at the edge of the taxiway or shoulder if provided, and the graded portion should not have an upward transverse slope exceeding :

- 2.5 per cent for strips of taxiways where code letter is C, D or E, and
- 3 per cent for strips of taxiways where code letter is A or B.

The upward slope being measured with reference to the transverse slope of the adjacent taxiway surface and not the horizontal. The downward transverse slope should not exceed 5 per cent measured with reference to the horizontal.

CHAPTER 9

Airport Aprons and Holding Bays

9.0 INTRODUCTION

An apron is an airside area on an airport for accomodating aircrafts for purposes of loading and unloading passengers, mail or cargo, fuelling, parking or maintenance. The apron is generally paved, but may occasionally be unpaved such as a turf parking apron for small aircrafts.

Aprons can be classified according to their main purpose and functions. This chapter describes characteristics of various types of aprons and aspects of their planning and design. Besides the aircraft stands, the associated apron taxiways, apron service roads and parking for ground service equipment is also included as a part of an apron system.

9.1 TYPES OF APRONS

The following are the various types of aprons.

(a) Terminal apron

The terminal apron is an area designed for aircraft manoeuvring and parking that is adjacent or readily accessible to passenger terminal facilities. From a terminal apron passengers board the aircraft. In addition terminal apron is also used for aircraft fuelling and mainteance as well as loading and unloading cargo, mail and baggage. Individual aircraft parking positions on the terminal are called as aircraft stands.

(b) Cargo apron

Aircrafts that carry only freight and mail may be provided a separate cargo apron adjacent to a cargo terminal building. The separation of cargo and passenger aircraft is desirable because of the different types of facilities each requires both on the apron and at the terminal.

(c) Parking apron

Parking apron is where aircrafts can park for extended periods of time. It may be required in addition to terminal aprons. Parking aprons may be used during crew layovers, or for light periodic servicing and maintenance of temporarily grounded aircrafts. Parking aprons are located as close to the terminal apron as possible.

(d) Service and hanger aprons

A service apron is an uncovered area adjacent to a repair hanger on which aircraft maintenance is carried out, while a hanger apron is an area on which aircraft move into and out of a storage hanger.

(e) General aviation apron

General aviation aircrafts, used for business or private flying, require several categories of aprons to support different general aviation activities.

(f) Transcient (Itinerant) apron

Transient general aviation aircraft use the itinerant (transient) aprons as temporary aircraft parking facility and as access to fuelling, servicing, and ground transportation. At aerodromes servicing only general aviation aircraft, the itenerant apron is usually adjacent to or an integral part of, a fixed-based operator area. The terminal apron will generally also set aside some area for transient general aviation aircraft.

(g) Based aircraft apron (tie downs)

General aviation aircraft based at an aerodrome required either hanger storage or a tiedown space in the open. Hangered aircrafts also need an apron in front of the building for manoeuvring. Open areas used for aircraft tie down may be paved, unpaved, or turfed depending on the size of aircraft and local weather and soil conditions.

(h) Other ground servicing aprons

Areas for servicing, fuelling or loading and unloading should also be provided as aprons as needed.

Not all of the apron types are required for every airport, but the need for them and their size should be estimated based on type and volume of forecast aircraft traffic at the airport.

9.2 PLANNING OF APRONS

9.2.1 Apron siting

Aprons are interrelated with the terminal complex and should be planned in connection with terminal building to achieve an optimum solution.

The following general objectives should be considered in siting aprons :

(i) Provide minimum taxiing distances between runways and aircraft stands for savings in fuel, time and maintenance.

(ii) Allow for freedom of aircraft movement to avoid unnecessary delay.

(iii) Reserve sufficient area for future expansion and change of technology.

(iv) Achieve maximum efficiency, operational safety and user convenience of each apron as well as the airport as a total system.

(v) Minimize adverse effects such as engine blast, noise, air pollution etc. on the apron and surrounding environment.

9.2.2 Size of aprons

The various factors which should be considered in arriving at the suitable size of an apron for its purpose and function are :

  (a) Number of aircraft stand at present and in future.

  (b) Aircraft mix, both present and future.

  (c) Aircraft dimensions and manoeuvring capabilities.

  (d) Aircraft parking configurations including shape of terminal and the surrounding area for development.

  (e) Clearance requirements between aircraft and aircraft building or other fixed objects.

  (f) Method of aircraft guidance onto the aircraft stand.

  (g) Aircraft ground servicing requirements i.e. vehicles versus fixed servicing installations etc.

  (h) Taxiway and service roads.

9.2.3 Aircraft parking configuration

In aircraft parking the method by which the aircraft will enter and leave the aircraft stand e.g. either under its own power (self-manoeuvring) or taxies in and push out (tractor assisted) should be considered.

Fig. 9.1 shows different parking configurations and Table 9.1 compares the main advantages and disadvantages of each cofiguration.

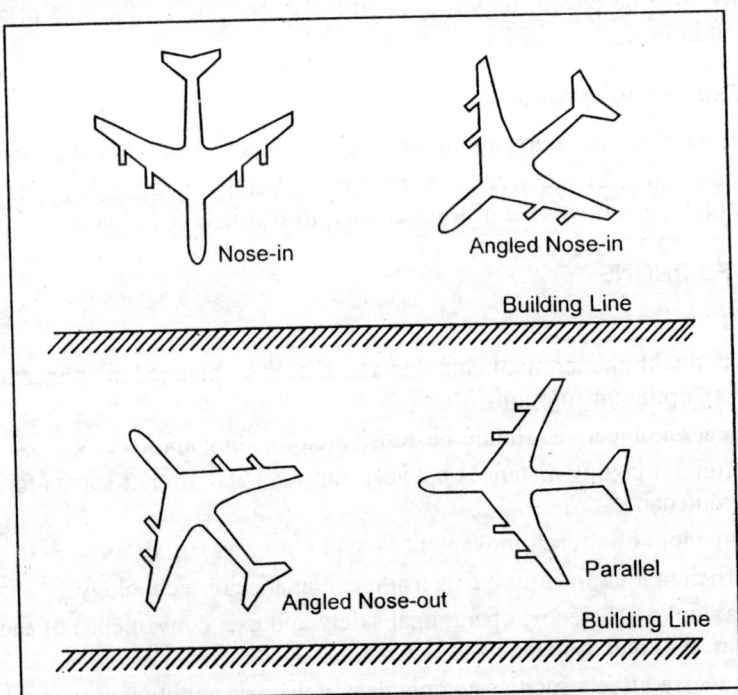

Fig. 9.1. Parking configurations of aircrafts (ICAO).

Table 9.1. Comparison of different aircraft configuration

	Nose-in (taxi in and push out)	Angled nose-in (in/out by own power)	Angled nose-out (in/out by own power)	Parallel (in/out by own power)
Advantages	• Requires smallest stand area for given aircraft • The effects of jet blast on equipment, personel and terminal are substantially less • Less noise while taxiing in because no turning is required • Reduces aircraft service time as ground equipment can be positioned prior to aircraft arrival and fewer removal requirement at aircraft departure • Easy to employ passenger loading bridge as door close to terminal building	• No requirement for tractor and so no additional power requirement	• No requirement for tractor and so less power requirement • The rear loading door is close to the terminal building	• Easiest manoeuvring for aircraft to taxi in/out • No requirement for tractor • Both the front and the rear doors are close to terminal building
Disadvantages	• Requires tractor for push-out operation and large power • Push-out operation requires time and skilled operator • Rear loading door is far away from trained builder	• Requires larger apron area than nose-in configuration • Relatively severe engine blast and noise is directed at terminal	• Requires larger apron area than angled nose-in configuration • Breakaway engine blast and noise are directed at terminal	• Requires largest apron area for given aircraft • Limits aircraft servicing activity at neighbouring stand when aircraft taxi in and out

As a general rule, nose-in-parking configurations are common at high traffic airports where the tractor cost is justified for more efficient use of limited apron area. Other parkings configurations are employed at low traffic airports where it is difficult to offset the tractor operation cost by savings in apron size.

The amount of apron area required for a particular aircraft varies largely with the parking configuration, and passenger/cargo handling concept, particularly for a passenger terminal apron, a nose-in parking configuration alongwith passenger loading bridges affords following advantages :

• Less apron area required.

• Less aircraft ground time due to efficient passenger handling and more efficiently positioned ground servicing equipment.

• Service road can be sited so as to reduce requirements to drive on aprons.

• Better passenger handling in terms of safety, convenience and comfort since passengers are

free from apron walking, climbing up/down stairs and adverse weather effects such as rain, snow, wind, heat etc.
- Substantially less adverse effects of jet blast noise and fumes of engines on ground equipment, personnel and terminal facilities.
- Greater security control of passengers on airside.

On the other hand it involves additional costs for purchasing and operating tractors and passenger loading bridges.

The present trend is towards nose-in/push out configuration with passenger loading bridges at high traffic volume airports.

9.3 DESIGN REQUIREMENTS OF APRONS

The design of any of the various apron types requires the evaluation of interrelated and often contradictory characteristics. Despite the distinct purposes of the different apron types, there are many general design characteristics relating to safety, efficiency, geometry, flexibility, and engineering that are common to all type. The following paragraphs give a brief description of these general design requirements.

9.3.1 Safety

Apron design should take into account safety procedures for aircraft manoeuvring on the apron. Safety in this context implies that aircraft maintain specified clearances and follow the established procedures to enter, move within, and depart from apron areas. Provision of services to aircraft parked on the apron should also incorporate safety procedures, especially regarding aircraft fuelling. Pavement should slope away from terminal buildings and other structures to prevent the spread fuel fires on the apron. Water outlets should be located at each stand position for routine hosing of the apron surface. Aircraft security should also be considered in locating the apron area where the aircraft can be protected from unauthorized person. This is accomplished by physically separating public access areas from any contact with the apron areas.

9.3.2 Efficiency

Apron design should contribute towards establishing a high degree of efficiency for aircraft movements and the dispensing of apron services. Freedom of movement, minimum taxi distances, and a minimum of delay for aircraft initiating movements on the apron are all measures of efficiency for any of the apron type. If the ultimate aircraft stand arrangement can be determined during the initial plan phase of the aerodrome, then utilities and services should be installed in fixed installations. Fuel lines and hydrants, compressed air hookups, and electrical power system must be carefully preplanned because these systems are often placed under the apron pavement. The high initial cost of these systems will be offset by the increased efficiency of the stand, which allows a greater utilization of the apron. Achieving these measures of efficiency will ensure the maximum economic value of the apron.

9.3.3 Geometry

The planning and design of any apron type are dependent upon number of geometric considerations. For example, the length and width of a land part available for apron development may preclude the choice of certain apron layout concepts. For a new aerodrome it may be possible to develop the most efficient arrangement, based upon the nature of the demand, and then to set aside an area of land ideally suited to the concept. However, expansion or addition of aprons at existing aerodrome will usually be less than ideal as a result of the limitations imposed by the shape size of available parcels. The overall area needed per aircraft stand includes the required for aircraft stand taxilanes as well as apron taxiways used in common with other aircraft stands. Therefore, the overall area needed for apron development is function not only of aircraft size, clearances, and parking method, but also of the geometric arrangement of aircraft stand taxilanes, other taxiways, blast fences, are used for the stationing of service vehicles, and roads for the movement of ground vehicles.

9.3.4 Flexibility

Planning for aprons should include an evaluation of the following flexibility characteristics :

(a) Range of aircraft sizes

The number and size of aircraft stands should be matched to the number and size of aircraft types expected to use the apron. A compromise must be developed between the extremes of :

- (a) using one size of aircraft stand large enough for the largest aircraft type; and
- (b) using as many different sized stands as there are aircraft types.

The first method is a highly inefficient use of area, while the second provides a low level of operating flexibility. For passenger terminal apron,a compromise solution that achieves adequate flexibility is to group the aircraft into two or three size classes and provide stands for a mix of these general sizes in proportion to the demand forecast. A greater number of general aviation parking space sizes can be used because the space may be leased and occupied by a single aircraft of known dimensions.

(b) Expansion capability

Another key element of a flexible apron system is allowance for expansion to meet future needs. To avoid undue restriction of the growth potential of a particular apron area, the apron should be designed in modular stages so that successive stages become integral additions to the existing apron with a minimum of disruption to ongoing activities.

9.3.5 Common design characteristics

Many technical design requirements for the construction of apron surfaces are common to all apron types. Several of these factors are described in the following paragraphs.

(a) Pavement

The choice of pavement surface is a function of aircraft mass, load distribution, soil conditions, and the relative cost of alternative materials. Reinforced concrete is routinely used at aerodromes serving the largest commercial aircraft where greater strength and durability is needed. As a minimum, most aerodromes require an asphalt (tarmac) surface to satisfy strength, drainage, and stabilization criteria, though turf and cement-stabilized aprons have been satisfactorily used in some locations. Reinforced concrete is usually more expensive to install than asphalt but is less expensive to maintain and usually lasts longer. In addition, concrete is relatively unaffected by spilled jet fuel, whereas asphalt surfaces are damaged if fuel remains on the surface for even short periods of time. This problem can be partially overcome by coating the asphalt with special sealants and by frequently washing off the pavement. Each part of the apron should be capable of withstanding the traffic of aircraft it is intended to serve, giving considerations to stationary and slow motion on aprons.

(b) Pavement slope

Slopes on an apron should be sufficient to prevent accumulation of water on the surface of the apron but should be kept as level as drainage requirements permit. Efficient storm drainage of large, paved apron areas is normally achieved by providing a steep pavement slope and numerous area drains. On aprons, however, too great a slope will create manoeuvrability problems for aircraft and service vehicles operating on the apron. Additionally, fuelling of aircraft requires nearly a level surface to achieve the proper fuel mass balance in the assorted aircraft storage tanks. The design of slopes and drains should direct spilled fuel away from building and apron service areas. In order to compromise the needs for drainage, manoeuvrability, and fuelling, apron slopes should be 0.5 to 1.0 per cent in the aircraft stand areas and no more than 1.5 per cent in the other apron areas.

(c) Jet blast and Propeller wash

The effects of extreme heat and air velocities from jet and propeller engines must be considered when planning apron areas and adjacent service roads and buildings. For some aircrafts, it may be necessary to provide greater aircraft to aircraft separations or erect blast fences between parking spaces to counteract these effects.

9.4 TERMINAL APRONS

Passenger terminal apron

The type of terminal apron parking layout best suited to a particular airport is a function of many inter-related criteria. Design of the terminal apron must be completely consistent with the choice of terminal design and vice versa. The best combination of apron and terminal building should be selected after comparing advantages and disadvantages of several alternative designs through an iterative procedure. The primary considerations that govern efficient design are the movement and physical characteristics of the aircraft to be served, the maneouvering, staging and location of ground service equipment and ground facilities, the dimensional relationship of parked aircraft to the terminal buildings and the safety, security and operational practices to apron control.

The volume of traffic using the terminal is an important factor in determining apron layout that is most efficient in serving particular terminal design. The number of aircraft stands at a passenger terminal apron depends upon passenger aircraft movements by aircraft type during peak hours and their gate occupancy time.

The number of aircraft stands at a passenger terminal may be estimated by the following formula :

$$S = \Sigma \left(\frac{Ti}{60} \times Ni \right) + \alpha$$

where

    S = Required number of aircraft stands

    Ti = Gate occupancy time in minutes of aircraft group i

    Ni = Number of arriving aircraft group i during peak hours

    α = Number of extra aircraft stands as spare.

The number of arriving aircraft can be obtained either by simply dividing the previously calculated passenger aircraft movements by two or by applying a heavy direction factor particular to the airport which may be in the order of 0.6 to 0.7 meaning that arriving aircraft represent 60 to 70% of the total peak hour arriving and departing aircraft movement, where the airport is planned to have different passenger terminals for domestic, national, international and foreign carrier, the above formula should be applied individually for each case.

9.5 PASSENGER LOADING METHODS

Passenger loading methods must be taken into account when planning the apron layout. Following methods are used for aircraft loadings :

    (i) Bridge loading
        (a) The stationary loading bridge
        (b) The apron drive loading bridge
   (ii) Movable steps
  (iii) Passenger transporters
  (iv) Aircraft contained steps

9.5.1 Bridge loading

Direct upper level loading is made possible by the development of the loading bridge, permitting the passenger to board the aircraft from the upper level of the terminal building. Two types of loading bridges are used :

(a) The stationary loading bridge

A short loading bridge which extends from a projection in the building as shown in Fig. 9.2. The aircraft parks nose-in alongside the projection, and stops with the aircraft front door opposite the bridge. The bridge extends a very short distance to the aircraft, allowing very little variation between the height of the aircraft main deck and the terminal floor.

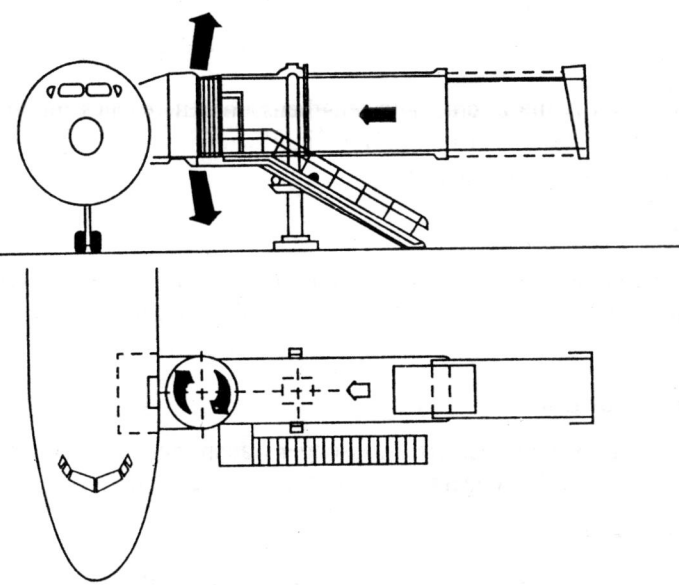

Fig. 9.2. Stationary loading bridge (ICAO).

(b) The apron drive loading bridge

In this type, one end of a telescoping gangway is hinged to the terminal building, the other end is supported by a steerable, powered dual wheel as shown in Fig. 9.3.

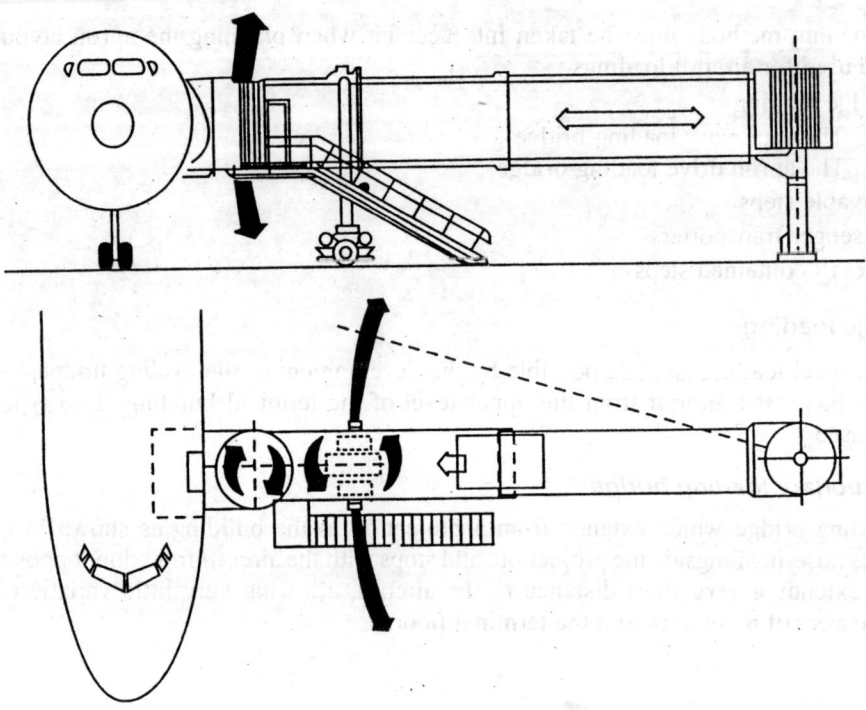

Fig. 9.3. Apron-drive bridge (ICAO).

The bridge pivots towards the aircraft and lengthens until it reaches the aircraft door. The end mating with the aircraft can be raised or lowered significantly, permitting aircraft of varying deck heights to be served from the loading bridge.

9.5.2 Movable steps

Movable steps are pushed or driven to the aircraft and set at door level. Passengers walk in the open on the apron or are driven by bus between the terminal and the aircraft and use the steps to board the aircraft.

9.5.3 Passenger transporters

Passengers board a bus or specially designed passenger transporters at the terminal building and are driven to remote aircraft stand. Passengers then use steps to board the aircraft.

9.5.4 Aircraft contained steps

Some aircrafts have self-contained steps. After stopping, the crew releases the self contained steps and passengers walk on the apron or are driven by bus between the aircraft and the terminal building.

9.6 TERMINAL APRON PARKING CONCEPTS

Apron arrangements are directly inter-related with the passenger terminal concept. Following paragraphs briefly describe the various concepts and their characteristics from the viewpoint of the apron.

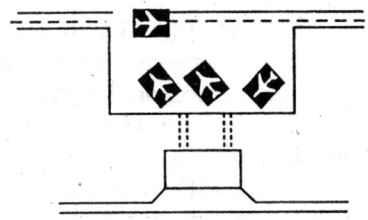

Fig. 9.4. Simple concept of parking (ICAO).

(i) Simple concept

This concept is applied at low traffic volume airports. Aircrafts are normally parked either angled nose-in or nose-out for self taxi-in or taxi-out (Fig. 9.4). Considerations should be given to providing adequate clearance between apron edge and terminal frontage facing airside to reduce the adverse effects of jet engine blast where this is not done, jet engine blast fence should be provided.

Apron expansion can be done incrementally in accordance with demands, causing little disruption of airport operation.

(ii) Linear concept

This concept is one of the advanced stages of a simple concept. Aircraft can be parked in an angled or parallel parking configuration. Nose-in parking affords relatively easy and simple manoeuvring for aircrafts taxiing into gate position.

With terminals requiring only a few aircraft stands, as at small airports, this system is commonly used. Linear system has also been used at airports with large number of stands, however, as the number of stands required increases, the passenger circulation between stands becomes more difficult and cost of the concourse related to each stand position increases. At busy traffic airports, it may become necessary to provide double apron taxiways to lessen the blocking of the taxiway, by push out operations. Fig. 9.5 shows the linear concept and its variations.

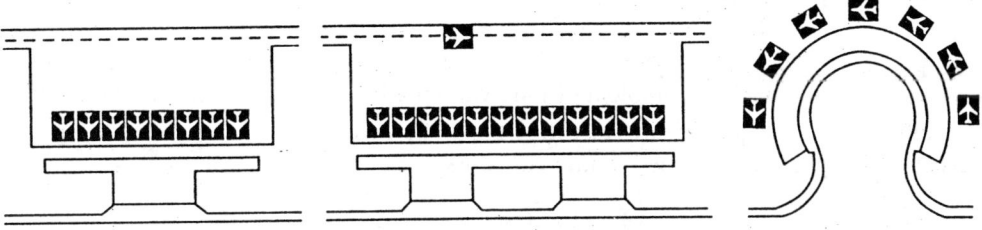

Fig. 9.5. Linear concept of barking and its variations (ICAO).

When apron size is planned from the outset to cater for the longest fuselage length, the linear concept has as much flexibility and expansibility as the simple concept and almost as much as the open apron concept.

(iii) Pier (Finger) concept

The use of pier concourses can greatly increase the apron area available to serve a given terminal core building. Fig. 9.6 shows several variations of this concept, according to the shape of the pier.

Aircrafts can be parked at gate positions on both sides of the piers, either angled, parallel or perpendicular (nose in). Where there is only a single pier, most advantages of the linear concept would apply for airside activities. When there are two or more piers, care must be taken to provide proper space between them. If each pier serves large number of gates, it may be necessary to provide double taxiways between piers to avoid conflicts between aircraft entering and leaving the gate positions.

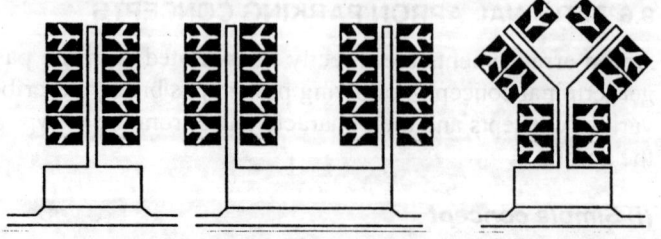

Fig. 9.6. Pier (finger) concept of parking (ICAO).

The walking distances from the ticket counter to the aircraft is another factor to be considered in pier concept of apron. This distance should be held to a maximum of 300 m. Both initial and final requirements should be considered in the planning phase of pier concept. If ultimate stand requirements do not exceed about twelve, a single pier should be used. If ultimate requirement is about 20, a double pier system may be more desirable. As the number of stand requirements exceeds thirty, a system of multiple piers, becomes more efficient. For better fitting of the available spaces piers of the shape 'Y' and 'T' as shown in Fig. 9.7 can be used, but they are considered less efficient because the distance of the farther positions are increased with unduly increase in the average walking distance.

(iv) Satellite concept

The satellite system terminal concept was developed to free the apron of obstructions and to permit more compact aircraft parking patterns. This concept consists of a satellite unitt, surrounded by aircraft gate positions, separated from the terminal. The passenger access to a satellite from the terminal is normally via an underground tunnel or elevated corridor to best utilise the apron space, as shown in Fig. 9.8.

Depending upon the shape of the satellite, the aircrafts are parked in radial, parallel or some other configuration around the satellite. For radial parking, push back operation is easy but requires large apron space. Use of this concept involves long walks between surface transportation and aircraft stands. The use of people-mover system such as moving sidewalks, trains etc. between the terminal building and the satellite can alleviate this problem of long walking, but at considerable additional costs.

The satellite concept is particularly effective if a large percentage of transfer passengers connect between different flights in the same satellite concourse.

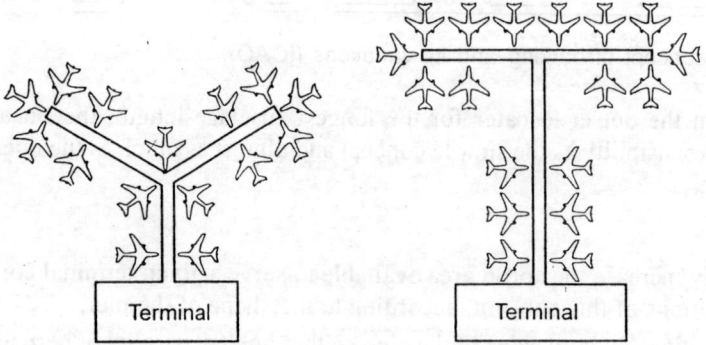

Fig. 9.7. 'Y' and 'T' type piers parking (ICAO).

Fig. 9.8. Satellite concept of parking (ICAO).

(v) Open apron (Transporter) concept

In the open apron system, aircrafts are parked away from the terminal building in rows as shown in Fig. 9.9.

Access to the aircraft with this system is by motor transporters. As the aprons may be ideally located for aircrafts close to the runway and remote from other structures, it provides advantages such as shorter overall taxiing distance, simple self-manoeuvring, ample flexibility and easy expansibility of aprons etc. However, as it requires transporting passengers, baggage and cargo for longer distances by transporters (buses/ mobile lounges), to and from the terminal, it can create traffic congestion problems on the air side.

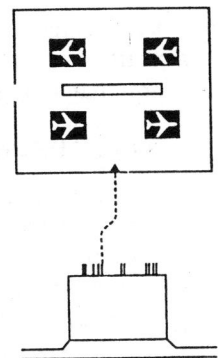

Fig. 9.9. Open-apron (transport) concept of parking (ICAO).

The movement of passengers to the waiting aircraft on foot, is considered unsafe, as it may be necessary to cross lanes of moving aircraft. This system is by far most flexible because additions or changes can be made to the apron area with little or no disruption to ongoing activities. In addition, if there are two taxiways serving each aircraft stand, one in front and the one behind the aircraft, the aircraft can manoeuvre solely under its own power. Though this form of aircraft parking is the simplest for the aircraft, but it has following disadvantages :

(a) The need for two taxiways increases the pavement area, and therefore, the cost.

(b) The apron service road cannot be located so that aircraft do not have to cross it.

(c) If a service road is located at the rear of the aircraft stands, the protection against breakaway blast can be given.

(d) Illumination and identification poles cannot be sited without increasing the aircraft stand separations.

(e) There are no areas suitable for parking apron equipment.

(f) The operating costs for some form of motor transportation can be excessive.

(vi) Hybrid concept

The hybrid concept means the combining of more than one of the above mentioned concepts. It is quite common to combine the transporter concept with one of the other concepts to cater to peak traffic. Fig. 9.10 shows a typical hybrid concept. Transporters take passengers to remote aprons/stands.

9.7 CARGO TERMINAL APRON

At the airports where the amount of air cargo is relatively small and mostly carried by passenger aircraft, there is no need to construct a cargo terminal exclusively for freight aircraft, and the cargo terminal building is best located close to the passenger terminal apron to minimise the travel distance, keeping in view the development of both activities in future.

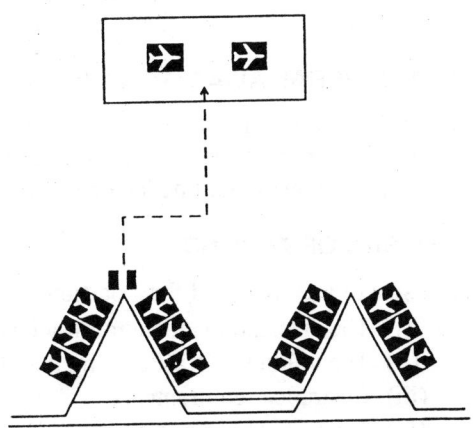

Fig. 9.10. Hybrid concept of parking (ICAO).

A separate cargo apron can be justified in view of the need, based on the air cargo forecasts. All cargo aircrafts are normally parked either parallel or nose-in, depending upon forecast volume and type of cargo handling system used.

9.8 PARKING APRON

At the airports, if aircrafts are obliged to be grounded for a long period of 6 to 8 hours or more, a parking apron may be justified. If such requirements of parking are not many and do not conflict with the peak hour periods of the aircraft, it may be possible for such aircrafts to remain at the terminals, however, as the number of such aircraft increases, it may be more economical to remove them from the passenger terminal apron and construct a separate parking apron.

The number of the required stands at a parking apron should be estimated based on the future needs and operating pattern of the airport. The parking apron should be located as close as is practical to the passenger terminal.

Sometimes, it may be economical to introduce transporters for an off-terminal parking apron. Such aprons as they are located remote from passenger terminal building are called 'remote parking aprons'.

9.9 SERVICE AND MAINTENANCE APRON

Maintenance and servicing of aircraft is an important preflight activity for safe and punctual operations.

The main categories of maintenance are line maintenance, airframe maintenance, power plant maintenance and component maintenance. The line maintenance can be carried out on a passenger apron and the airlines can schedule aircraft so that other types of maintenance are completed at their home base. Thus not all airports need to have a major maintenance area or apron.

For an airport which serves as the base for an airline, a maintenance terminal including hanger, workshop, storage and apron will be required. In addition to maintenance apron, it may be required to provide an engine test-run area with facilities to reduce engine blast and noise.

It is generally recommended that maintenance terminal aprons be located at a fairly remote area from the passenger apron.

Fixed aircraft servicing installations reduce apron congestion and permit shorter service times. The items of possible installation include hydrant fuelling, fixed ground power, potable/non-potable water supply, compressed air and airconditioning.

9.10 GENERAL AVIATION APRON

When an airport is also intended to serve general aviation airport, a general aviation terminal including a separate apron and other related facilities may be required. The general aviation terminal and its aprons, however, should be located as to minimise conflict with the scheduled aircraft operation.

9.11 SIZE OF APRONS

The size of area required for a particular apron layout depends upon the following factors :
   (i) The size and manoeuvrability characteristics of the aircrafts using the apron.
  (ii) The volume of traffic using the apron.
 (iii) Clearance requirements.
 (iv) Type of ingress and egress to the aircraft stand.
  (v) Basic terminal layout and parking system used.

(vi) Aircraft ground activity requirements.

(vii) Taxiway and service roads

9.11.1 Aircraft size and manoeuvrability

The size and manoeuvrability of the mix traffic expected to use apron must be known before a detailed design is undertaken. Fig. 9.11 shows the dimensions needed for stand space and Table 9.2 lists the values of these dimensions for several common type of aircrafts.

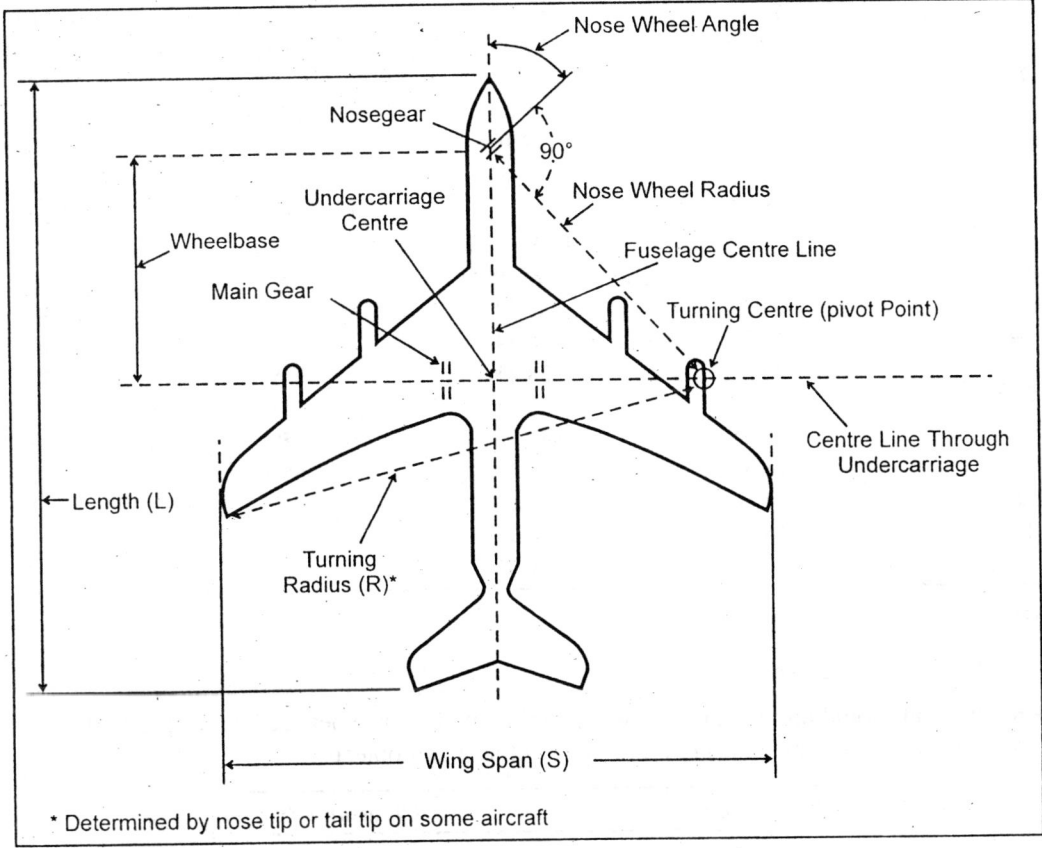

* Determined by nose tip or tail tip on some aircraft

Fig. 9.11. Dimensions for sizing aircraft stand spacing (ICAO).

The overall aircraft size i.e. the length and wing span can be used as the starting point in establishing the overall apron area requirement for an airport. The manoeuvrability characteristics of an aircraft depend upon the turning radius (R).

9.11.2 Volume of traffic

Apron areas should be planned for development as per the growth of volume of traffic. The number and size of aircraft stand position needed for any type of apron will depend upon the future forecasts during the period of planning. Aprons should be able to provide for reasonable peak activity with a minimum amount of delay.

Table 9.2. Selected aircraft dimensions

Aircraft type	Length (m)	Wing span (m)	Nose wheel angle	Turning radius (m)
A-300B-B2	46.70	44.80	50°	38.80[a]
B-727-100	40.59	32.92	75°	21.90[c]
B-727-200	46.68	32.92	75°	25.00[c]
B-737-100	28.65	28.35	70°	18.40[a]
B-737-200	30.58	28.35	70°	18.70[a]
B-747	70.40	59.64	60°	60.20[a]
B-757	47.32	37.95	60°	27.90[a]
B-767	48.51	47.63	60°	36.00[a]
BAC 111-400	28.50	27.00	65°	21.30[a]
Caravelle	36.70	34.30	45°	29.00[a]
Concorde	62.10	25.50	50°	30.10[c]
DC-8-40/50	45.95	43.41	70°	29.20[a]
DC-8-61/63	57.12	43.41/45.2	70°	32.70[c]
DC-9-10/20	31.82	27.25/28.5	75°	17.80[c]
DC-9-30	36.36	28.44	75°	20.40[c]
DC-9-40	38.28	28.44	75°	21.40[c]
DC-9-50	40.72	28.45	75°	22.50[c]
DC-9-80	45.02	32.85	75°	25.10[b]
DC-10-10	55.55	47.35	65°	35.60[a]
DC-10-30	55.35	50.39	65°	37.30[a]
DC-10-40	55.54	50.39	65°	36.00[a]
L-1011	54.15	47.34	60°	35.59[a]
Vickers Viscount 800	26.10	28.60	50°	21.60[a]

a = to wing tip, b = to nose, c = to tail.

9.11.3 Clearance requirement

An aircraft stand should provide the following minimum clearance between aircraft using the stand as well as between aircraft and adjacent buildings or other fixed objects :

Code letter	Clearance (m)
A	3.0
B	3.0
C	4.5
D	7.5
E	7.5

The clearance for code letters D and E can be reduced between the terminal (including passenger loading bridges) and the nose of an aircraft. These clearance may be increased as needed to ensure safe operation on the apron.

Location of aircraft stand taxilane and apron taxiways should provide separation distance between the centre line of these taxiways and an aircraft at the stand with following minimum distances :

Code letter	Minimum separation distances	
	Aircraft stand, taxilane centreline to object (m)	Apron taxiway centre line to object (m)
A	12.0	13.5
B	16.5	19.5
C	24.5	28.5
D	36.0	42.5
E	40.0	46.5

9.11.4 Type of ingres and egress

There are several methods used by aircrafts to enter or leave an aircraft stand. It may enter or leave an aircraft stand. It may enter or leave its position under its own power or it may be towed in and towed out or it may enter a position under its own power and be towed out. Apron size requirement should take into account these methods of ingress and egress as described below.

(a) Self-manoeuvring method

In this procedure an aircraft enters and leaves the aircraft stand under its own power without recourse to a tractor for any part of manoeuvre. Fig. 9.12 shows the area required for aircraft manoeuvring into and out of an aircraft stand position angled to the terminal building.

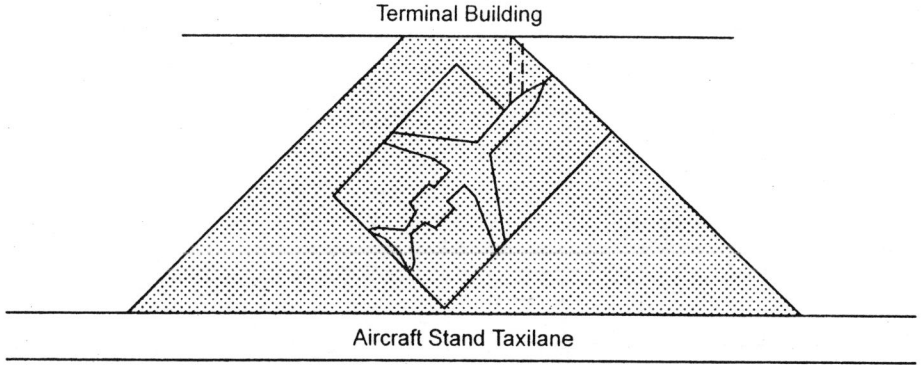

Terminal Building

Aircraft Stand Taxilane

Fig. 9.12. Self manoeuvring method (ICAO).

This requirement of area shall vary with the angle used during entry and exit manoeuvres. The normal manoeuvre of taxiing into and out of an aircraft stand which adjoins the terminal building or pier involves an 180° turn. The radius of this turn and geometry of the aircraft determine the aircraft stand spacing.

This method of parking requires more pavement area than tractor assisted methods, but there is the saving of the equipment and personnel required for tractor operation.

(b) Tractor assisted method

In this method ingress and egress requires the use of a tractor and tow bar. The most common procedure is the taxi-in, and push-out method, but aircraft can also be towed in and out in other combination.

Use of tractors allow a much closer spacing of aircraft stands, reducing both apron and terminal space required to accommodate a high volume of terminal aircraft parking. Fig. 9.13 shows the area required for aircraft that taxi-in and push-out perpendicular to the terminal building.

This procedure results in more efficient use of apron space than self manoeuvring procedure. This is a simple manoeuvre which can be made without creating excessive engine blast problems. It take an average of 3 to 4 minutes from the beginning of the push back until the tractor is disconnected and the aircraft is moving under its own power. The push-out operation requires skill and practice on part of the driver.

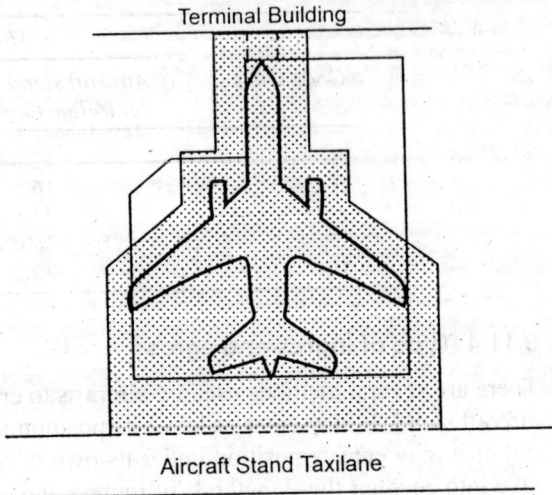

Fig. 9.13. Tractor assisted method (ICAO).

(c) Stand spacing

In the simplest case for aircraft that taxi in perpendicular to the terminal building and push-out straight back as shown in Fig. 9.14, the minimum stand spacing (D) simply equals the wing span (S) plus the required clearance (C).

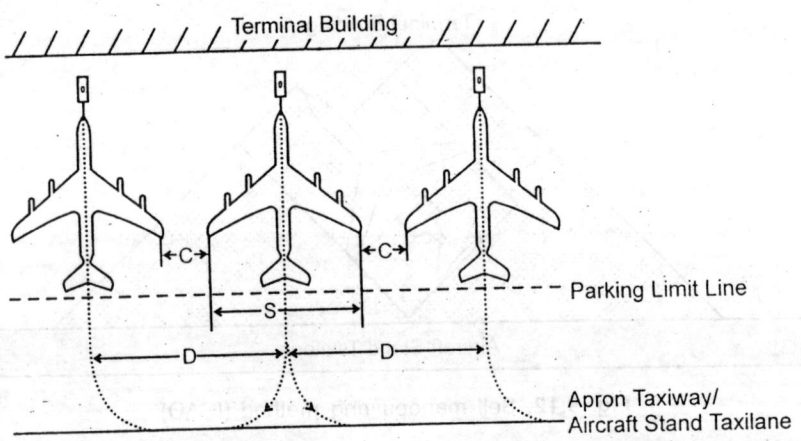

Fig. 9.14. Stand-spacing for taxi-in, push-out procedure (ICAO).

For other ingress and egress procedures or for other parking angles, the geometry is more complex and requires a detailed analysis for determining stand spacing. For example Fig. 9.15 shows the stand spacing for a self manoeuvring aircraft stand, which depends upon the angle the aircraft can comfortably manoeuvre into a stand position with other aircrafts parked in the adjacent positions.

9.11.5 Basic terminal layout and parking system

The basic parking systems, affecting size of the aprons have already been described in previous articles.

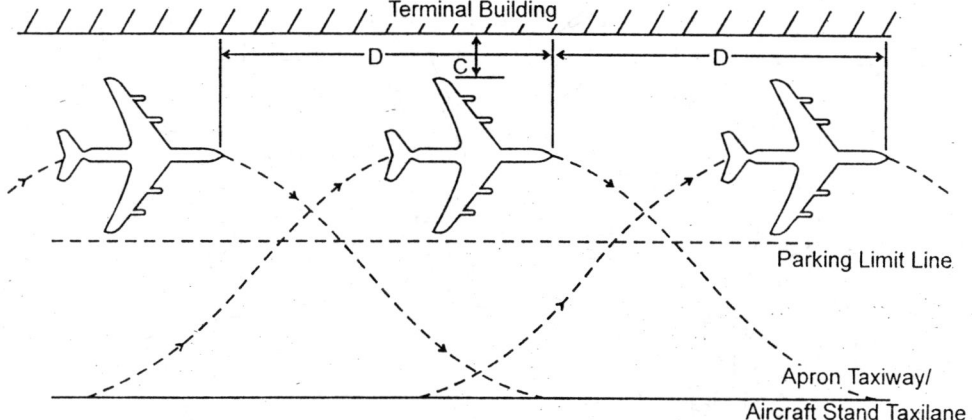

Fig. 9.15. Stand spacing for self manoeuvering procedure (ICAO).

9.11.6 Aircraft ground servicing

Passenger aircrafts need services during the time it is parked in a stand position. The normal services include toilet service, gallery service, baggage handling, potable water service, fuelling, airconditioning, oxygen, aircraft towing, electric power supply and starting air. Most of these functions have a vehicle and/or equipment associated with them or have some type of fixed installation established to conduct these services.

Fig. 9.16 shows a typical ground service layout plan for a medium sized aircraft. The areas to the right of the aircraft nose forward of the wing is often used as a prepositioned service area to store vehicles and equipment.

9.11.7 Taxiways and service roads

The total area needed for an apron includes not only the individual aircraft stand, but also the area required for apron taxiways, aircraft stand taxilane, and service roads needed to access the aircraft stands and provide necessary support services.

Location of these facilities will depend upon the terminal arrangement, runway locations and locations of off-apron services such as flight kitchen, fuel forms etc.

Aircraft stand taxilanes branch off of apron taxiways, which in turn are generally located on the edge of the apron pavement.

The space needed for service roads must be considered during the overall apron planning. They are usually located either adjacent and parallel to the terminal building or on the airside of the aircraft stand parallel to the aircraft stand taxilane.

The width of the service roads will depend upon anticipated level of traffic. Adequate clearance must be available under the loading bridges for the largest vehicle expected to use the road.

Overall apron planning should also take into account manoeuvering and storage areas for ground equipment.

9.12 Apron safety

In planning the location and design of aprons, the need to maintain security of operations from possible

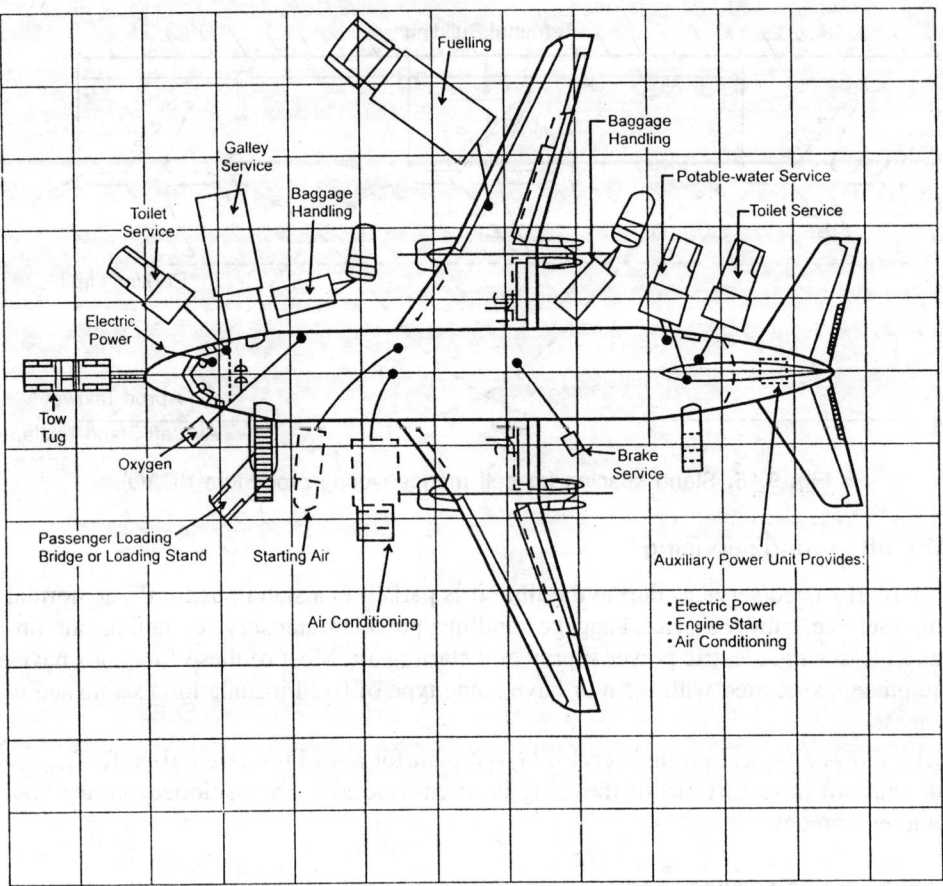

Fig. 9.16. Typical ground service layout (ICAO).

sabotage or armed aggression should be considered in areas where this may be a problem. This will require control of public access to apron, such as through doors in passenger building and any other barriers preventing the public from having ready access to the apron.

9.13 Holding bays and bypasses

Procedures for aircraft traffic control state that departures shall normally be cleared in the order in which they are ready for take-off, except that deviations may be made from this order of priority to facilitate the maximum number of departures with the least average delay. As the operations at an airport increase the deviation from departure sequence may be required. At high activity level, airports with single taxiways and no holding bays or other bypasses, provide airport controllers with no oportunity to change the sequence of departures once the aircraft has left the apron. In particular, at airports with large apron areas it is often difficult to arrange for aeroplane to leave the apron in such a way that they will arrive at the end of the runway in the sequence required by air traffic services unit.

Holding bays and bypasses, are therefore, provided where traffic volumes are high. The provision of an adequate number of holding bays and bypasses, based upon the analysis of current and future hourly aircraft departure demand, will allow a large degree of flexibility in controlling the departure sequence, to overcome delays and thus increase the capacity of an airport.

In addition holding bays and other bypasses allow :

(a) departure of certain aeroplanes to be delayed due to unforseen circumstances without delaying the following aeroplane.

(b) engine runups for piston type aircrafts or use as a VOR aerodrome check point.

(c) aircraft to carry out the pre-flight altimeter checks etc.

9.13.1 Types of bypasses

The taxiway features which allow an aircraft to bypass a preceeding aircraft are of following three types:

(i) Holding bays

(ii) Dual taxiways, and

(iii) Dual runway entrances.

(i) Holding bays

Holding bay is an area where aircraft can be held or by passed. Fig. 9.17 shows a typical example of a holding bay. There may be different configurations of holding bays some examples of which are shown in Fig.9.18. When a holding bay is used, aircraft can, on the basis of their priority take-off in any order. Holding bay allows aircraft to leave and independently re-enter the departure stream. Fig. 9.17 is a typical design of the pavement area for a holding bay located at the holding point for a precision approach runway where the code number is 3 or 4 and incorporates an aircraft wing-tip to wing-tip clearance of 12 m. Holding bay designs for other runway types or locations along the taxiway will have proportional dimensional requirements.

(ii) Dual taxiways

Dual taxiway is a second taxiway or a taxiway bypass to the normal parallel taxiway. Some examples are illustrated in Fig. 9.19.Dual taxiway or taxiway bypass can only achieve relative departure priority by separating the departure stream into two parts.

Taxi passes can be constructed at a relatively low cost, but provide only a small amount of flexibility to alter the departure sequence. A full length dual taxiway is an expensive alternative and can be justified at very busy airports where there is a clear need for two-directional movement parallel to the runway.

(iii) Dual runway entrances

Dual runway entrances is a deplication of the taxiway entrance to the runway. Some typical examples are shown in Fig. 9.20.

The dual runway entrance reduces the take-off run available for aircraft using the entrance not located at the extremity of the runway. The remaining take-off length should be adequate for the take-off operation. A dual runway entrance also makes it possible to bypass an aircraft delayed on another entrance taxiway or even at the extremity of the runway. The use of dual entrances in combination with dual taxiways will give a degree of flexibility comparable with that obtained with a well designed holding bay.

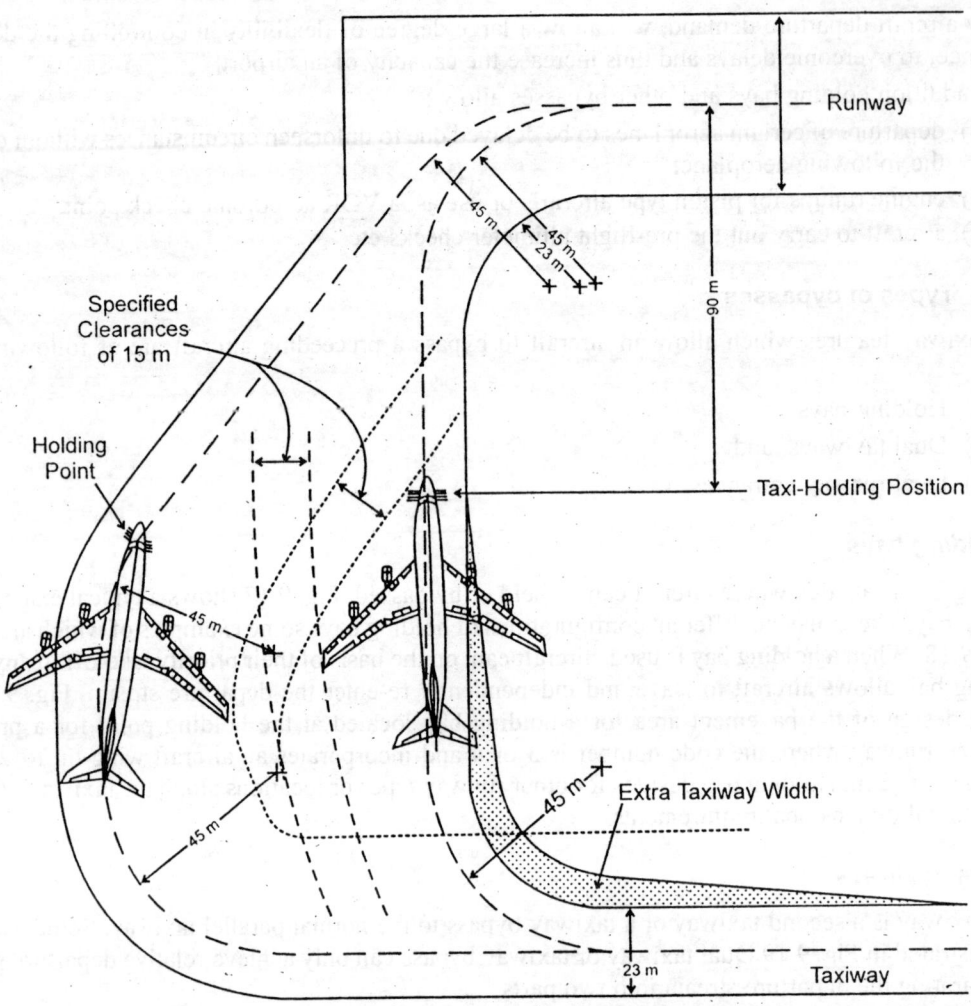

Fig. 9.17. Detailed example of holding bay (ICAO).

Oblique entrances permit entry at same speed but they make it more difficult for the pilot to see aircraft approaching to land. They are also more expensive to provide because of larger paved area required.

For a given airport, the best choice between above three methods depends upon the geometry of the existing runway/taxiway system and the nature of the aircraft demand, the economic considerations will often be decisive when choosing between the three types. These three types can also be used in various combinations to optimise the efficient surface movement of aircraft to the threshold.

Regardless of the type of bypass used, minimum centre line to centre line separation between taxiway and runway must be maintained, as specified for the type of runway served.

The design used should provide at least one entrance to the beginning of the runway usable for take-off so that aircraft requiring the entire take-off run available may easily align themselves for take-off without significant loss of runway length.

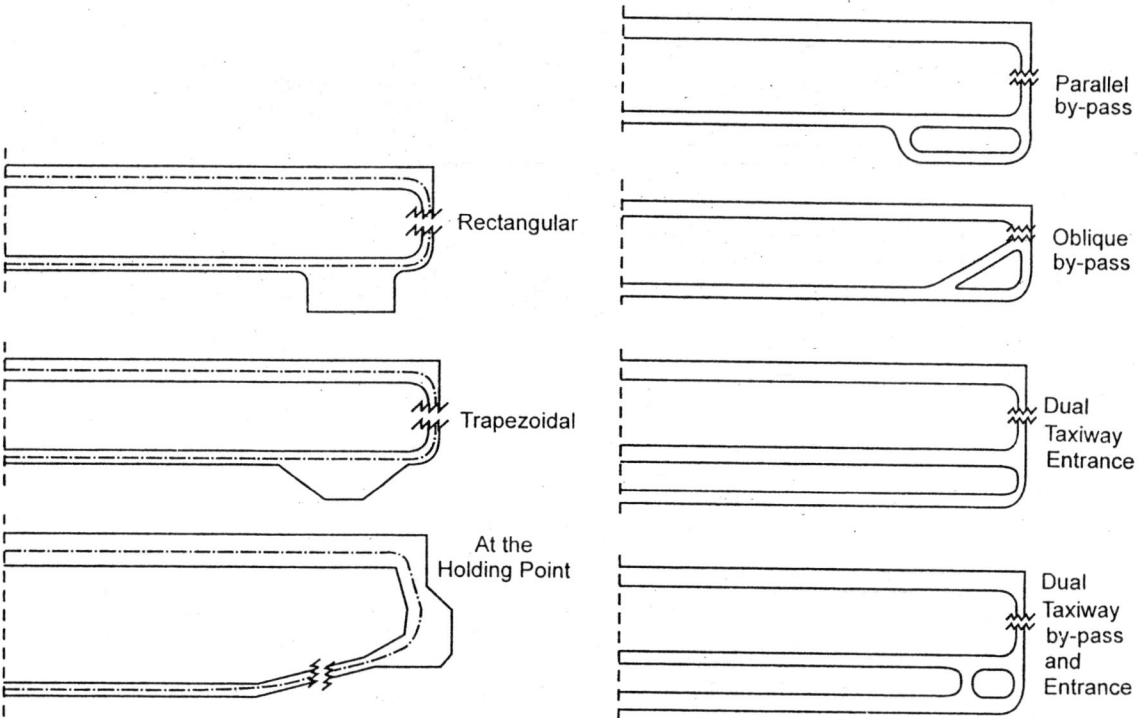

Fig. 9.18. Examples of holding bay configurations (ICAO).

Fig. 9.19. Examples of dual taxiways (ICAO).

9.13.2 Size and location of holding bays

The space required for a holding bay depends upon the number of aircraft positions to be provided, the size of aircraft to be accommodated, and the frequency of their utilisation. The dimensions must allow for sufficient space between aircraft to enable them to manoeuvre independently.

In general, the wing tip clearance between a parked aircraft and one moving along the taxiway should not be less than that given in the following table.

Code letter	Wing tip clearance
A	4.5 m
B	5.25 m
C	7.5 m
D or E	12 m

When used to allow flexible departure sequencing, the most advantageous location for a holding bay is adjacent to the taxiway serving the runway end. Other locations along the taxiway may be suitable for performing pre-flight checks or engine run ups or as a holding point for aircraft awaiting departure clearance.

Criteria for the location of holding bays with respect to the runway are given below :

The distance between a holding bay or a taxi holding position and the centre line of a runway shall

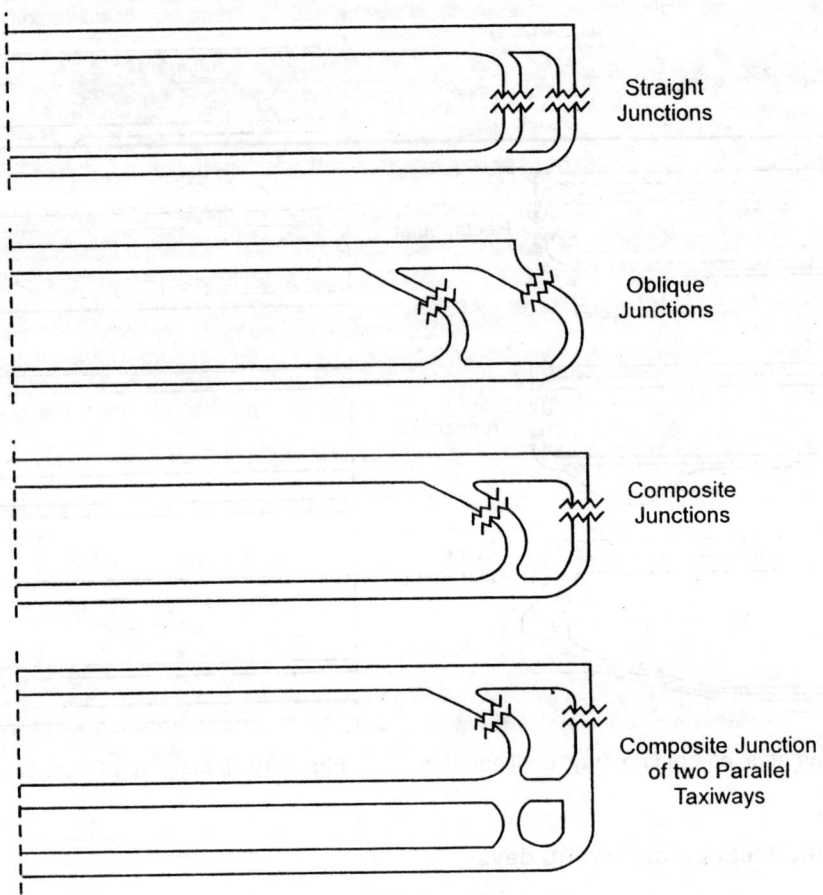

Fig. 9.20. Examples of dual runway entrances (ICAO).

be as given in Table 9.3. In the case of a precision approach runway it should be such that holding aircraft will not interfere with the operation of radio aids.

At elevation greater than 700 m the distance of 90 m specified in above table for precision approach runway code number 4 should be increased as follows :

    (a) up to an elevation of 2000 m : 1 m for every 100 m in excess of 700 m,

    (b) elevation in excess of 2000 m and up to 4000 m, 13 m plus 1.5 m for every 100 m in excess of 2000 m, and

    (c) elevation in excess of 4000 m and up to 5000 m, 43 m plus 2 m for every 100 m in excess of 4000 m.

If the holding bay or taxi holding position for a precision approach runway code number 4 is at a greater elevation compared to the threshold, this distance of 90 m as specified in above table should be further increased 5 m for every metre the bay or position is higher than the threshold.

The distance of 90 m for code 3 or 4 based on an aircraft with a tail height of 20 m, a distance from the nose to the highest part of the tail of 52.7 m and a nose height of 10 m, holding at an angle of 45° or

Table 9.3. Minimum distance from the runway centre line to a holding bay or taxi-holding position

Type of runway	Code number			
	1	2	3	4
Non-instrument	30 m	40 m	75 m	75 m
Non-precision approach	40 m	40 m	75 m	75 m
Precision approach category I	60 m[b]	60 m[b]	90 m[a,b]	90 m[a,b]
Precision approach categories II and III	–	–	90 m[a,b]	90 m[a,b]
Take-off runway	30 m	40 m	75 m	75 m

a. If a holding bay or taxi-holding position is at a lower elevation compared to the threshold the distance may be decreased 5 m for every metre the bay or holding position is lower than the threshold, contingent upon not infringing the transitional surface.
b. This distance may need to be increased to avoid interference with radio aids; for a precision approach runway category III the increase may be of the order of 50 m.

more with respect to the runway centre line, being clear of the obstacle free zone and not accountable for the calculation of obstacle clearance altitude/height.

The distance of 60 m for code number 1 or 2 is based on an aircraft with a tail height of 8 m, a distance from the nose to the highest part of the tail of 24.6 m and a nose to the highest part of the tail of 24.6 m and a nose height of 5.2 m holding at an angle of 45° or more with respect to the runway centre line, being clear of the obstacle free zone.

9.14 SEGREGATION OF AIRCRAFTS AND VEHICLES

Aircrafts and ground vehicles interact in their movement on the runways, taxiways and aprons.

On the apron areas ground vehicles are used for loading/unloading of passengers, baggage, cargo and mail, gallery services, fuelling service, provision of compressed air for engine starting, aircraft maintenance, electric power and airconditioning etc.Emergency and security vehicles also move on apron areas.

Ground vehicle activities that occur on areas other than aprons include the following :

Emergency operations : Rescue and fire fighting equipment, which may be required at any point of the airport or runway approach areas.

Security operations : Vehicles used for the patrol of boundary lines and restricted areas.

Aerodrome maintenance and construction : Activities like repair of pavement, navigational aids, and grass mowing etc. require use of ground vehicles.

The interaction should be minimised in the planning phase by proper segregation of air and ground traffic to minimise collisions and increase safety and efficiency of aircraft movement. Due to different physical characteristics of airports it is not possible to develop specific design criteria for segregation, however, following measures and concepts could be used for segregation.

9.14.1 Methods of achieving segregation

The degree of segregation which can be achieved depends largely on the available space. More is the available space on aprons, easier it is to segregate. The other factor on which degree of segregation depends is dimension and other characteristics of aircrafts (wingspan, manoeuvarability, jet blast etc.) and nature of ground vehicles.

The need to keep the volume of ground traffic should be minimised. All ground vehicles not required in the movement area should be excluded. Measure should be taken to prevent unauthorised access of public vehicles in the movement area, by providing fences, gates and other security systems.

Airside service roads should be planned to eliminate or lessen the necessity for the use of runways and taxiways by ground vehicles. Such roads may provide direct access to navigational aids, construction areas, etc. For terminals with loading bridges, airside roads may beneath the immovable part of the loading bridge. Fig. 9.21 shows examples of airside service roads used on aprons. Following considerations should be kept in mind.

- Airside service roads should not cross runways and taxiways. Tunnels may be used beneath runways and taxiways to avoid such crossings.
- Need for provision of emergency access roads for used by rescue and fire fighting vehicles to various areas of aerodrome and in particular to the approach areas up to 1000 m from the threshold should be realised and planned for.

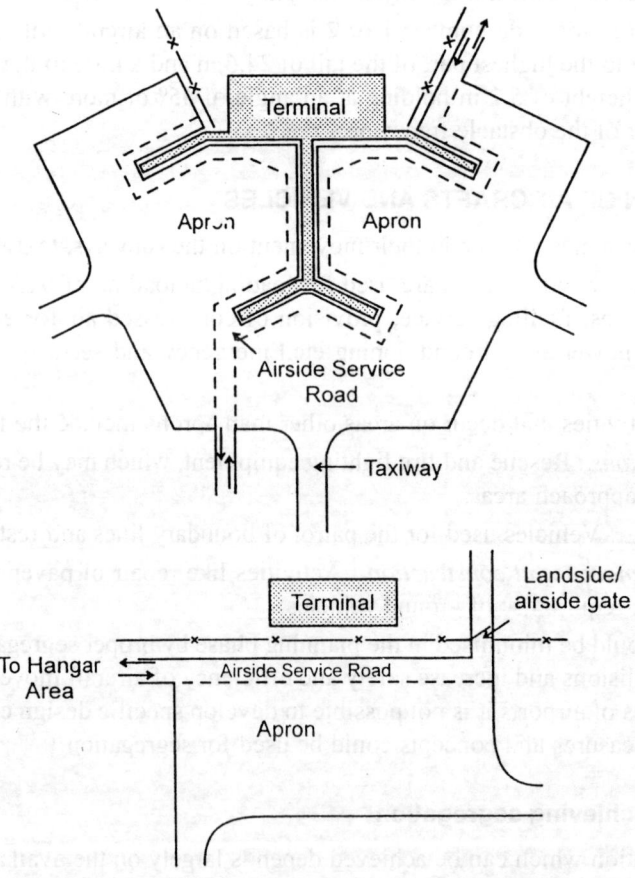

Fig. 9.21. Examples of airside service roads (ICAO).

- Service roads to navigational aids should be planned in such a way to minimise interference to the function of the aids.
- If an access road has to cross an approach area it should be located so that vehicles travelling on them are not obstacles to airport operations.
- The airside service road system must be designed to account for local security measure. Ground vehicle movement should be strictly controlled if they affect surface movement of aircraft on runways and taxiways.

  Signs, signals and two way radio communications can be used to control the movement.
- Many apron service vehicles can be eliminated by using fixed service installations either set within the apron or within terminal building adjacent to the aircraft stands. For example provision of hydrant fuelling system, compressed air outlet, static power supply, drainage outlets, drinking water pipes etc. Loading bridges for passenger loading and unloading also reduces use of ground vehicles for this purpose. The initial cost of these fixed installations may be high and they may not provide required flexibility for different aircraft types. Initial proper planning may reduce these disadvantages of fixed service installations.
- Point markings should be used to facilitate the segregationof traffic on apron. Markings can be used to provide guidance to assist pilots in manoeuvring their aircraft safety and expeditiously on aprons. Markings may be used to designate safety limits for placement of equipment on apron e.g., wing tip clearance lines etc.

CHAPTER 10

Structural Design of Airport Pavements

10.1 INTRODUCTION

Airport pavements provide adequate support for the loads imposed by aircrafts using the airport, and produce a firm, stable, smooth, all-year, all-weather surface free from dust or other particles that may be blown or picked up by propeller wash or jet blast. In addition, it must possess sufficient inherent stability to withstand, without damage, the abrasive action of traffic, adverse weather conditions, and other damaging elements.

Types of pavements used for airports are 'flexible' and 'rigid'. The design and evaluation techniques are similar to highway pavements. A flexible pavement consists of one or more layers of material named as surface, base and sub-base course resting on a prepared subgrade layer. A rigid pavement consists of a slab of portland cement concrete. The layer directly under the slab is called sub-base layer as shown in Fig. 10.1. The main funcion of a pavement is to bear heavy loads of aircrafts. Stresses are transmitted over larger area of the underlying soil causing deformations within elastic range.

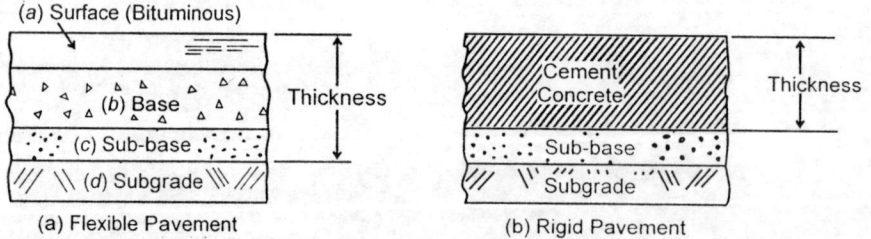

Fig. 10.1. Typical airport pavements. (a) Flexible pavement, (b) Rigid pavement.

Structural design involve, working out the total thickness of pavement to protect the subgrade soil from excessive deformation.

(a) **Surface course or wearing course** includes hot mix bitumen, sand-bituminous mixtures, sprayed bituminous surface treatment or portland cement concrete. It must prevent penetration of surface water to the base course, provide a smooth, well bonded surface free from loose particles, resist shear stresses, and furnish a texture of non-skid qualities without causing undue wear on tires.

(b) **Base course** consists of variety of different materials which may be treated or non-treated such as crushed or uncrushed aggregates, mixed with stabilisers such as cement,bitumen etc. It is principal structural component of the flexible pavement, which distributes the imposed wheel loads to pavement foundation, the subbase and/or subgrade. It should prevent failure in the subgrade, withstand stresses produced in the base, resist vertical pressure resulting in distortion of the course.

(c) **Sub-base course** consists of a graular material, a stabilised granular material or a stabilised soil. Its function is similar to base course. It is subjected to lower loading intensities.

(d) **Subgrade** depending upon the soil type and ground water conditions geotextile - a permeable flexible textile material layer is sometimes used to provide separation between pavement aggregate and the underlying subgrade. Subgrade is subjected to lower stresses than surface, base and sub-base courses. Compaction control is important for subgrade to resist shear deformation.

Though basic principles of design of highways and airport pavements are common, several differences exist, notable among them are :

- Magnitude of applied loads.
- Tire-pressures.
- Number of repetitions of load applied during the entire life of pavement.
- Impact of loads.
- Geometric section of the pavement.
- Increased attention towards drainage.
- Design life and considerations for stage construction.

10.2 CONSIDERATIONS IN STRUCTURAL DESIGN OF PAVEMENTS

Design considerations associated with both aircraft and pavement must be recognised to produce satisfactory design. Subgrade soil ultimately provides support for the pavement and imposed loads. The pavement serves to distribute the imposed loads to the subgrade over an area greater than that of the tire contact area. The greater the thickness of pavement, the greater is the area over which the load on the subgrade is distributed. Weaker or unstable the subgrade soil, the greater is the required area of load distribution and consequently the greater is the required thickness of pavement.

It is impossible to cover entire field of structural design of airport pavement in a book of this size, however, the basic principles particularly important to the airport paving engineers are highlighted. Consultation of ICAO and FAA publications should be done for an exhaustive treatise on the subject.

The structural design of airport pavements consists of determining both the overall pavement thickness and thickness of component parts of the pavement. There are number of factors which influence the thickness of pavement. These include magnitude and character of aircraft loads to be supported, volume and concentration of traffic, quality of subgrade and materials comprising of pavement structure.

10.2.1 Equivalent single wheel load (design wheel load)

In design we must consider the vertical stress imposed by group of stationary wheels on one landing gear (assembly) at different depths beneath a pavement. It is apparent that, at the surface stresses will result from an individual wheel, whilst at great depth they will be identical to those imposed by one large wheel with the loading equivalent to the total load of all wheel together.

The stress distribution depends upon physical characteristics of the pavement layers. In rigid pavement, most of the stresses will be dissipated through concrete slab, the stresses in subbase and subgrade being relatively small, whilst in a flexible pavement the stresses in subbase and subgrade at similar depths will be higher, although they will depend, to some extent, upon the stiffness parameters of the pavement layer. However, in practical terms all flexible pavements are classified on the assumption that they have the same stiffness-characteristics.

It follows, therefore, that any flexible or rigid pavement design will have different strengths at different depths and these are expressed in terms of equivalent single wheel load ESWL (or LCN, which is a function of ESWL) at different depths. Additionally a concrete pavement can be classified in terms of its 'radius of relative stiffness' (which is a function of its depth).

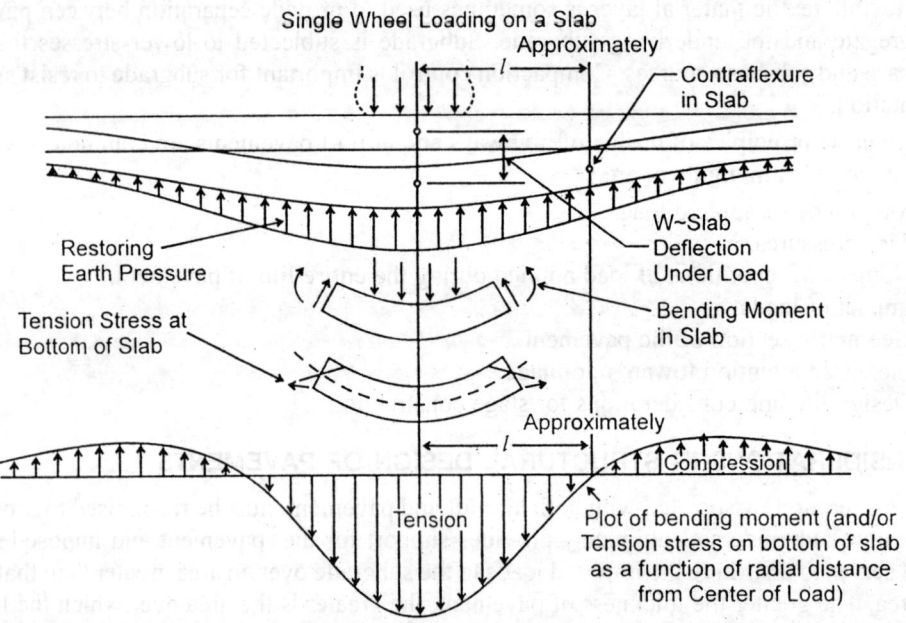

Fig. 10.2. Physical concepts of Westergaard's Radius of Relative Stiffness (l) (FAA).

Design considerations associated with aircraft and the pavement must be recognised and addressed in pavement design process. Aircraft manufacturers provide charts showing stresses produced, by their aircrafts at different loading conditions and tire pressures.

ESWL on flexible pavement of a given gear load is defined as single wheel which operating at the same tire (contact) pressure as the wheels in the assembly, produces critical effect in a particular pavement of surfacing, base, sub-base and subgrade, equivalent to those produced by the group of wheels. With a single wheel load under carriage it is thus constant, while with multi-wheeled under carriages it will increase with depth.

ESWL on rigid pavement is defined as the single isolated wheel load which, operating at the same tire (contact) pressure as the wheels in an assembly, produces stresses in the concrete equivalent to those produced by the whole assembly. In case of multi-wheeled under carriages, it will also, increase with depth of the concrete.

10.3 FUNDAMENTAL CONSIDERATIONS FOR FLEXIBLE PAVEMENTS

There are a number of factors which influence the thickness of pavement required to provide satisfactory service, including the following :

(a) Load

The pavement design methods are based on gross weight of the aircraft. For design purposes the pavement is designed for maximum take-off weight of the aircraft. The design procedure assumes 95 per cent of the gross weight is carried by the main landing gears and 5 per cent as carried by the nose gear. Wide body aircrafts require special attention, due to their different landing gear assemblies, special design curves are available for these aircrafts.

(b) Landing gear type and geometry

The gear type and configuration, dictate how the aircraft weight is distributed to the pavement and determine pavement response to aircraft loading. Gear configuration, tire contact areas, tire pressure affect the pavement required. Design curves are developed making reasonable assumptions for distribution of aircraft gross weight.

(i) *Single gear aircraft* : No special assumptions are required.

(ii) *Dual gear aircrafts* : Spacing between the dual wheels, dimensions between centre line of the tires, are used in load distribution, for design of pavement.

(iii) *Dual tandem gear aircraft* : Studies indicated that dual wheel spacing, and a tandem spacing for lighter aircraft and a dual wheel spacing and tendom spacing, for heavier aircrafts, are used in design of pavements, for load distribution.

(iv) *Wide body aircrafts* : Aircrafts like B-747, D-10 have large differences in gross-weights and gear geometrics, and therefore, separate design curves are prepared for wide body aircrafts.

(c) Stress distribution

For a single wheel, bulb of vertical stresses can be computed using Boussinesq's equation as shown in Fig. 10.3. Fig. 10.3(a) shows the zone of equal stress while Fig. 10.3(b) shows how they dissipate with

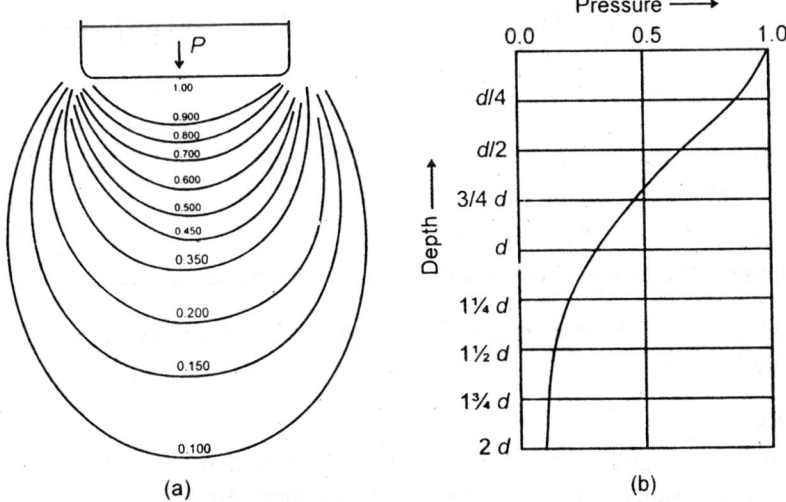

Fig. 10.3. Stress distribution beneath a single wheel.

depth. For a given load as tire pressure is reduced, the stresses are reduced proportionately, however, this will require large area of contact (assumed to be circular) thus limiting the reduction. At infinite depth, the stresses become independent of contact pressure. With increasing depth, the stress distribution rapidly dissipates, enabling progressively weaker materials to be employed.

Fig. 10.4 illustrates the effect with a twin wheel gear assembly. Fig. 10.4(a) shows the two pressure bulbs and their superimposition and Fig. 10.4(b) the resultant pattern of maximum vertical stress dissipation. This will occur, near the surface, under each wheel, but at depth under the geometric centre of the wheels. The rate of stress dissipation with depth is not the same as under the single wheel and will depend on spacing of the wheels.

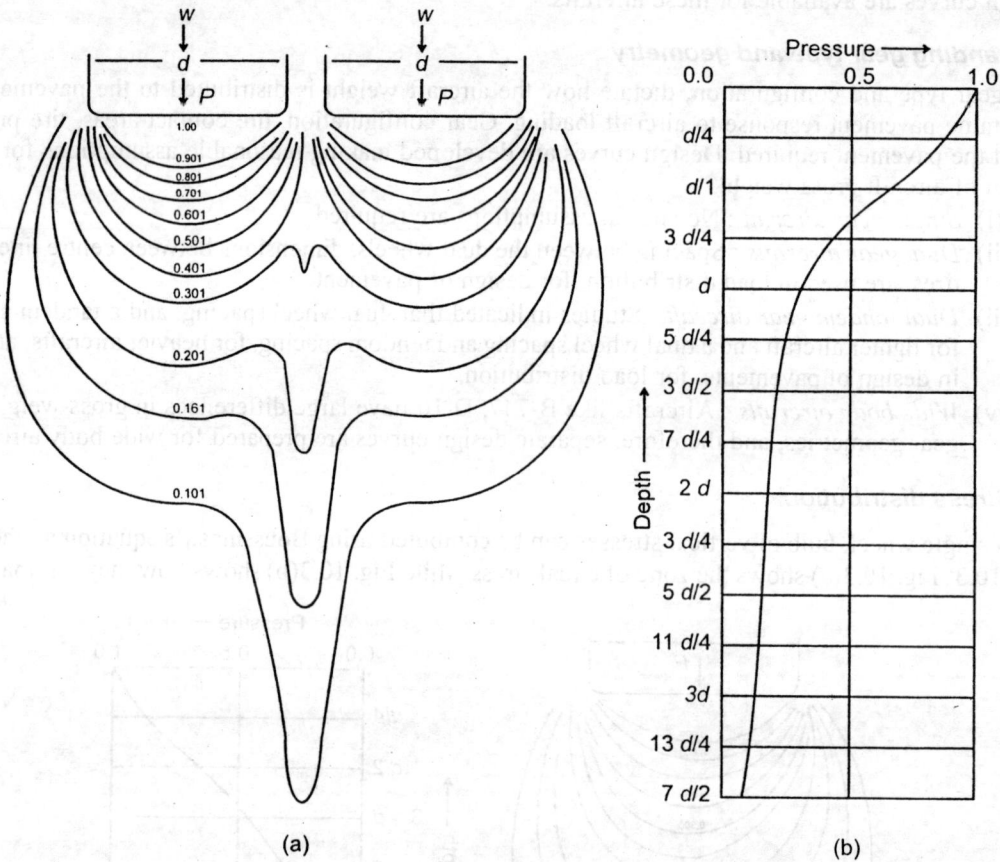

(a)   (b)

Fig. 10.4. Stress distribution beneath double wheel same pressure and individual loadings, as in Fig. 10.3.

Most commercial aircrafts have more than 1 wheel whilst the larger have 4 wheels in a twin-tandem configuration, which has a very complicated rate of stress distribution.

(d) Tire-pressure

Tire pressure may be 75 to 200 psi (0.52 to 1.38 MPa) or more depending on gear configuration and gross weight. As tire pressure asserts less influence on pavement stresses as gross weight increases and assumed 200 psi (1.38 MPa) may safely exceed if other parameters are not exceeded.

(e) Traffic volume

Forecasts of annual departures by aircraft type, are to be done for pavement design. Consultations with concerned authorities, study of airport master plan, and other statistical information is consulted for forecasting.

(f) Design aircraft

The design aircraft is selected on the basis of one requiring the greatest pavement thickness which may not necessarily be the heaviest one.

(g) Determination of equivalent annual departures by the design aircrafts

As the traffic forecast is a mixture of a variety of aircraft having different landing gear types and different weights, the effects of all traffic must be accounted for in terms of the 'design aircraft'.

First all aircrafts must be converted to the same landing gear type as the design aircraft. The following conversion factors are used to convert from one landing gear type to another (Table 10.1).

Table 10.1. Conversion factors for landing gears

To convert from	To	Multiply departures by
Single wheel	Dual wheel	0.8
Single wheel	Dual tandem	0.5
Dual wheel	Dual tandem	0.6
Double dual tandem	Dual tandem	1.0
Dual tandem	Single wheel	2.0
Dual tandem	Dual wheel	1.7
Dual wheel	Single wheel	1.3
Double dual tandem	Dual wheel	1.7

Secondly after the aircrafts have been grouped into same landing gear configuration, the conversion to equivalent annual departures of the 'design aircraft' is determined by the following formula :

$$\text{Log } R_1 = \text{Log } R_2 \times \left(\frac{W_2}{W_1} \right)^{1/2}$$

where R_1 = Equivalent annual departures by the design aircraft
R_2 = Annual departures expressed in design aircraft landing gear
W_1 = Wheel load of the design aircraft
W_2 = Wheel load of the aircraft in question

The above procedure is relative rating which compares different aircrafts to a common design aircraft.

Special considerations are needed for wide body aircrafts. After equivalent annual departures are determined, the design proceeds using appropriate design curves.

(h) Quality of materials used in pavement layers and subgrade

The quality of material used in surface, base and sub-base, is evaluated by different methods, suitable to requirement of design method in particular. California Bearing Ratio test (CBR) and plate bearing load tests are conducted on materials in the flexible pavements while flexural strength of concrete, elastic modulus and Poisson's ratio are determined as strength parameters in rigid pavements.

Quality of subgrade support is more important in flexible pavement design as it influences the pavement design. In rigid pavement subgrade support does not play that important role.

(i) Climatic conditions

The design of airport pavement must consider climatic conditions which will act on the pavement during its construction and service life. Severe cold and hot temperatures cause warping stresses, expansions and contraction of joints in the concrete slab, and develop subgrade restraint.

In frost areas, control of pavement deformation resulting from frost action is necessary by using non-frost-susceptible material or limiting frost penetration into subgrade, or by some other treatments of frost protection.

Example 10.1

An airport pavement is to be designed for the following forecast traffic :

Aircraft	Gear type	Forecast annual departures	Wheel load (kg)	Maximum take-off weight (kg)
727–100	Dual	3760	17240	72600
727–200	Dual	9080	20520	86500
707–320B	Dual tandem	3050	17160	148500
DC–9–30	Dual	5800	11630	49000
CV–880	Dual tandem	400	9940	83948
737–200	Dual	2650	12440	52440
L–1011–100	Dual tandem	1710	16160	204120
747–100	Double dual tandem	85	16160	317800

(a) Determine design aircraft.
(b) Group forecast traffic landing gear of design aircraft.
(c) Convert aircraft to equivalent annual departures of the design aircraft.
(d) Determine the annual departures and weight for designing the pavement.

Solution

(a) Pavement thickness is determined for each aircraft in the above forecast using appropriate design curves. The pavement input data i.e., CBR, K-value, flexural strengths etc. are kept same for all aircrafts. Aircraft weights and departures must correspond to particular aircraft as shown in the above forecast.

After doing this exercise, it will be observed that 727–200 aircraft requires greatest pavement thickness, therefore, the design aircraft is 727–200, which is equipped with dual wheel landing gear system.

(b) All the forecasted aircrafts are converted to dual wheel landing aircrafts as shown in table below.

(c) Equivalent annual departure of design aircraft are also shown in table below :

S. No.	Aircraft	Dual gear departure	Wheel load (kg)	Wheel load of design aircraft (kg)	Equivalent annual departures design aircraft (R_1) from formula
1.	727–100	3760	17240	20520	1891
2.	727–200	9080	20520	20520	9080
3.	707–320B	3050 × 1.7 = 5185	17610	20520	9080
4.	DC–9–30	5800	11630	20520	682
5.	CV–880	400 × 1.7 = 680	9940	20520	94
6.	737–200	2650	12440	20520	463
7.	L–1011–100	1710 × 1.7 = 2907	16160*	20520	1184
8.	747–100	85 × 1.7 = 145	16160*	20520	83
	Total				16241

* Wheel loads for wide body aircrafts (L–1011–100 anmd 747–100) are taken as the wheel load for 136100 kg aircraft for annual departure calculations.

(d) For this example, therefore, pavement will be designed for 16000 annual departures of a dual wheel aircraft weighing 86500 kg the maximum take-off weight of design aircraft. The design should, however, provide for the heaviest aircraft in the traffic mixture when considering depth of compaction, thickness of a bituminous surface etc.

10.4 CALIFORNIA BEARING RATIO (CBR) METHOD OF FLEXIBLE PAVEMENT DESIGN

Among the various methods of flexible pavement design the CBR design method is most popular and practical. It is basically an empirical method, however, a great deal of research has been done with the method and reliable correlations have been developed. The structural design of pavement consists of determining both the overall pavement thickness and the thickness of components of the pavement (surface, base, sub-base).

California Bearing Ratio (CBR) test is basically a penetration test conducted at a uniform rate of strain. The force required to produce a given penetration in the material under test is compared to the force required to produce the same penetration in a standard crushed lime stone. The result is expressed as a ratio of the two forces. Laboratory and field CBR tests are performed in accordance with procedure laid down in testing manuals.

10.4.1 Design CBR

Subgrade soils are usually variable and selection of design CBR value requires some judgement. As a general rule of thumb the design CBR should be equal to or less than 85% of all the CBR values. This corresponds to a design value of one standard deviation below the mean. Sometimes it may be more economical to remove and replace a weak layer than designing for it. Local conditions dictate best treatment for improving subgrade.

10.4.2 Equivalent Single Wheel Load (ESWL)

To calculate single wheel load, for multi wheel gear loads, graphical methods of plotting on log-log paper as shown in Fig. 10.6 are used. From this curve the ESWL-depth relationship can be obtained with sufficient accuracy. To construct Fig. 10.6, it is necessary to know the load per wheel (W) tonnes, the tire pressure (μ) MPa, the spacing between dual wheels (S_t) meters, in the case of dual tandem gears, the diagonal spacing (S_d). The contact area is assumed to be elliptical and d is the thickness and P is single wheel load. Fig. 10.5 shows the above dimensions.

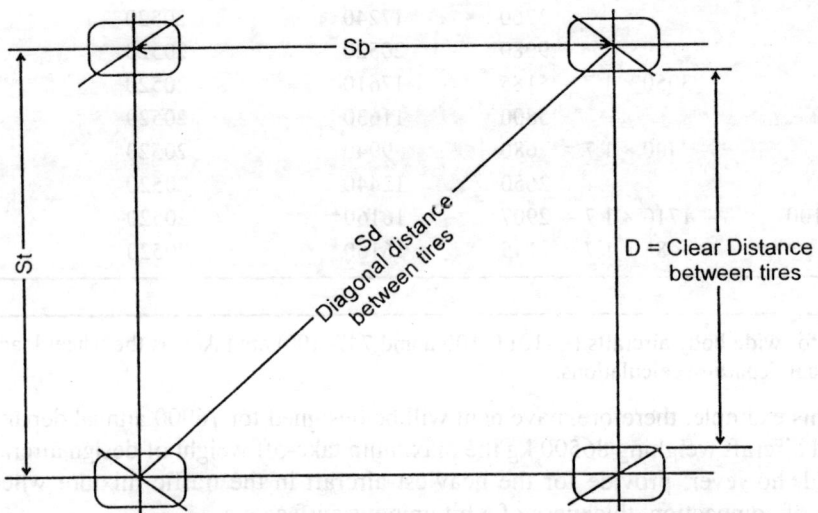

Fig. 10.5. Dual-in-tandem landing gear assembly.

To compute the equivalent single wheel load (ESWL) for a dual assembly, two points are plotted with coordinates as (D/2, P) and (2 S_d, 2 P) (Viscount aircraft, Boeing aircraft). For dual-in-tandem landing gear assembly where the number of wheels is increased from two to four the coordinates of two points A and B are taken as (D/2, P) and ($2S_d$, 4P). The definition of terms for a typical boeing and viscount land gear assembly is shown in Fig. 10.6.

For different gear configurations, separate flexible pavement design curves for several gear configurations have been prepared. One typical design chart is shown in Fig. 10.7 to illustrate a numerical example.

Example 10.2

Calculate the equivalent single wheel load at various depths for a viscount dual wheel assembly aircraft of 6.82 tonne gross wheel load. The clear distance between the tires (D) is 200 mm and distance centre to centre of tires (St) is 480 mm.

Solution

Plot two points as shown in Fig. 10.6 (dotted lines) with coordinates as (D/2, P) and ($2S_t$, 2P) point A and B. The values of these coordinate as shown are (100, 6.82) and (960, 13.64). The dotted straight line is drawn connecting these two points.

For this plot, ESWL can be read for different depths, beneath surface for given viscount, shown by dotted lines.

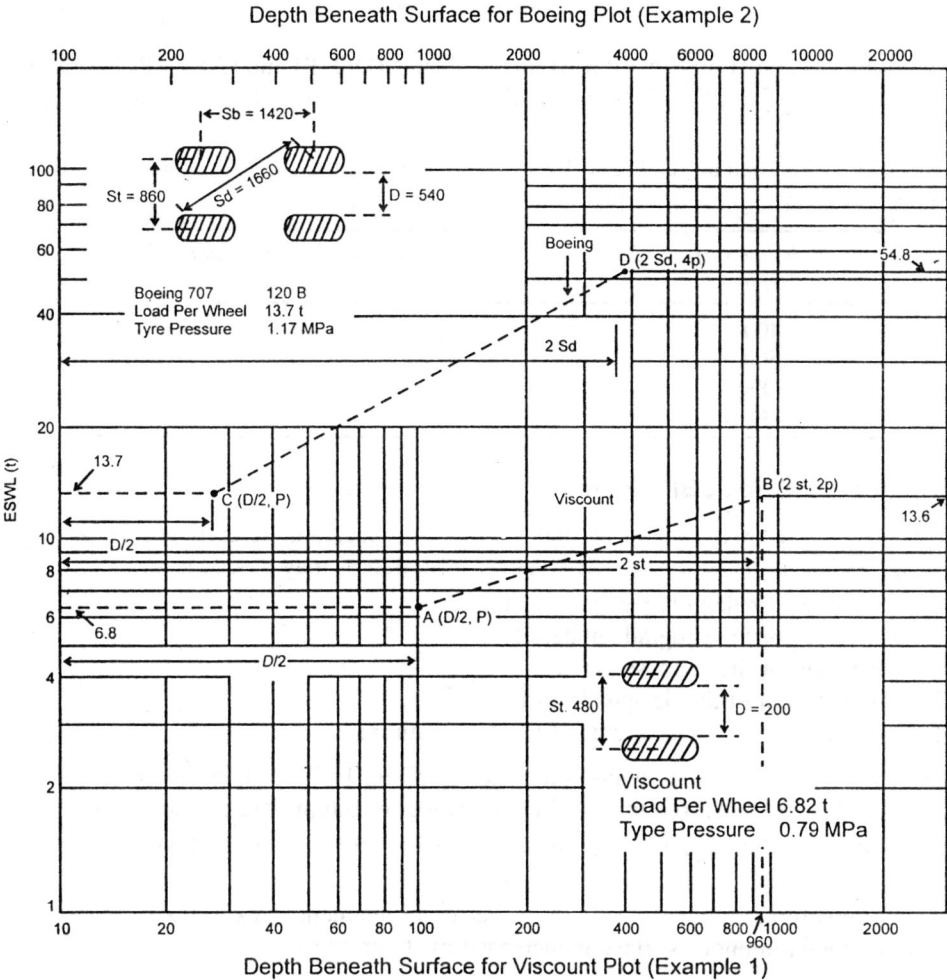

Depth Beneath Surface for Boeing Plot (Example 2)

Depth Beneath Surface for Viscount Plot (Example 1)

Fig. 10.6. Method of determination of ESWL at varying depths for given aircraft (multiple wheeled under carriage).

Depth below surface (mm)	Equivalent single wheel load (ESWL) (tonnes)
100	6.8
200	8
400	10
600	12.5
and so on	

Example 10.3

For a Boeing (120–B) with dual in tandem gears, calculate the equivalent single wheel load, at different depths beneath the surface. Load per wheel is 13.7 tonnes. The diagonal spacing between wheels (S_d) = 1660 mm and clear distance between the wheel D is 540 mm.

Solution

Plot the two points as shown in Fig. 10.6 with coordinates (D/2, P) and (2 S_d, 4P) i.e. point C and D with numerical values of as (27, 13.7) and (3320, 54.8). The straight line is drawn connecting those points.

From this plot ESWL can be read for different depths below the surface for the given Boeing (120–B) aircraft, as shown in Fig., using depth values given on the top.

Depth below surface (mm)	Equivalent single wheel load (ESWL) (tonnes)
400	17.5
600	20.0
2000	40.0
and so on.	

10.4.3 Design procedure : CBR method

As shown in Fig. 10.6, ESWL varies with depth for multi-wheeled undercarriages. Data shown can be plotted as Fig. 10.7 then the necessary CBR values at all depths may be deduced.

CBR design curves for flexible pavements require :
- A CBR value for the subgrade material.
- A CBR value for the subbase material.
- The gross weight of the design aircraft.
- The number of annual departures of the design aircraft.

Design curves for some gear assemblies are shown in Fig. 10.8 to 10.16. These curves indicate the total thickness of the pavement required and the thickness of bituminous surfacing.

Fig. 10.18 shows the minimum thickness of base course for a given total pavement thickness and CBR values.

For annual departures in excess of 25000, the total pavement thickness should be increased as given below and the bituminous surface is increased by 3 cm.

Annual departure level	Percentage of 25000 departure thickness
50000	104
100000	108
150000	110
200000	112

The design curves, determine the total thickness 'T' of the critical areas. For non-critical areas thickness factor of 0.9T may be used.

Example 10.4

A flexible pavement is to be designed for a dual gear aircraft having a gross mass of 34000 kg and 6000 annual equivalent departures of the design aircraft. Design CBR values for the sub-base and sub-grade are 20 and 6 respectively.

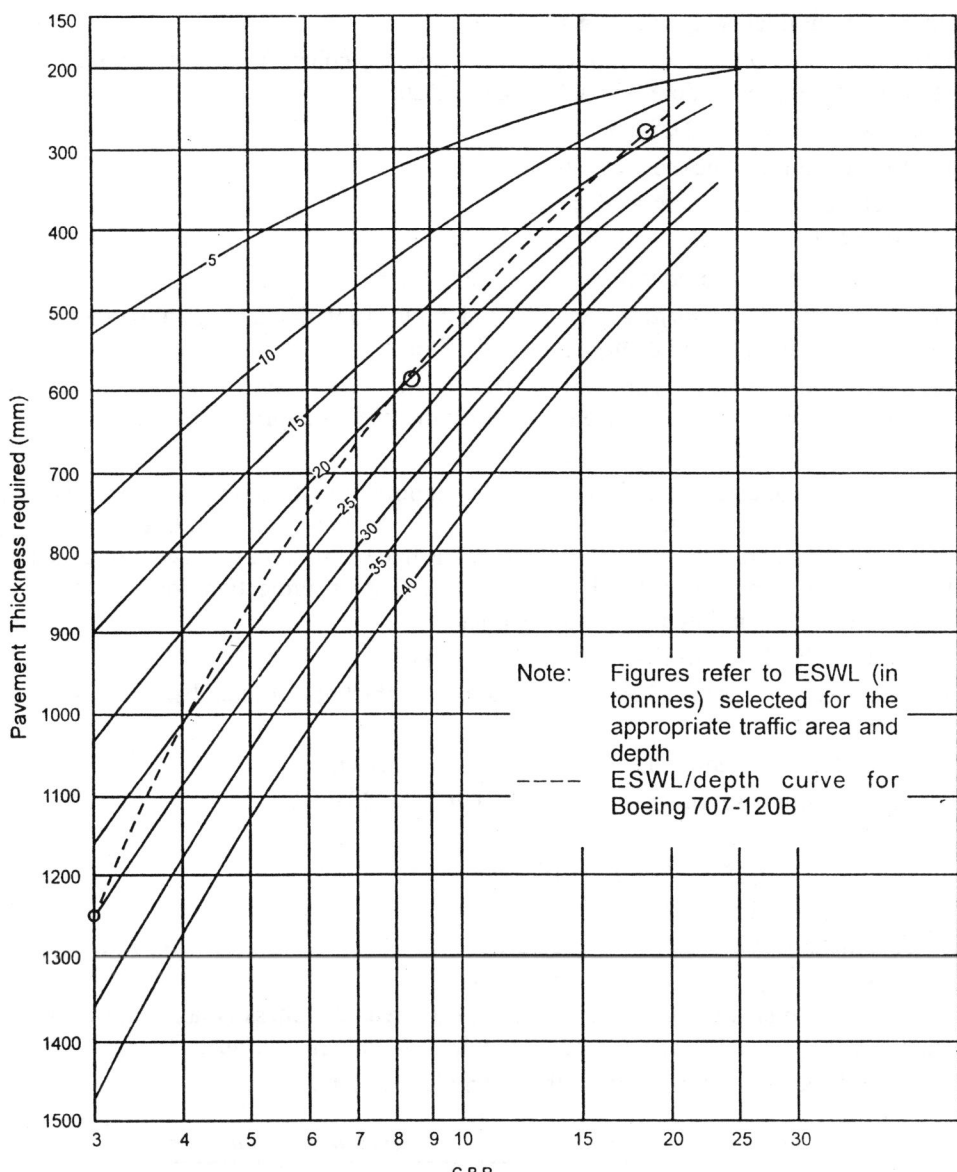

Fig. 10.7. CBR Design Curves for varying ESWL (t) small aircrafts.

Solution

(a) *Total pavement thickness* : For the dual gear aircraft refer to Fig. 10.9. Enter the upper abacissa with subgrade value 6. Project vertically downward to the gross mass of the design aircraft 34000 kg. At the point of intersection of the vertical projection and the aircraft gross weight, make a horizontal projection to the equivalent annual departures 6000. From the point of intersection of the horizontal projection and annual departure level, make a vertical projection down to lower abscissa **and read the total pavement thickness.**

Total pavement thickness = 51.2 cms.

(b) *Thickness of sub-base course* : Thickness of sub-base course can be determined from Fig. 10.9, using similar procedure with CBR value of 20.

Thickness of sub-base = 21.8 cms.

Therefore, combined thickness of bituminous

Surface and base course over = 21.8 cm

20 CBR base is

Sub-base thickness required = 51.2 – 21.8 = 29.40 cm.

(c) *Thickness of bituminous surface* : As indicated in Fig. 10.9, thickness of bituminous surfacing is 10 cm in initial areas and 8 cm in non-critical areas.

(d) *Thickness of base course* : Thickness of base-course can be determined by substracting the thickness of bituminous surface from combined thickness of surface and base material.

Thickness of base course = 21.8 – 10 = 11.8 cms.

Thickness so calculated should be compared with minimum thickness of base course required as shown in Fig. 10.17. Enter the left ordinate, with the total pavement thickness of 51.2 cm. Make a horizontal projection to the subgrade CBR line 6. From the intersection of the horizontal projection and the subgrade CBR line, make a vertical projection down to the lower abscissa and read the minimum base course thickness.

The minimum thickness of base required = 15 cms.

The extra thickness base is taken out of the sub-base thickness, rather than increasing the total pavement thickness.

Additional thickness required = 15 cm – 11.8 cm = 3.2 cm.

Thus reduced sub-base thickness = 29.40 – 3.20 = 26.20 cm.

Final thickness will be "T"

  (i) Thickness of surface course = 10 cms

  (ii) Thickness of base course = 15 cms

  (iii) Thickness of sub-base = 26.2 cm

      Total   51.2 cms

(e) *Thickness of non-critical areas* : For non-critical areas thickness is taken 0.9 of the critical pavement base and sub-base thickness + required bituminous surface.

Thus final thickness requirements in this example would be :

	Thickness requirement	
	Critical areas (cms)	Non-critical areas (cms)
Bituminous surface course	10	8
Base course	15	13
Sub-base course	27	24
Total	52	45

Since design aircraft weighs less than 45,300 kg, stabilised base and sub-base are not required, but may be used if local conditions require.

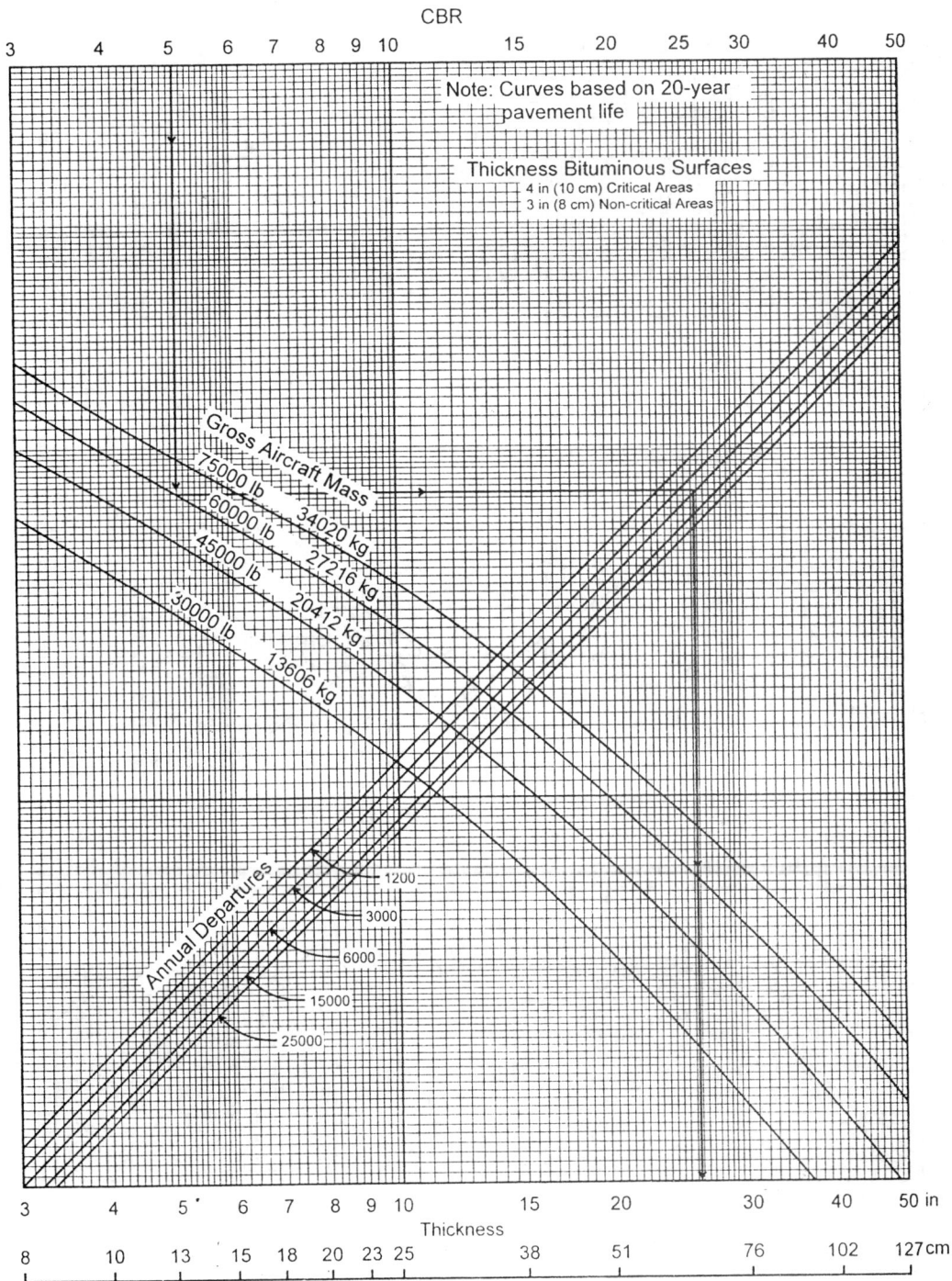

Fig. 10.8. Flexibla pavement design curves for critical areas, single wheel gear (ICAO).

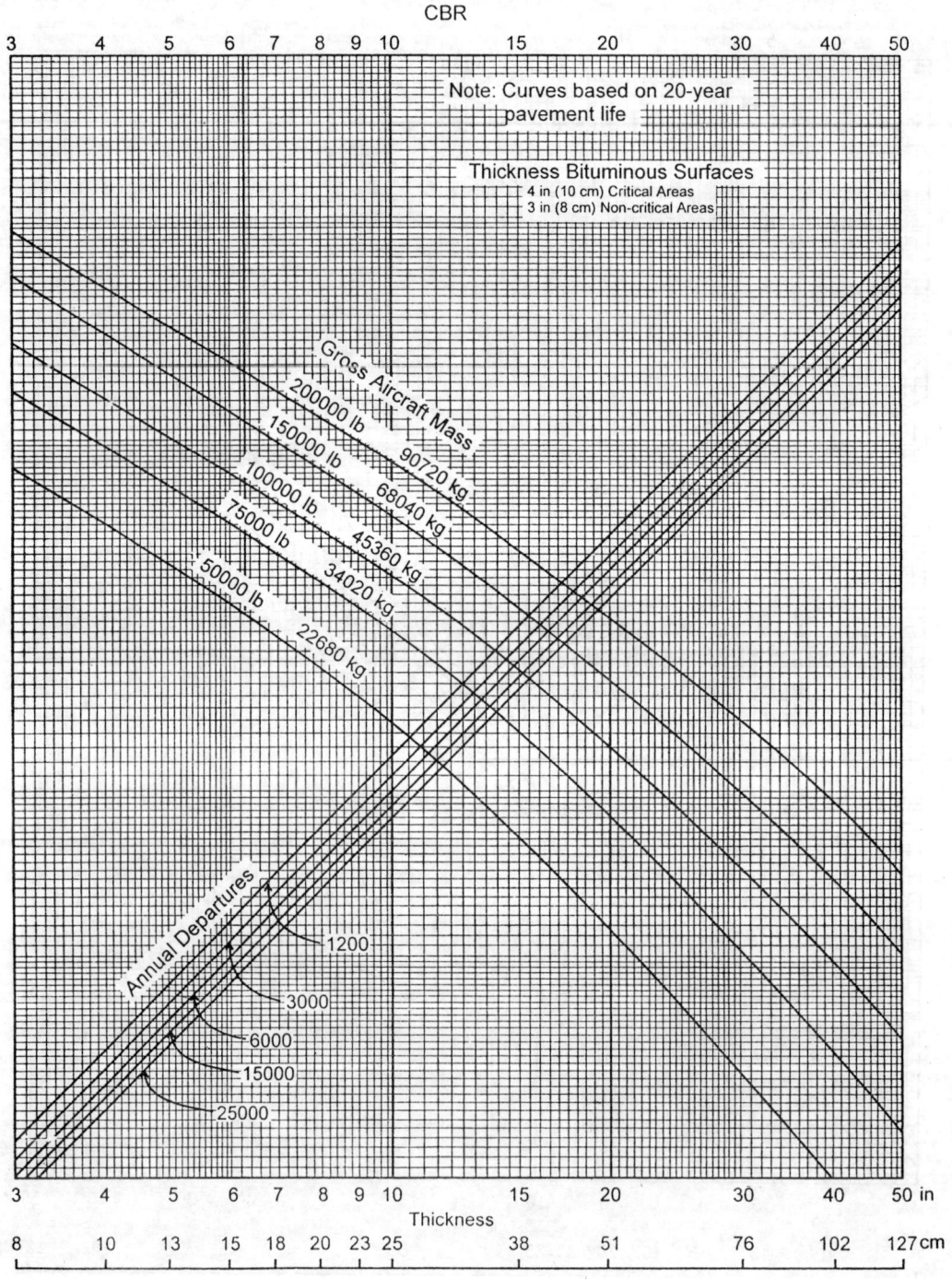

Fig. 10.9. Flexible pavement design curves for critical areas, dual wheel gear (ICAO).

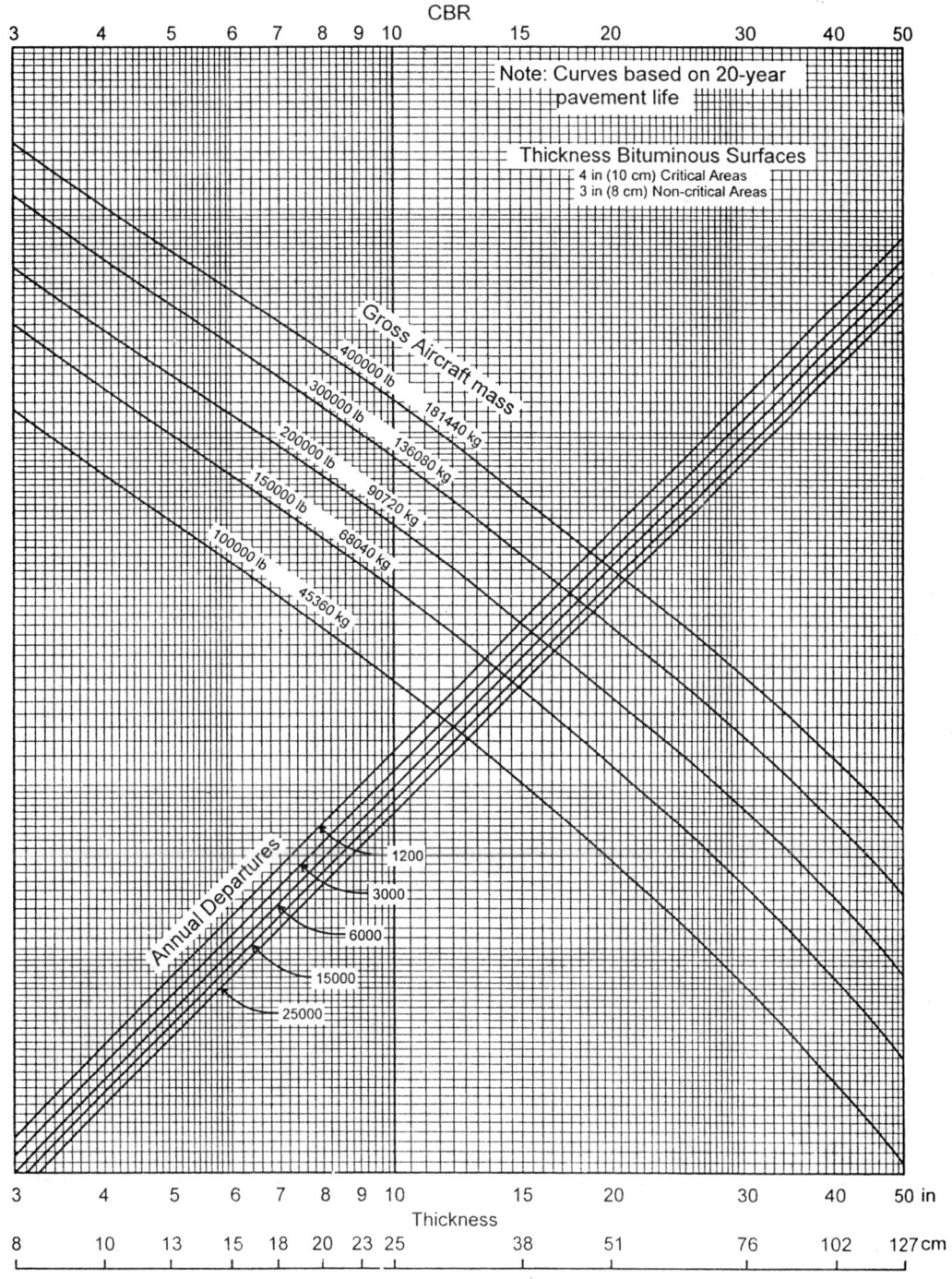

Fig. 10.10. Flexible pavement design curves for critical areas, dual tandem gear (ICAO).

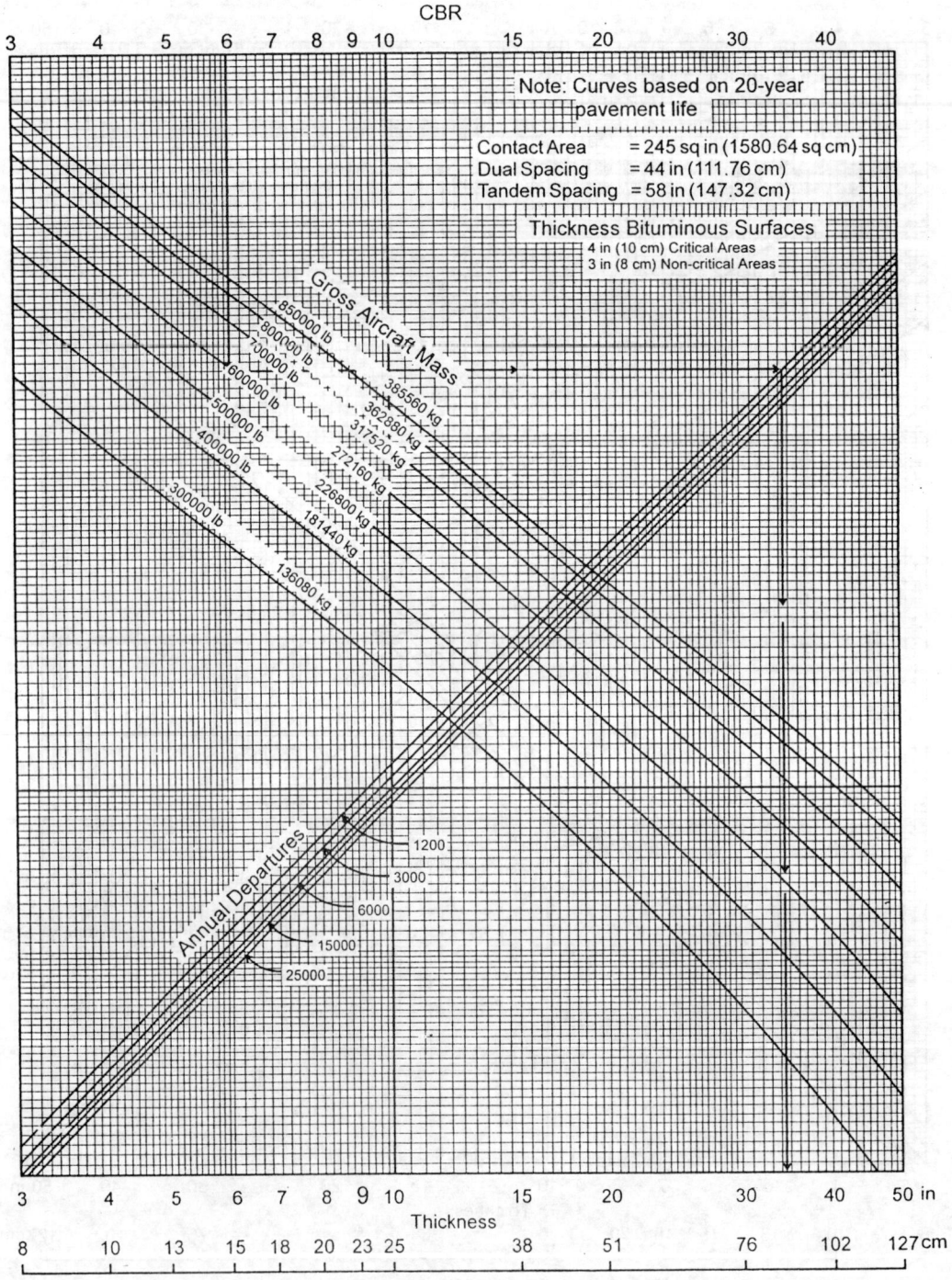

CBR

Note: Curves based on 20-year pavement life

Contact Area	= 245 sq in (1580.64 sq cm)
Dual Spacing	= 44 in (111.76 cm)
Tandem Spacing	= 58 in (147.32 cm)

Thickness Bituminous Surfaces
4 in (10 cm) Critical Areas
3 in (8 cm) Non-critical Areas

Gross Aircraft Mass

850000 lb — 365560 kg
800000 lb — 362880 kg
700000 lb — 317520 kg
600000 lb — 272160 kg
500000 lb — 226800 kg
400000 lb — 181440 kg
300000 lb — 136080 kg

Annual Departures

1200
3000
6000
15000
25000

Thickness

3 4 5 6 7 8 9 10 15 20 30 40 50 in
8 10 13 15 18 20 23 25 38 51 76 102 127 cm

Fig. 10.11. Flexible pavement design curves for critical areas, B747–100, SR, 200 B, C, F (ICAO).

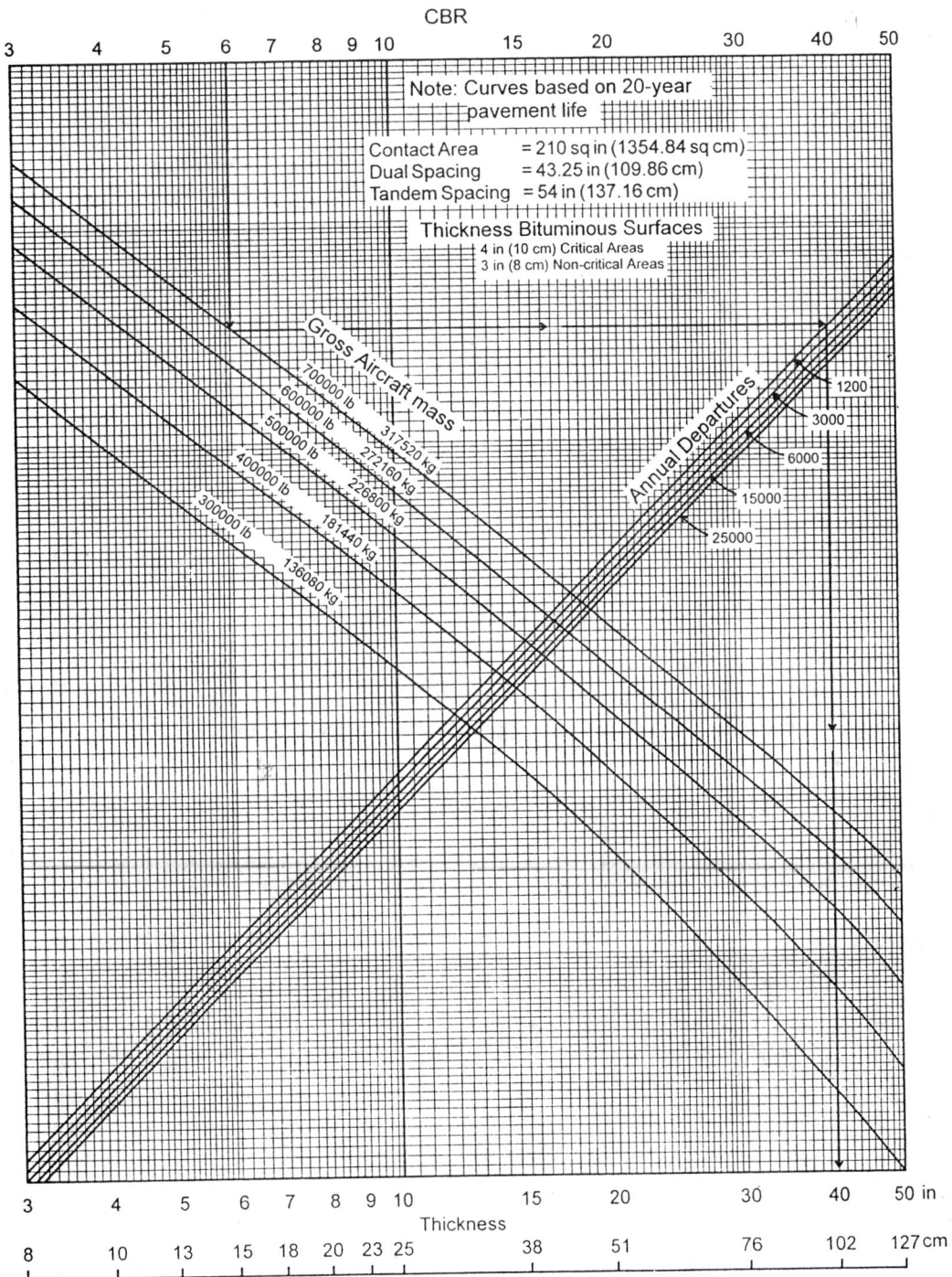

CBR

Note: Curves based on 20-year pavement life

Contact Area = 210 sq in (1354.84 sq cm)
Dual Spacing = 43.25 in (109.86 cm)
Tandem Spacing = 54 in (137.16 cm)

Thickness Bituminous Surfaces
4 in (10 cm) Critical Areas
3 in (8 cm) Non-critical Areas

Gross Aircraft mass

700000 lb   317520 kg
600000 lb   272160 kg
500000 lb   226800 kg
400000 lb   181440 kg
300000 lb   136080 kg

Annual Departures

1200
3000
6000
15000
25000

Thickness

Fig. 10.12. Flexible pavement design curves for critical areas, B747–SP (ICAO).

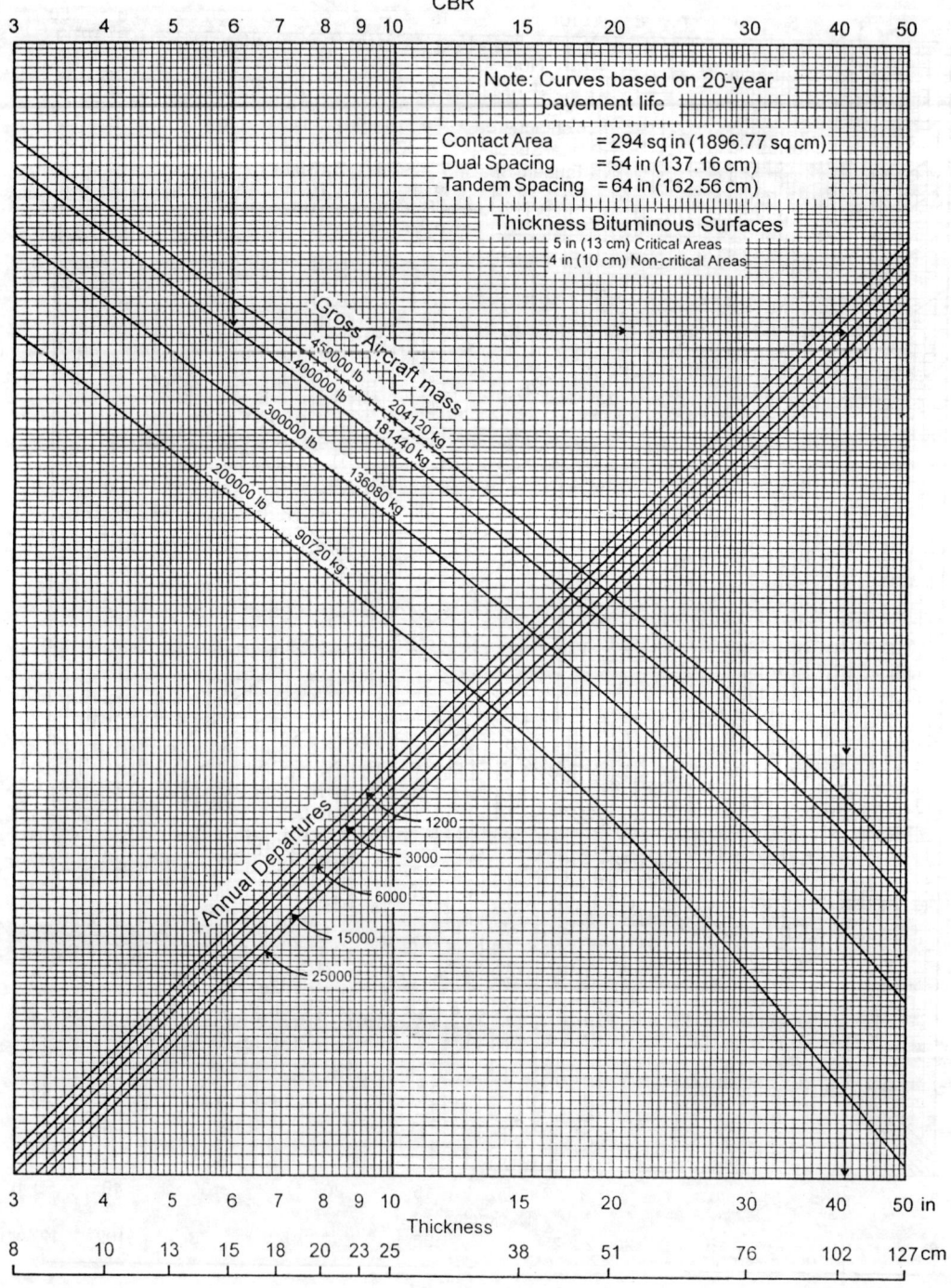

Fig. 10.13. Flexible pavement design curves for critical areas, DC10–10, 10CF (ICAO).

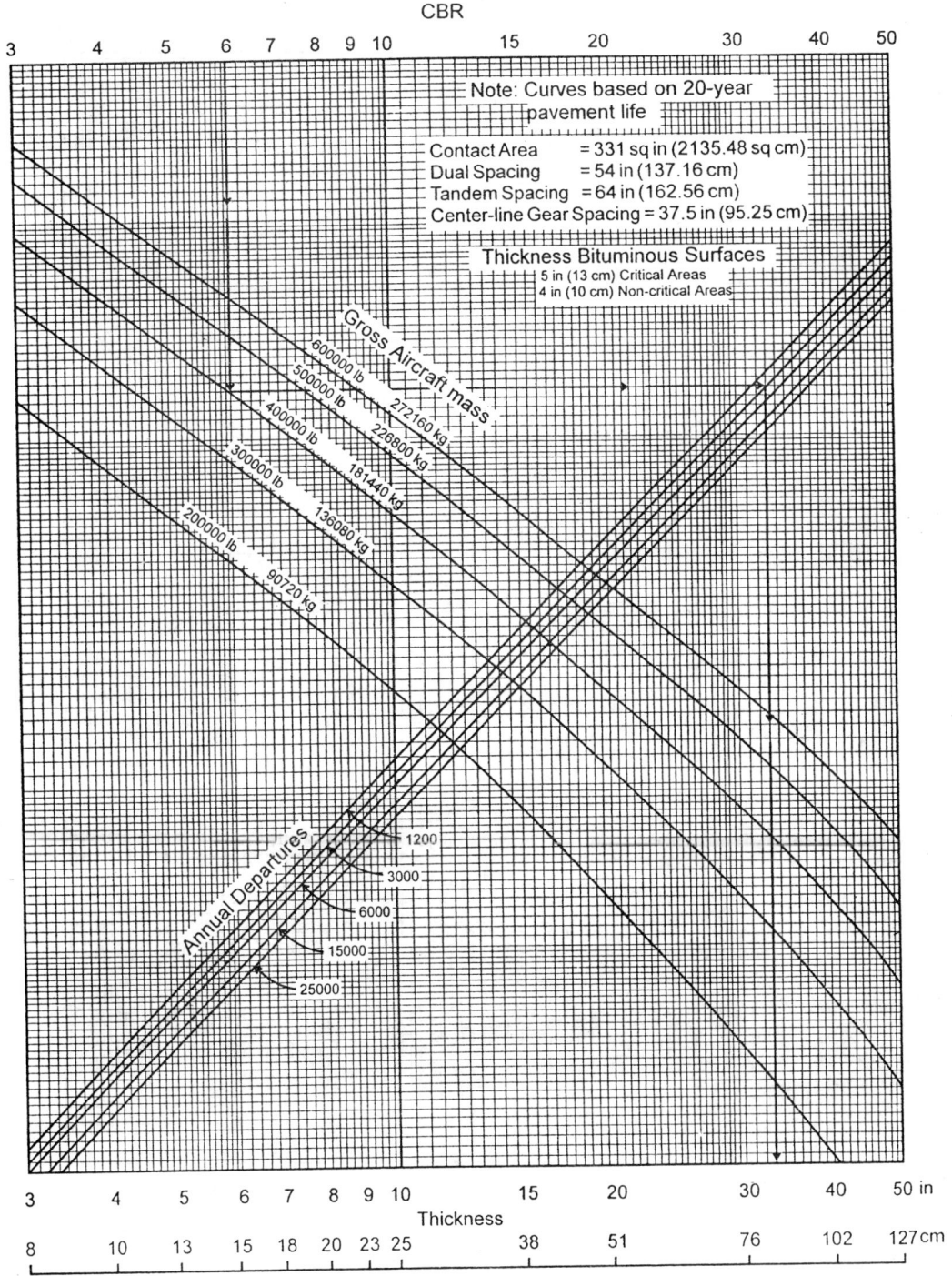

CBR

Note: Curves based on 20-year pavement life

Contact Area = 331 sq in (2135.48 sq cm)
Dual Spacing = 54 in (137.16 cm)
Tandem Spacing = 64 in (162.56 cm)
Center-line Gear Spacing = 37.5 in (95.25 cm)

Thickness Bituminous Surfaces
5 in (13 cm) Critical Areas
4 in (10 cm) Non-critical Areas

Gross Aircraft mass

600000 lb 272160 kg
500000 lb 226800 kg
400000 lb 181440 kg
300000 lb 136080 kg
200000 lb 90720 kg

Annual Departures
1200
3000
6000
15000
25000

Thickness

Fig. 10.14. Flexible pavement design curves for critical areas, DC10–30, 30CF, 40, 40CF (ICAO).

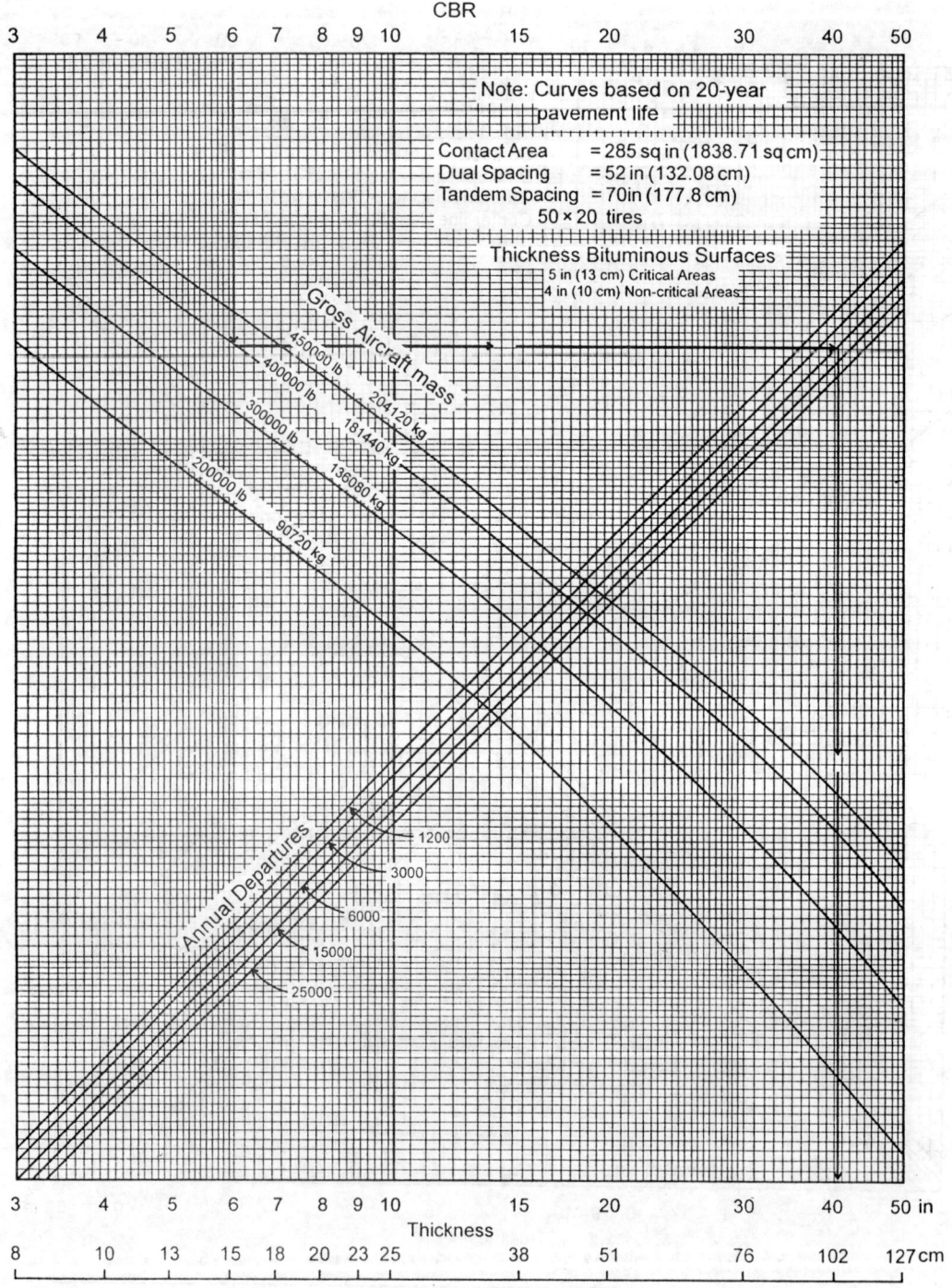

Fig. i0.15. Flexible pavement design curves for critical areas, L–1011, 100 (ICAO).

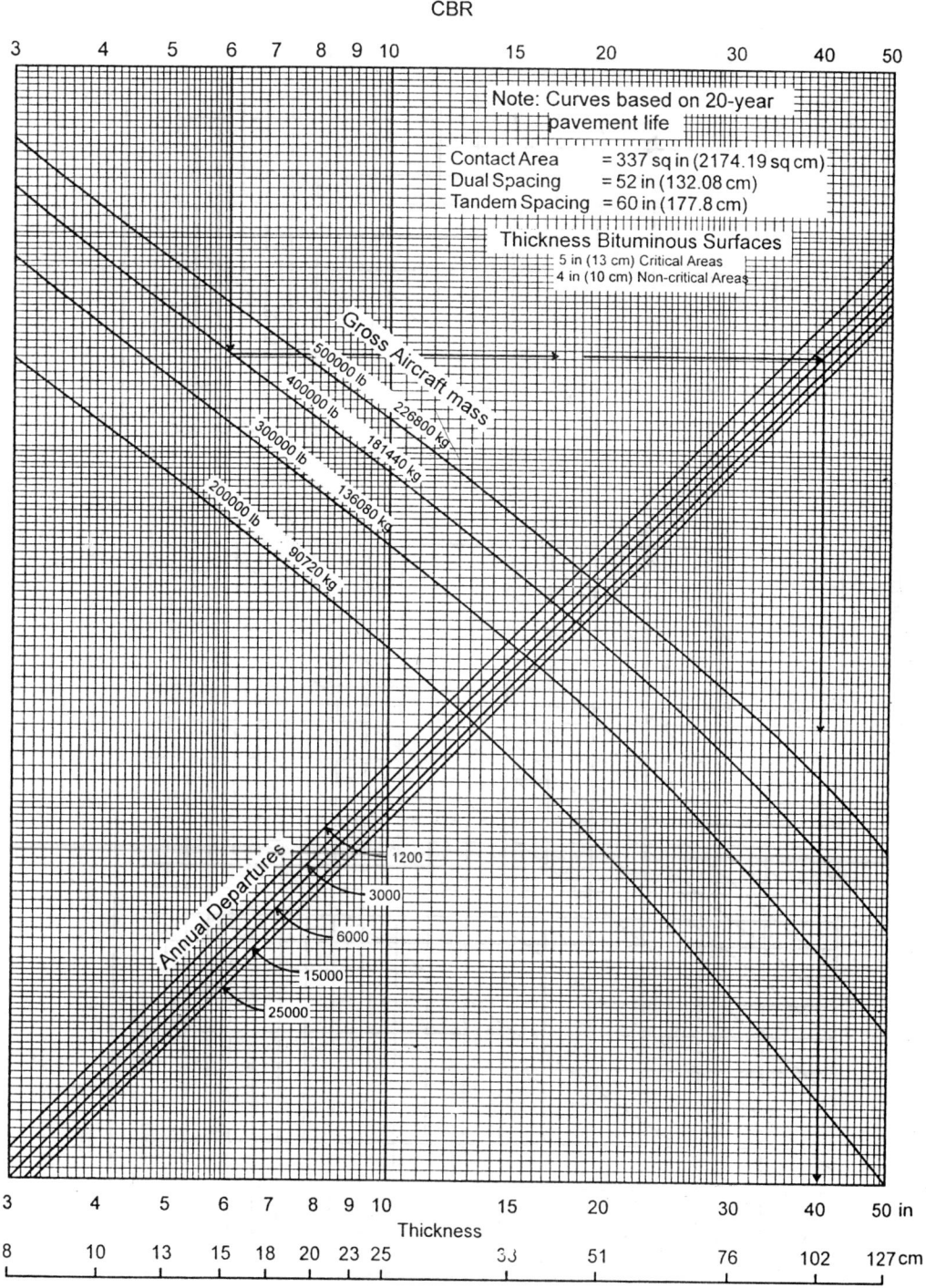

CBR

Fig. 10.16. Flexible pavement design curves for critical areas, L–1011–100, 200 (ICAO).

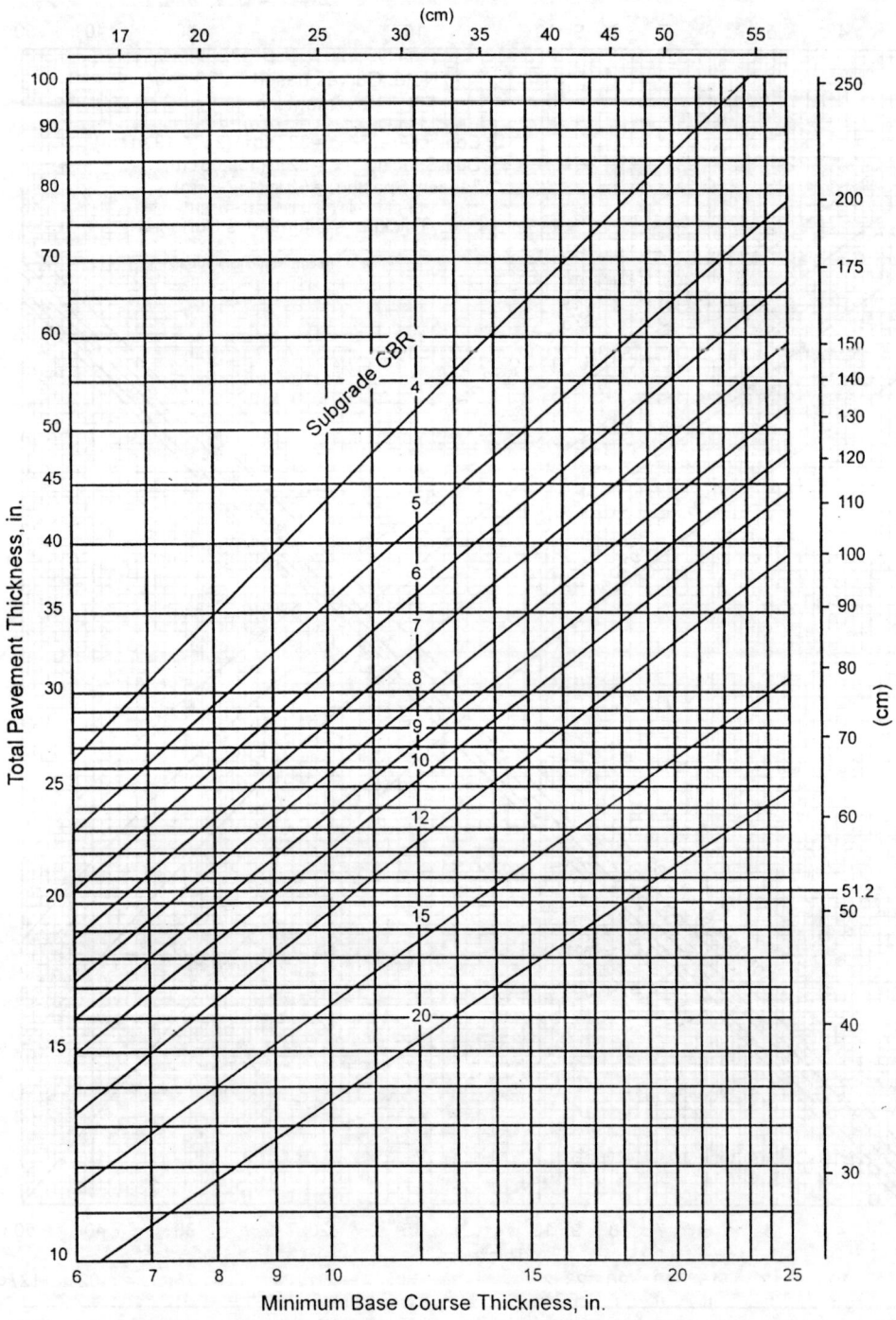

Fig. 10.17. Minimum base course thickness requirements (ICAO).

10.5 FUNDAMENTAL CONSIDERATIONS FOR RIGID PAVEMENTS

For airports, Westergaard developed equations for stresses and deflections for interior and edge of the concrete slab.

$$Sc = 0.316 \frac{W}{h^2} \left[4 \log_{10} \frac{1}{b} + 1.0693 \right]$$

where

Sc = flexural stress at the centre of the slab in lbs/sq.in.
W = the applied load in lbs.
h = the thickness of the slab in inches
E = the modulus of elasticity of the concrete in lbs/sq.in.
b = radius of contact area
μ = Poisson's ratio
l = the radius of relative stiffness in inches

$$\text{l is given by } l = 4 \sqrt{\frac{Eh^3}{12(1 - \mu^2) K}}$$

where

K = the modulus of subgrade reaction in lbs/sq.in/in.

In S.I. units where h and l are in mm
E = in MPa
K = MPa/m

$$L = 4 \sqrt{\frac{1000 \, Eh^3}{12(1 - \mu^2) K}}$$

The required thickness of the concrete slab thus chiefly depends on the applied load (W), and its radius of relative stiffness (l). The modulus of subgrade reaction K has theoretically relatively little effect in ultimate slab design thickness.

Design practice is to use a cement stabilised sub-base which will not only provide a working platform for placing the concrete, but will also tend to control subgrade movement and will strengthen it.

Westergaard assumed the pavement slab to be a thin plate resting on a special subgrade which is considered elastic in vertical direction only. Thus the reaction is proportional to the deflection of the subgrade.

$$R = K.Z$$

where

R = vertical reaction proportional to the amount of deflection
K = a soil constant i.e. modulus of subgrade reaction
Z = deflection

Assumptions involved in the Westergaard's theory are :
• concrete slab is homogeneous
• slab is isotropic elastic solid

- wheel load of the aircraft is distributed over an elliptical area
- slab rests on subgrade which is considered as a dense liquid

Though these assumptions donot satisfy theory in strict sense the result have compared reasonably with observations. For a single wheel load the equivalent single wheel load (ESWL) will be constant at all depths, whilst in other cases the ESWL will vary with l - the radius of relative stiffness, which itself varies principally with slab thickness.

10.5.1 Foundation for Rigid Pavement (Sub-base)

Material which supports a rigid pavement is termed as sub-base. Granular and treated material is placed over the existing soil as subbase. The purposes of sub-base in a rigid pavement are :

- It prevents pumping (ejection of water near joints and cracks).
- Adverse action due to swell and shrinkage in soils is avoided.
- In frost areas, damage to pavement is prevented.
- Supporting capacity of existing soil is improved.
- Continuous, smooth and stable platform for concrete slab is provided.

For this purpose treated sub-bases using portland cement or bitumen, are used particularly for heavily trafficked airport pavements. Granular or stabilised when used on the top of subgrade soil, there will be an increase in 'K' value. If it is not feasible to test increased 'K' value of such a construction, an estimate can be made from Fig. 10.28. Such figures can be developed for different type sub-bases.

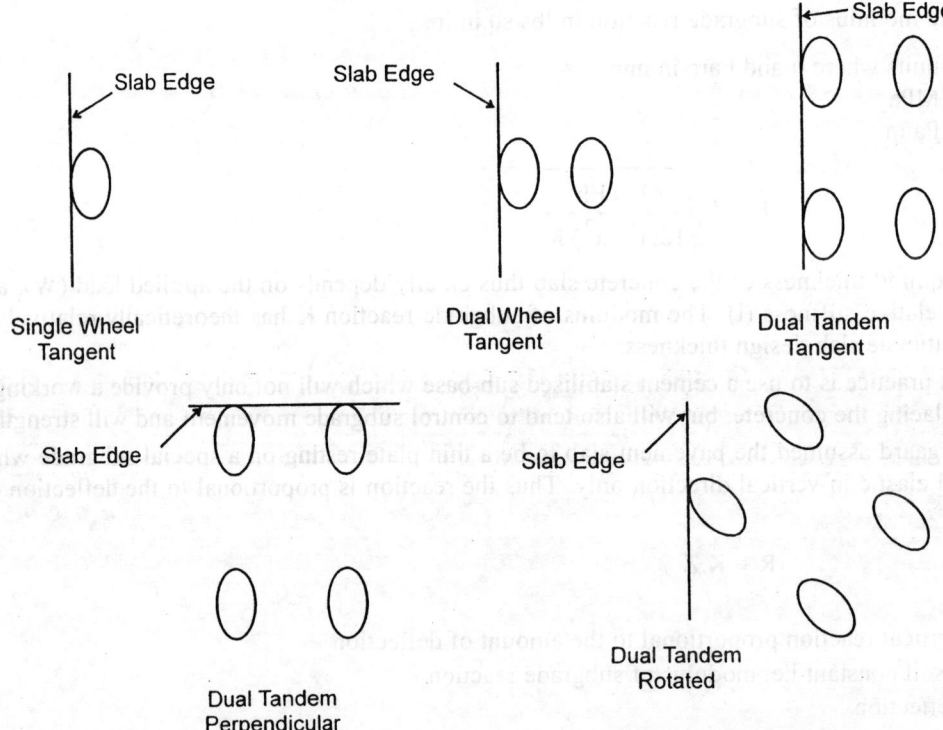

Fig. 10.18. Gear assembly positions for rigid pavement design curves.

10.5.2 Design Procedure for Rigid Pavements

Design curves have been prepared for rigid pavements similar to those for flexible pavements i.e., separate curves for single, dual and dual tandom landing gear assemblies and separate design curves for wide-body jet aircraft, as shown in Fig. 10.19 to 10.27. These curves are based on a jointed edge loading assumption where the load is tangent to the joint as shown in Fig. 10.18.

10.5.3 Design Inputs : For Rigid Pavement Design

For using the design curves, to know the thickness of the concrete slab, following design parameters are required :
  (a) Concrete flexural strength
  (b) Subgrade modulus (K-value)
  (c) Gross weight of the design aircraft
  (d) Annual departure of the design aircraft

(a) Concrete flexural strength

The required thickness of concrete pavement depends on the strength of the concrete used in the pavement. Concrete strength is assessed by the flexural strength of the concrete. A laboratory test is performed to determine concrete flexural strength. Normally 90 days flexural strength is used for design. It is assumed that 90 day flexural strength of concrete will be 10 per cent higher than the 28-day strength.

(b) Subgrade modulus (K-value)

K value, the subgrade modulus indicates the bearing value of the supporting material. It is determined by field plate bearing test. It is made on representative foundation material which is to support the pavement. The usual procedure is to apply loads by means of a hydraulic jack through a jacking frame onto a steel plate and plotting a load-versus-deformation curve. It is expressed as kg/cm^3 or Mn/m^3. K value is corrected for saturated condition, bending of plate. Consolidation test is conducted for saturation correction.

(c) Gross weight of the aircraft

Each design curve is for a particular gross weight of the aircraft. The design curves are grouped in accordance with main landing gear assembly type. Interpolation, if required, can be done between the range of gross weights shown on the design curves.

(d) Annual departure of design aircraft

The annual departures of design aircraft are computed, as illustrated before for the flexible pavements.

  For using the design curves, the left ordinate showing concrete flexural strength is first entered. A horizontal projection is made until it intersects with the appropriate foundation modulus line. A vertical projection is made from the intersection point to the appropriate gross weight of the design aircraft. A horizontal projection is then made to the right ordinate showing annual departures. The pavement thickness is read from the annual departure line. The pavement thickness, thus obtained is thickness of the concrete pavement only, exclusive of the sub-base. An example is included to illustrate the use of design curves.

  A minimum thickness of 10 cm of sub-base is required under all rigid pavements. If economical

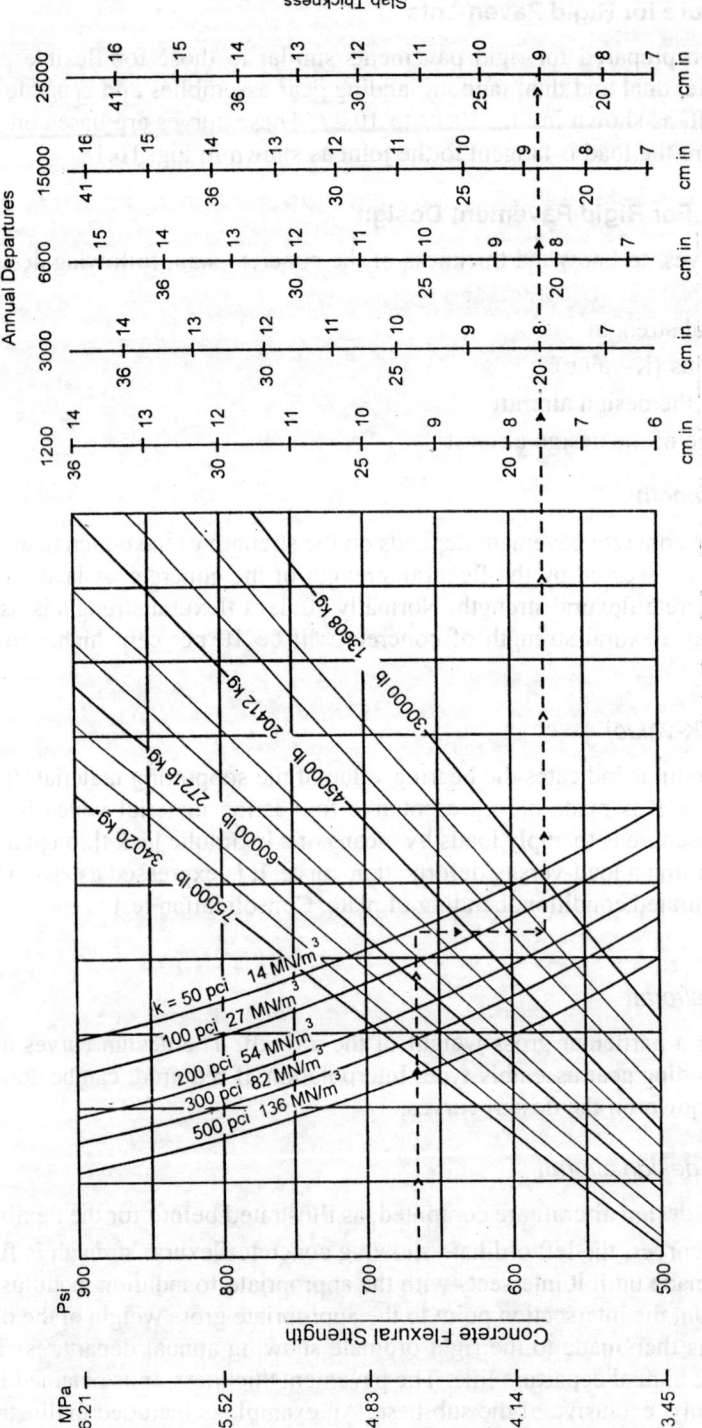

Fig. 10.19. Rigid pavement design curves - single wheel gear (ICAO).

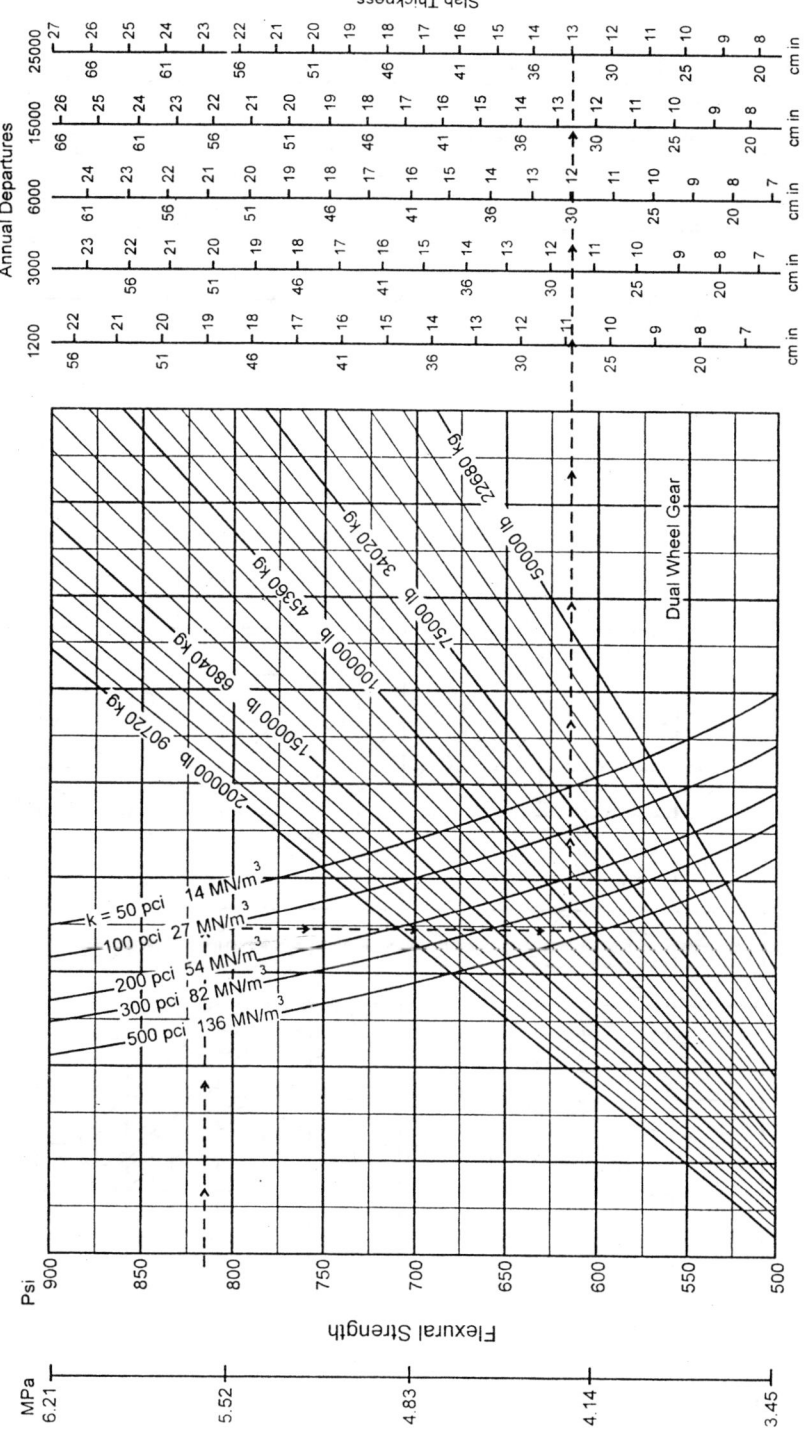

Fig. 10.20. Rigid pavement design curves - dual wheel gear (ICAO).

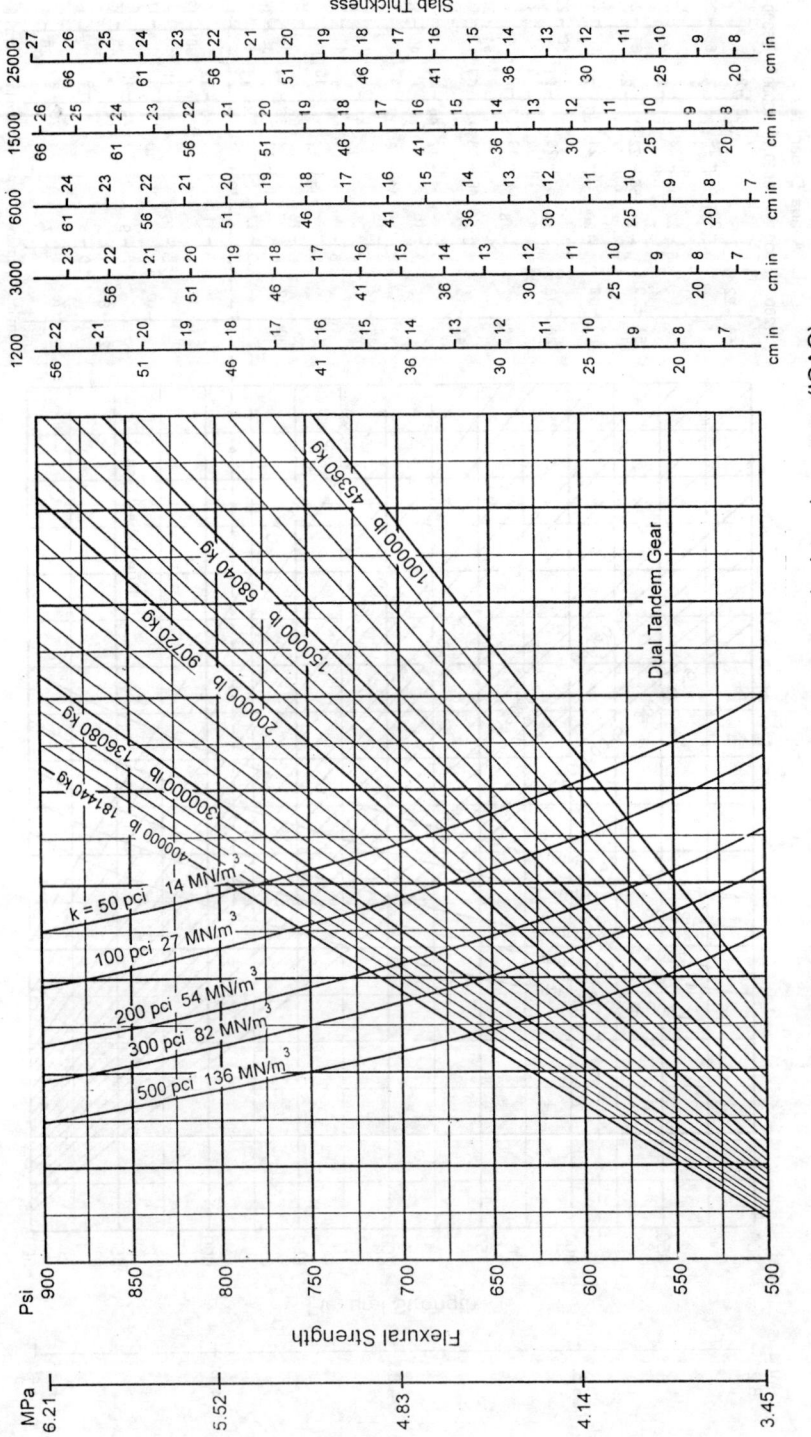

Fig. 10.21. Rigid pavement design curves - dual tandem gear (ICAO).

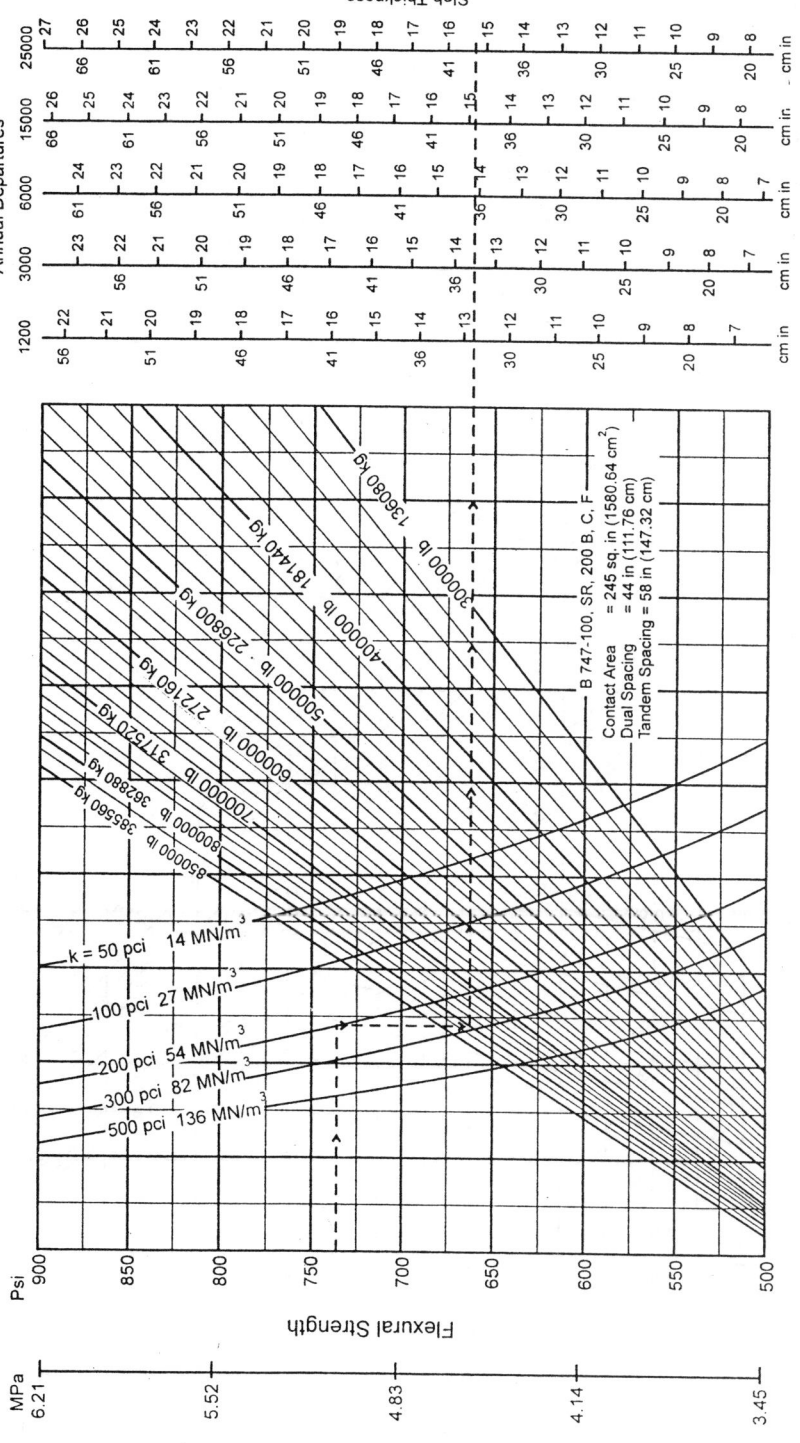

Fig. 10.22. Rigid pavement design curves - B-747-100, SR, 200 B, C, F (ICAO).

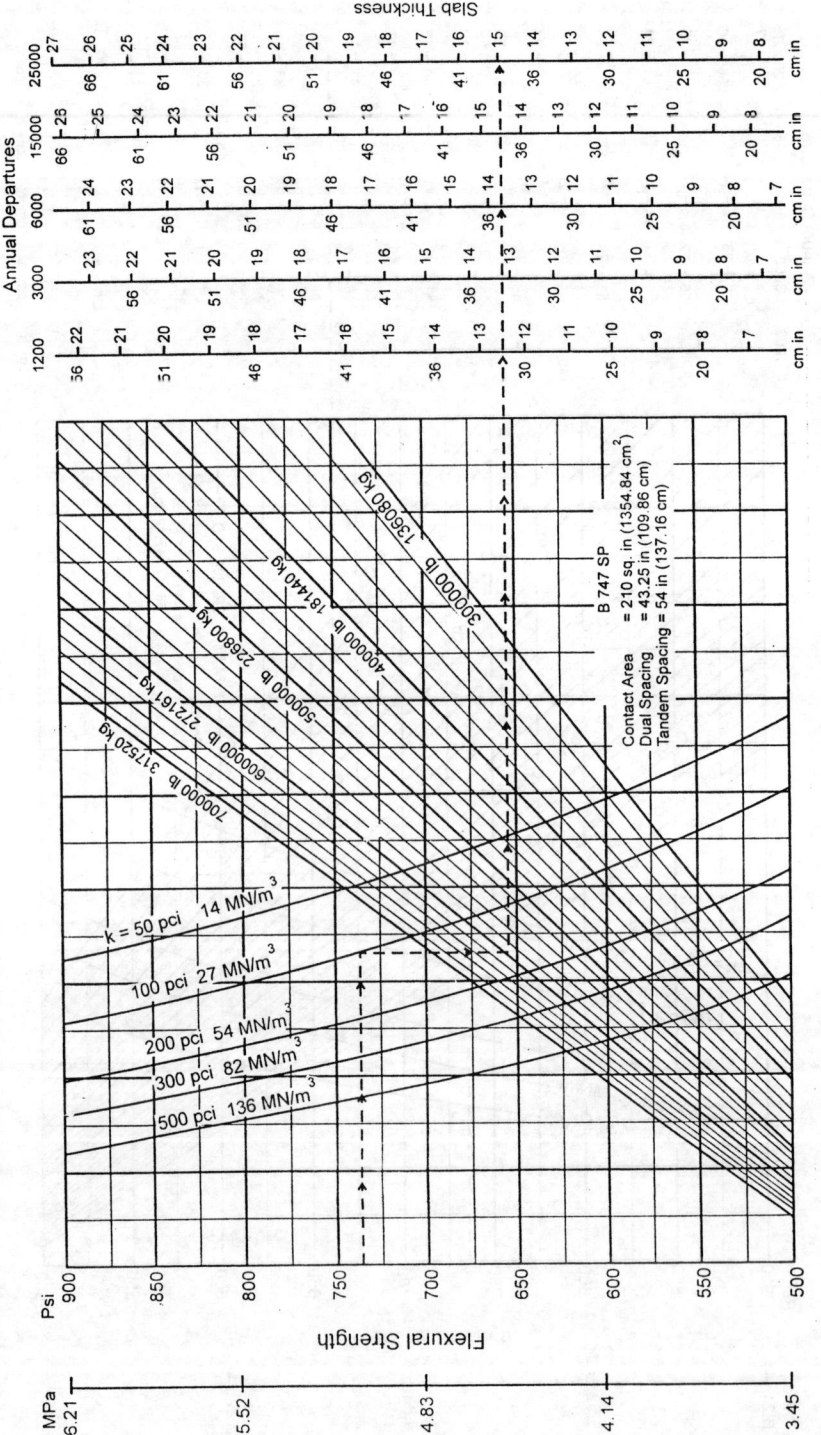

Fig. 10.23. Rigid pavement design curves - B-747-SP (ICAO).

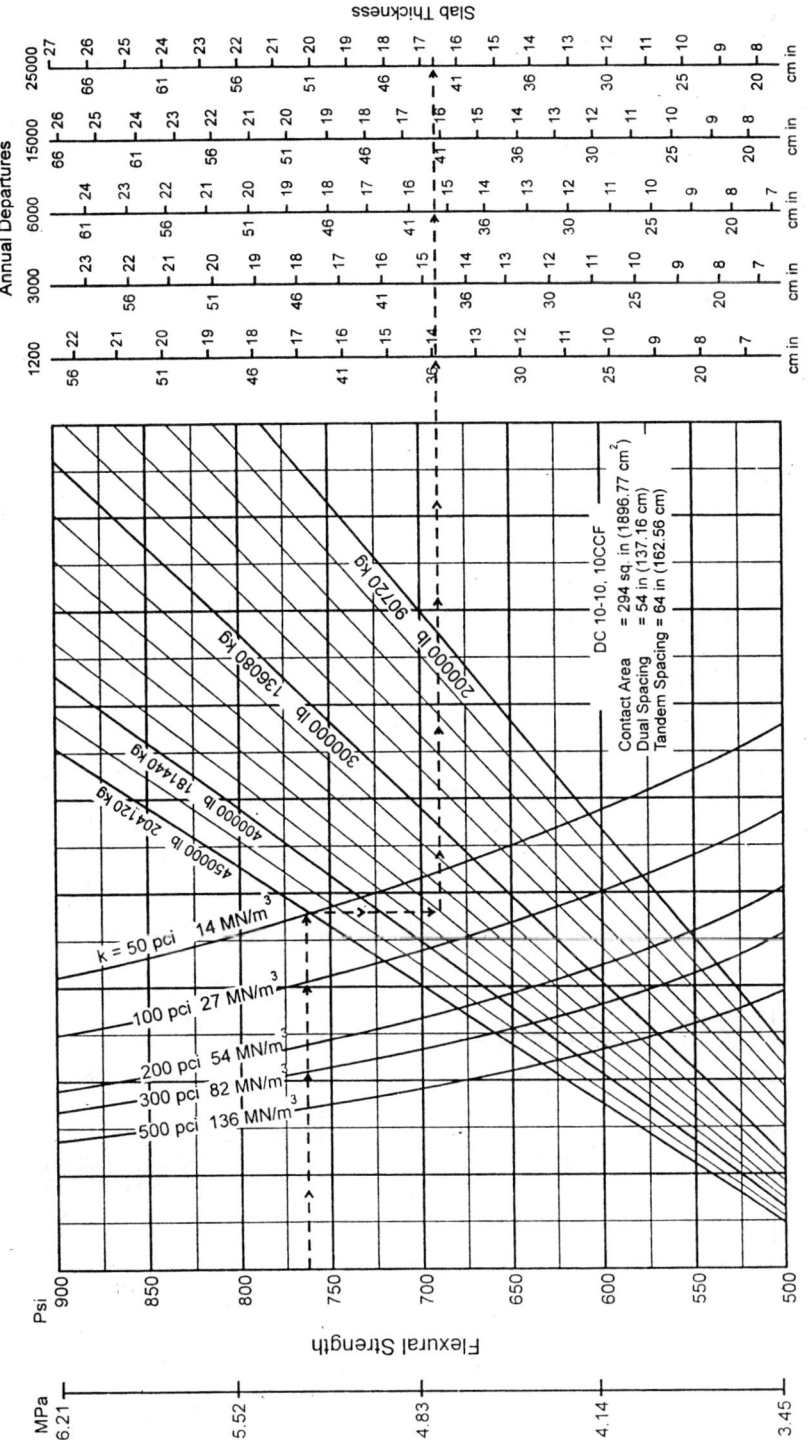

Fig. 10.24. Rigid pavement design curves - DC 10-10, 10CF (ICAO).

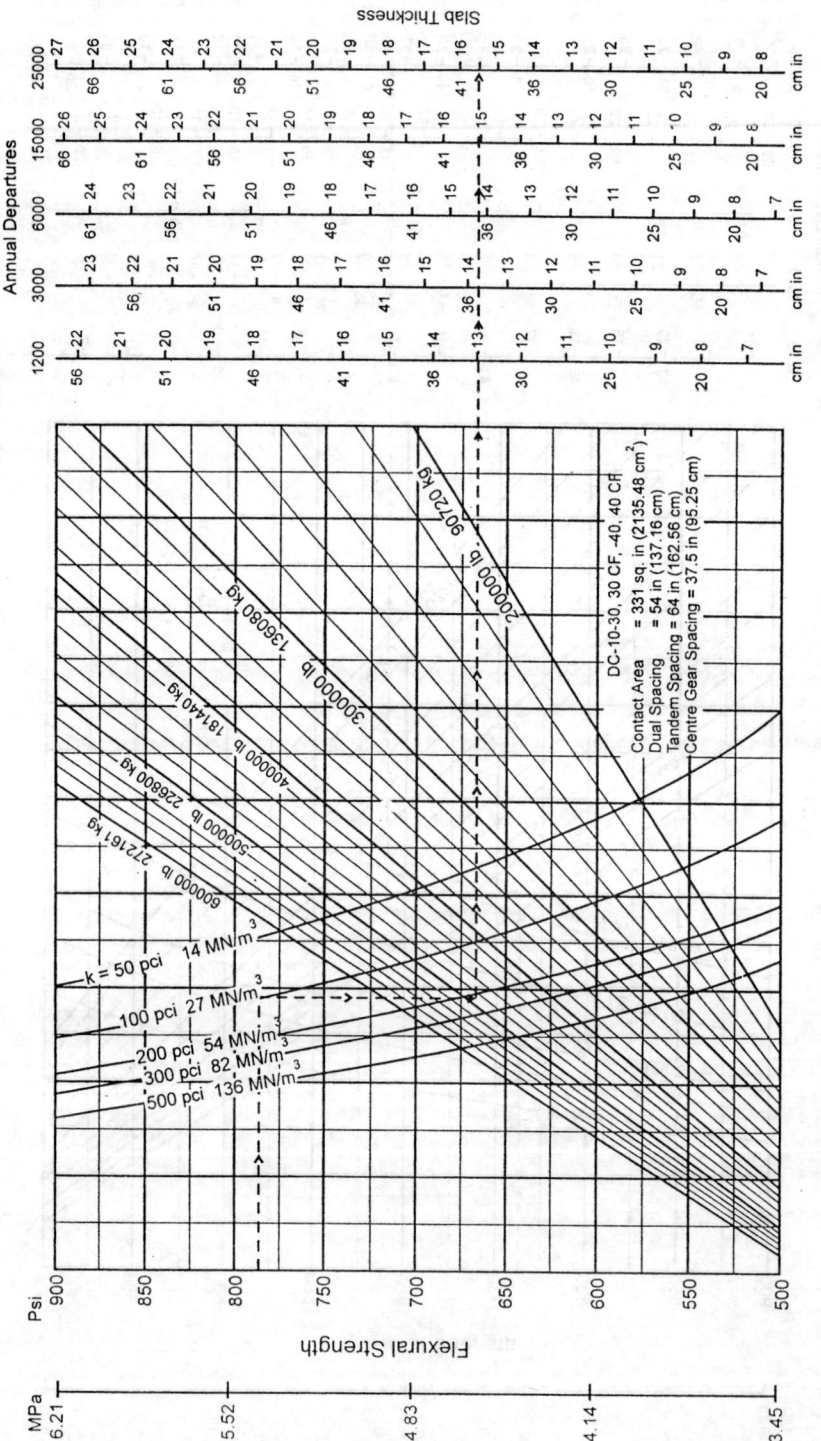

Fig. 10.25. Rigid pavement design curves - DC 10-30, 30CF, 40, 40GF (ICAO).

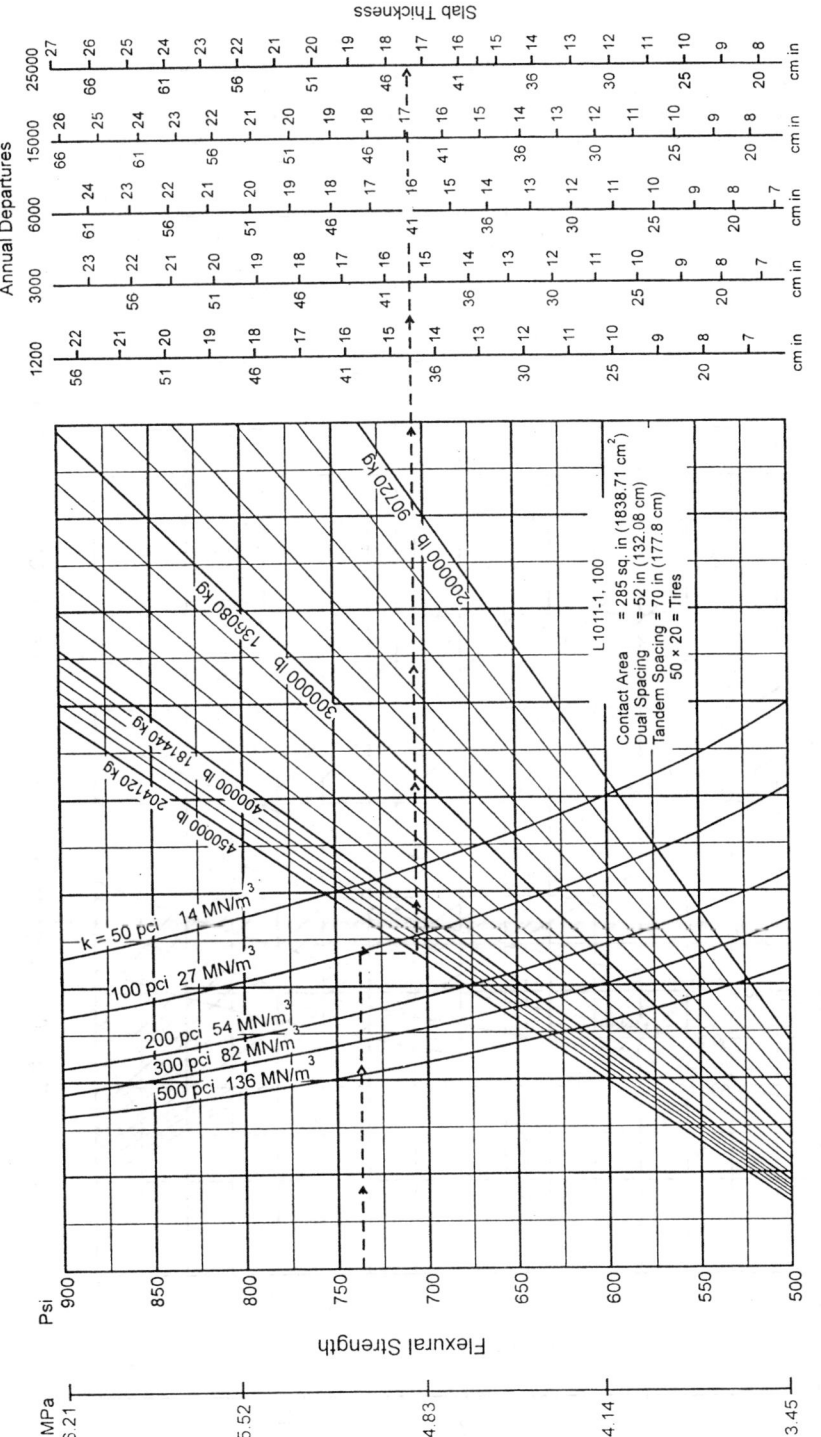

Fig. 10.26. Rigid pavement design curves - L-1011-1, 100 (ICAO).

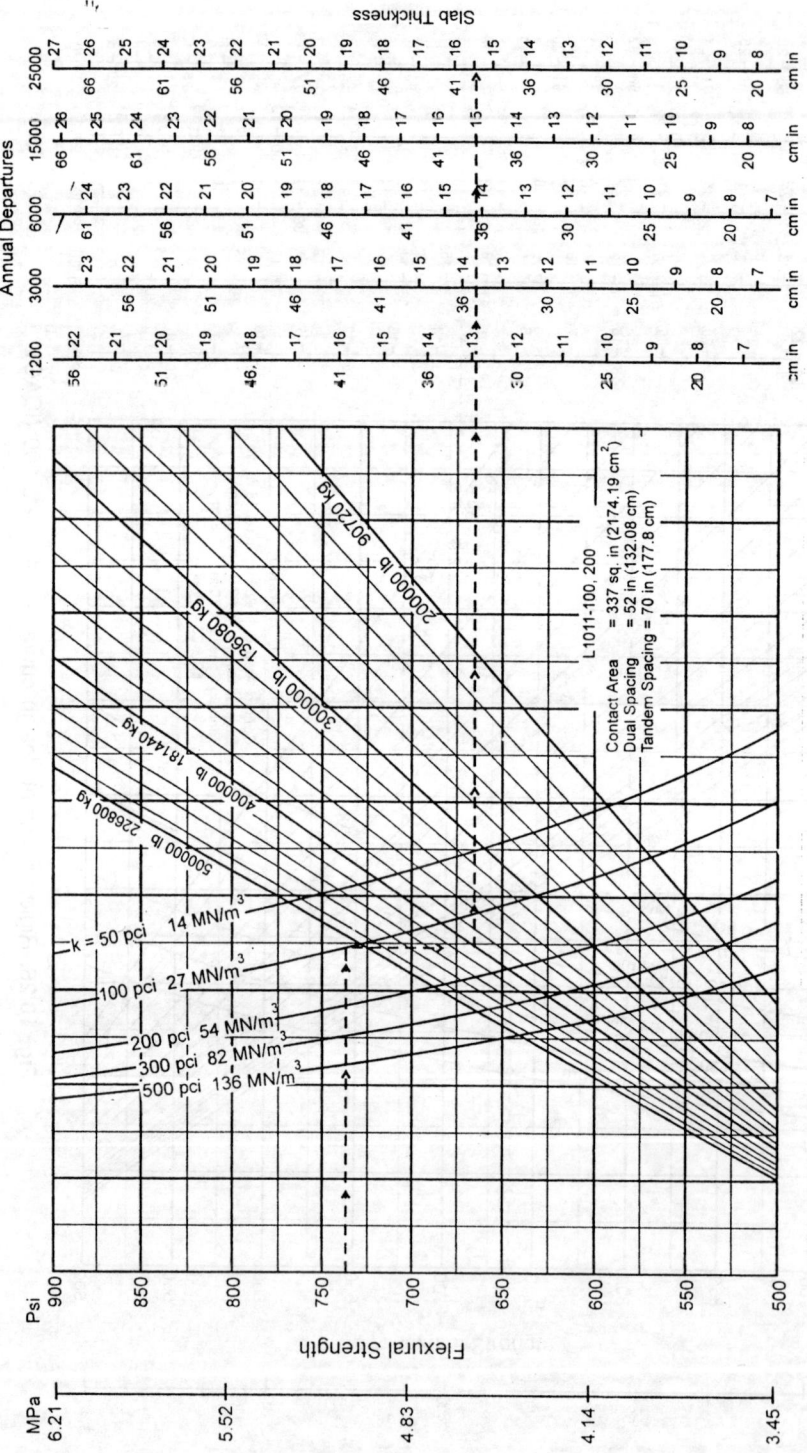

Fig. 10.27. Rigid pavement design curves - L-1011-100, 200 (ICAO).

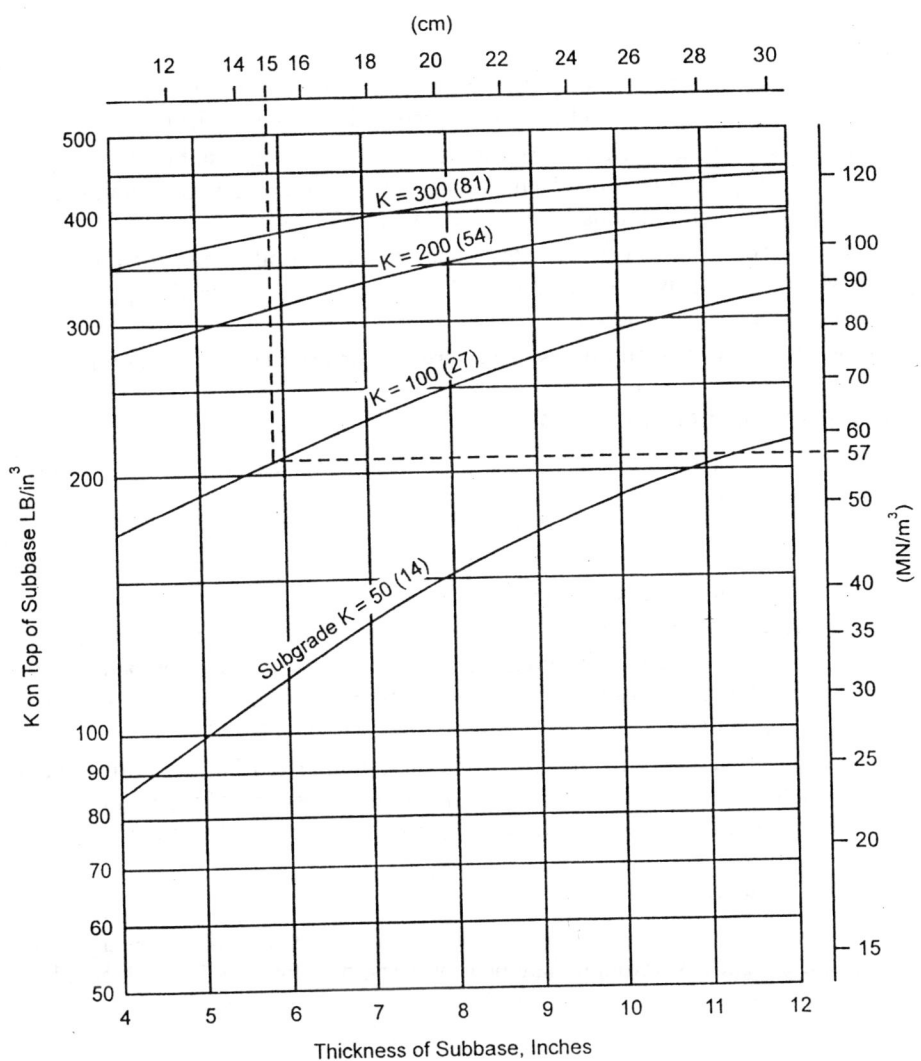

Fig. 10.28. Effect of stabilized sub-base on subgrade modulus (ICAO).

thicker sub-base than 10 cm can be used to reduce required thickness of concrete, as modulus of soil reaction will be increased. Suitable materials for sub-base are gravel, bituminous layer, crushed aggregates, soil cement etc.

Example 10.5

A rigid pavement is to be designed for dual tandem aircraft having a gross weight of 160,000 kg and for 6000 annual equivalent departures of the design aircraft. The equivalent annual departures of 6000 includes 1200 annual departures of B-747 aircrafts weighing 350,000 kg gross weight.

The subgrade modulus is 25 MN/m^3 with poor drainage. Concrete mix designs indicate that a flexural strength of 4.5 MN/m^2 can be obtained with locally available aggregates.

Solution

The required thickness of pavement is determined by assuming and trying several thickness of sub-bases, so that modulus of subgrade reaction is increased and pavement thickness is reduced.

In this example, as the gross weight of the aircraft is 160,000 kg, a stabilised sub-base should be used. Let us try a sub-base thickness of 15 cm. Using the Fig. 10.28, the 15 cm thickness of sub-base would increase the foundation modulus from 25 MN/m³ to 57 MN/m³.

Now using the Fig. 10.21 the dual tandem design curve, for flexural strength of 4.5 MN/m², subgrade modulus of 57 MN/m³, gross weight of 160,000 kg and equivalent annual departures of 6000, the thickness of concrete pavement is 42 cm.

Similarly other thicknesses of sub-bases can be tried to determine most economical section.

10.5.4 Other procedures of pavement design

Other design procedures such as those based on layered elastic analysis, Burmister theory etc. developed by Asphalt Institute, Portland Cement Association etc. are also utilised to determine pavement thicknesses. These procedures might give different design thicknesses due to different basic assumptions. Because of thickness variations, the evaluation of existing pavement should be done using the same method which was employed in the design.

Empirical, semi-empirical and analytical design methods for airfield pavements based on available materials, conditions of loading, climatic parameters and existing soils are developed by different countries to meet the challenges of heavier and wider aircrafts. Suitable design models, curves, based on computer based solutions of stress, strain and deformations are developed.

Practices followed for design in some countries are summarised below. Further details will be available in the technical manuals on design and evaluation of airports of the particular country. The practices outlined here do not cover all the essential design data or detailed design analysis.

10.5.4.1 Canadian practice

In the early days, Norman McLeod of Canadian department of transport, based on large scale testing of highway and airfield pavements, through plate bearing tests, developed a design method. He repeated plate bearing test with different sizes of plate diameter, and recommended the following empirical design equation :

$$T = K \log \frac{P}{S}$$

where
 T = thickness of gravel material cm
 P = gross wheel load kg
 S = subgrade support kg/cm²
 K = base course constant

This method related plate diameter and subgrade modulus (K) values and converted wheel loads to equivalent wheel load, as illustrated earlier.

(a) Flexible pavement design curves

A flexible pavement design curve for a given aircraft is a plot of pavement thickness required to support the aircraft loading as a function of subgrade bearing strength. The equation utilised to generate the design curve is :

$$S = (ESWL) (C_1 \ 10^{-c_2 t})$$

where

$\quad$ S = subgrade bearing strength KN

$\quad$ ESWL = equivalent single wheel load of the design aircraft loading KN

$\quad$ t = pavement equivalent granular thickness cm

$\quad$ C_1, C_2 = factors depending on contact areas of ESWL

The steps involved in determination of flexible pavement thickness are :
(a) Determination design loading, for the pavement on the basis of traffic studies and projections.
(b) Determination subgrade bearing strength through plate test.
(c) Determination from the curve the pavement-equivalent granular thickness, requirement for design load.
(d) Determination of pavement thickness required for partial frost protection.

Pavement thickness provided will be greater between the two obtained in (c) or (d) above. Granular thickness determined in (c) is converted to actual pavement thickness, through equivalency factors for pavement construction material.

(b) Rigid pavement design curves

A rigid pavement design curve for a given aircraft is a plot of concrete slab thickness required to support the aircraft loading as a function of the bearing modulus of the surface on which the slab rests. Slab thickness required to support aircraft loading is based on (limiting to 2.75 MPa) the flexural stress occurring at the bottom of the slab directly under the centre of one tire of the aircraft gear. The stress calculations are carried out according to the Westergaard analysis for interior slab loading conditions, using a computer programme.

The bearing modulus is based on the load in mega Newtons which will produce a deflection of 1.25 mm when the load is applied through rigid circular plate 762 mm in diameter.

10.5.4.2 French practice

"Optimised design" design which takes into account all aircraft types having significant effect on the pavement and another design "general design" a design for specific load which the pavement must support are undertaken. General design is used in the absence of accurate data.

(a) Designing flexible pavement

The design of a flexible pavement involves two stages :
(i) Collection of data :
$\quad$ – traffic data (load, movements).
$\quad$ – characteristics of the natural soil.
(ii) Calculation of the thickness which is also done in two stages :
$\quad$ – the determination of an "equivalent pavement thickness" using optimised or general design methods.

– selection of pavement structure which provides an equivalent thickness corresponding to or greater than the thickness determined above.

CBR value is used to determine bearing strength of subgrade. Graphs are then used to determine the pavement thickness.

(b) Designing rigid pavement

The design of rigid pavement involves the following two stages :
- (i) Collection of data :
    - traffic (load, movement).
    - characteristics of the subgrade and of the hydraulic cement concrete.
- (ii) Calculation of the thickness of the concrete slab (non-reinforced and non-prestressed).

Flexural breaking strength is measured on prismatic specimen after 90 days. Plate bearing test is carried in situ to determine modulus of subgrade reaction K.

Thickness of the concrete slab is determined by the graphs. The optimised design method takes into account a number of actual movements over a fixed pavement life 10 years (flexible) and 20 years (rigid).

In France, two systems of runway bearing strength are used :
- (i) the method based on a typical undercarriage leg.
- (ii) the ACN-PCN method.

10.5.4.3 United Kingdom practice

United Kingdom practice is to design for unlimited operational use by a given aircraft taking into account the loading resulting from interaction of adjacent leading gear wheel assemblies where applicable. The aircraft is designated as "the design aircraft" for the pavement.

While number of computer programmes based on plate theory, multilayer elastic theory and finite analysis are available, a system called Reference Construction Classification system (RCC) has been developed from the British load classification number (LCN) and load classification group (LCG).

For aircraft reaction on rigid pavements, a simple two layer model is used. The model is analysed by Westergaard centre case theory. A range of equivalency factors appropriate to the relative strengths of indigenous construction materials are adopted to convert between theoretical model construction depths and actual pavement thickness.

Aircraft reaction on flexible pavements follows the same basic pattern adopted for rigid pavement design. In this case four pavement model is analysed using US Corps of Engineers, California Bearing Ratio (CBR) method. This includes Boussinesq deflection factors. Equivalency factors are used to relate materials and layer thicknesses to the theoretical model to calculate construction depth of pavements.

10.6 REINFORCED CONCRETE PAVEMENT

Reinforcement in concrete pavements keeps the cracks tightly closed. Such interlocking provides structural integrity and improves pavement performance. By keeping the cracks tightly closed, the steel minimises the infiltration of external material, debris etc. into the crack. Steel reinforcement allows longer joint spacing and thus reduces the cost due to fewer joints. Decision between plain or reinforced concrete pavement is done, basically on cost benefits between the two.

10.6.1 Type and spacing of reinforcement

Reinforced pavements use either bars or welded wires with end and side covers to provide complete reinforcement throughout the slab panel. End laps should be a minimum of 31 cms, but not less than 30 times the diameter of the longitudinal wire or bar.

Side laps should be a minimum of 15 cm but not less than 20 times the diameter of the transverse wire or bar. End and side clearance should be a minimum of 15 cm and a minimum of 5 cm to allow for nearly complete reinforcement and yet achieve adequate concrete cover.

Longitudinal members should be spaced not less than 10 cm, not more than 31 cm apart. Transverse members should be spaced not less than 10 cm, nor more than 61 cm apart.

10.6.2 Amount of reinforcement

The steel area required for a reinforced concrete pavement is determined from the subgrade drag formula and coefficient of friction combined. The resultant formula is expressed as follows :

$$As = \frac{0.64 \, L \, \sqrt{Lt}}{f_s}$$

where

  As = area of steel, square centimeters per metre

   L = length or width of slab in metres

   t = thickness of slab in millimetres

  f_s = allowable tensile stress in meganewtons for square metre

In this formula the weight of slab is assumed to be 23.6 Mn per m². Allowable tensile stress in steel varies with type and grade of steel. Allowable tensile stress is taken as two-thirds of the yield strength of the steel.

The minimum percentage of steel reinforcement should be 0.05 per cent.

$$\text{Percentage of steel} = \frac{\text{Area of steel (As)}}{\text{Area of concrete per unit of length/width}} \times 100$$

The maximum allowable slab length is 23 metres.

10.6.3 Continuously reinforced concrete pavement

A continuously reinforced concrete pavement (CRCP) is a portland cement concrete pavement with continuous longitudinal steel reinforcement and no intermediate transverse expansion or concentration joints. The advantage of continuously reinforced concrete pavement is the elimination of transverse joints which are costly to construct, require periodic resealing and are often a maintenance problem. They provide a very smooth riding surface. Continuously reinforced concrete pavements normally contain 0.5 to 1.0 per cent longitudinal steel reinforcement.

Foundation support requirements, thickness design etc. are the same as for plain concrete.

10.6.4 Prestressed Concrete Pavement

Prestressed concrete pavement have been used in airport applications in Europe and to a limited extent in USA. Prestressed concrete pavements are usually post-tensioned with high strength steel strands. These pavement are considerably thinner than plain, reinforced or continuously reinforced concrete pavements, yet provide high load carrying capacity. Slab length of 120-150 m are generally used.

When properly designed and constructed any pavement type (rigid, flexible, composite etc.) can provide a satisfactory pavement for any type of aircraft, however, some designs are more economical than others and still provide satisfactory performance. The engineer must, therefore, work out the rationale and economic aspects for the design selected, based on evaluation of several design alternatives. Life-cycle cost analysis should be performed to select best and cheapest type of pavement, keeping inview the operational constraints, funding limitations, future expansion etc.

10.7 AIRCRAFT AND AIRPORT PAVEMENT CLASSIFICATION SYSTEM

ICAO have adopted a pavement classification system for reporting aerodrome pavement bearing strength (PCN) and an aircraft classification system (ACN) for aircrafts of mass greater than 5700 kg. This system reports a "pavement classification number" (PCN) which indicates that an aircraft with an "aircraft classification number" (ACN) equal to or less than PCN can operate on the pavement subject to any limitation of tire pressure.

ACN : A number expressing the relative effect of an aircraft on a pavement for a specified standard subgrade strength, in terms of a standard single wheel load. It is the load carried by a single wheel in terms of 500 kg at a constant tire pressure of 1.25 MPa, by the pavement required for the aircraft using standard CBR method for flexible pavement and interior loading for rigid placement.

PCN : A number expressing the bearing strength of a pavement for unrestricted operations, in terms of a standard single wheel load. ACN of aircraft which the pavement is capable of carrying for unrestricted movements.

The ACN-PCN method reports pavement strength on a continuous scale and the same scale is used to measure the load ratings of both aircraft and pavement. The lower end of scale is zero and there is no upper limit.

To facilitate the use of method, aircraft manufacturers publish, in the document following details :

- Characteristics of their aircraft.
- ACN computed at two different masses.
  - maximum apron mass.
  - representative operating empty mass.
- Both on rigid and flexible pavements.
- For 4 standard subgrade strength categories.

Table 10.2 shows ACN numbers of some selected aircrafts. Static mass is shown in table, no allowance is made for dynamic effects.

For each pavement following information is reported in ACN-PCN method :

- Pavement type
- Subgrade category
- Maximum tire pressure allowable
- Pavement evaluation method used

PCN for light aircrafts those with a maximum gross take-off weight less than 5700 kg is reported in terms of maximum aircraft allowable weight and tire pressure (maximum allowable).

The concept of the single wheel load was adopted to determine the ACN so that interaction between the landing gear and pavement could be evaluated without reference to the pavement thickness.

In the table for ACN of aircrafts the subgrade strength is classified as high, medium, low and ultra low based upon whether modulus of subgrade reaction K is equal to 150, 80, 40 or 20 MN/m^3 for rigid pavements and the CBR value of 15, 10, 6 or 3 for flexible pavements. A flow chart illustrating the procedure for determining the aircraft classification number for pavements is shown in Fig. 10.29.

Two mathematical models, one for rigid pavement : the Westergaard solution for a loaded elastic plate of an interior load situation and another for flexible pavement : the Boussinesq solution for stresses and displacements in a homogeneous isotropic elastic surface loading are used in ACN-PCN method. Two computer programmes are developed using these mathematical models, which are used for all new type of aircrafts to calculate ACN/PCN.

The manufacturers publish pavement thickness requirement charts for their aircrafts, and the ACN.

Example 10.7

The following information related to runway pavement is available :

PCN for the pavement = 80
Subgrade category = medium strength
Tire pressure limitation = none.
Pavement type = rigid

Determine whether the pavement can accept the following aircrafts at the indicated operating masses and tire pressures.

	Mass	Tire pressure
Airbus A 300 Model B4 at	165000 kg	1.29 MPa
Boeing 747-400 at	395987 kg	1.41 MPa
Concorde at	185066 kg	1.26 MPa
DC-10 30/40 at	268816 kg	1.21 MPa

Solution

Referring to Table 10.2 of ACN for selected aircrafts are 55, 63, 71 and 55 respectively.

Since pavement in question has a PCN of 80, it can accept all these aircrafts.

Example 10.8

Find the ACN of Boeing 767-300 ER at 160000 kg on a flexible pavement resting on a medium strength subgrade (CBR 10). The tire pressure of the main wheels is 1.38 MPa.

Solution

Referring to Table 10.2 of ACN for selected aircrafts by proportioning :

$$\text{ACN} = 57 - \frac{185520 - 160000}{(185520 - 88470)} \times (57 - 22)$$

$$= 57 - \frac{25520}{97050} \times 32 = 57 - 8.32 = 48.48 \approx 49$$

10.8 JOINTS IN RIGID (CONCRETE) PAVEMENTS

Variation in temperature and moisture content causes volume changes and slab warping resulting in significant stresses. Joints are, therefore, placed in concrete pavements to permit expansion and contraction of the pavement, to relieve stresses due to curling and friction and to facilitate construction.

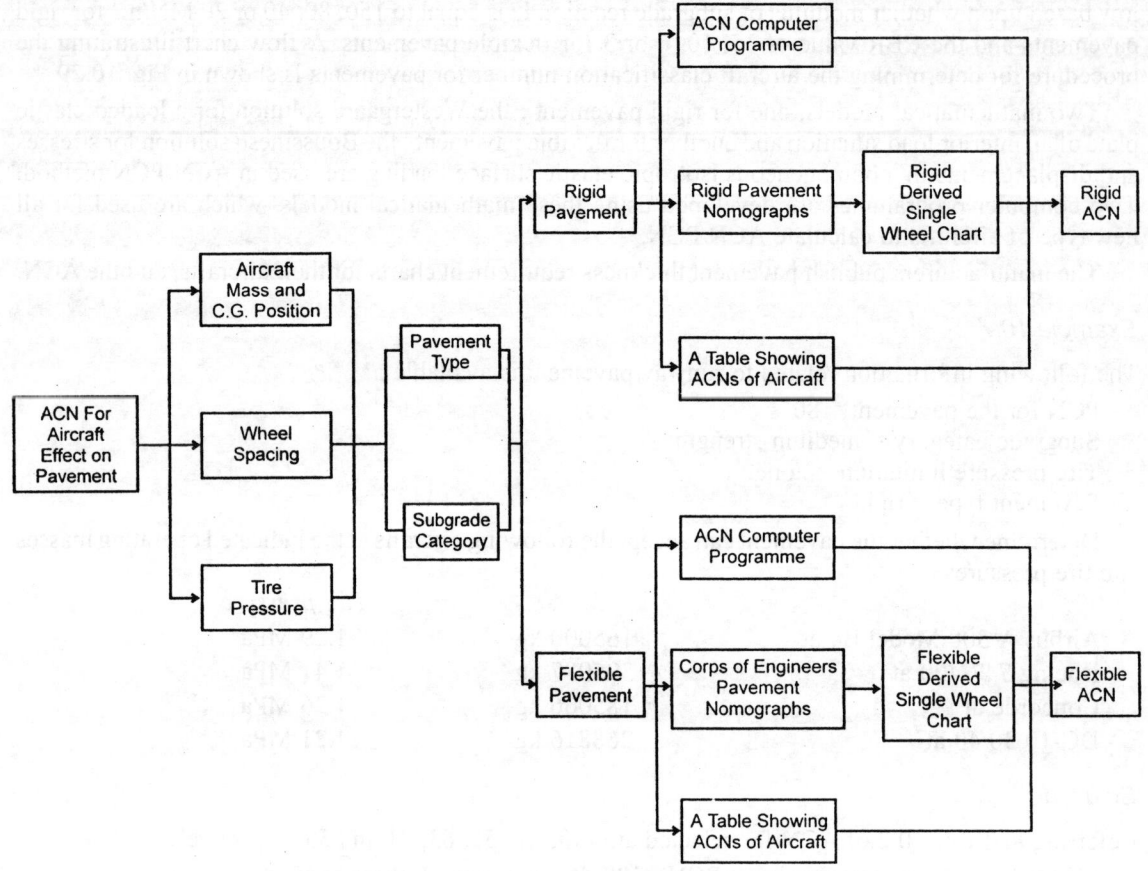

Fig. 10.29. Flow chart for determination of ACN (ICAO).

10.8.1 Type of joints

There are three type of joints : expansion, contraction and construction joints.

(i) Expansion joints

The functions of an expansion joint is to isolate intersecting pavements and to isolate structures from the pavement, thus provide space for expansion of the pavement. This prevents development of very high compressive stresses, which can cause pavement to buckle.

There are two types of expension joints :

Type A is used when load transfer across the joint is required. This joint contains 19 mm non-extruding compressible material and is provided with dowel bars for load transfer.

Type B is used when conditions preclude the use of load transfer devices which span across the joint, such as where pavement abuts a structure. These joints are formed by increasing the thickness of the pavement along the edge slab. No dowel bars are provided.

Expansion joints are shown in Fig. 10.30.

Table 10.2. ACN for selected aircrafts on rigid and flexible pavements

Aircraft	All-up mass (kg)	Load on one main gear leg (%)	Tire pressure MPa	ACN for rigid pavement subgrade MN/m³				ACN for flexible pavement			CBR
				High	Medium	Low	Ultra low	High	Medium	Low	Very low
				150	80	40	20	15	10	6	3
Airbus A300B4	165000 / 88505	47.0	1.29	46 / 17	55 / 20	64 / 25	73 / 29	49 / 20	56 / 21	68 / 25	84 / 36
Airbus 300-600R	170000 / 85033	47.4	1.35	49 / 17	58 / 19	68 / 23	78 / 28	52 / 19	58 / 20	71 / 23	89 / 34
Airbus 310-200	142000 / 75961	46.7	1.33	37 / 15	44 / 17	52 / 20	60 / 23	40 / 17	44 / 18	54 / 20	70 / 27
Airbus 310-300	157000 / 78900	47.4	1.49	45 / 14	54 / 15	63 / 18	71 / 22	47 / 15	53 / 15	64 / 16	81 / 25
BAC I-II Series 500	47400 / 24757	47.5	1.08	32 / 15	34 / 16	35 / 16	36 / 17	29 / 13	30 / 13	33 / 15	35 / 17
Boeing 707 320/420	143335 / 64682	46.0	1.24	36 / 13	43 / 14	52 / 17	59 / 20	40 / 15	44 / 15	54 / 17	69 / 22
Boeing 727-100	77110 / 41322	47.6	1.14	46 / 22	48 / 23	51 / 25	53 / 26	41 / 20	43 / 20	49 / 27	54 / 26
Boeing 727-200	95254 / 45677	46.5	1.19	58 / 24	61 / 25	64 / 27	67 / 29	52 / 22	55 / 22	62 / 25	66 / 29
Boeing 737-200	58332 / 29620	46.0	1.25	34 / 15	36 / 16	38 / 17	39 / 18	30 / 14	31 / 14	35 / 15	39 / 17
Boeing 737-300	61462 / 32904	45.9	1.14	35 / 17	37 / 18	39 / 19	41 / 20	31 / 15	33 / 16	37 / 17	41 / 20
Boeing 737-400	64864 / 33643	46.9	1.44	41 / 19	43 / 20	45 / 21	47 / 22	35 / 16	37 / 17	41 / 18	45 / 21
Boeing 747-400	395987 / 178459	23.4	1.41	53 / 19	63 / 21	75 / 25	85 / 29	57 / 21	64 / 22	79 / 25	101 / 32
Boeing 757-200	109316 / 60260	45.2	1.17	27 / 12	32 / 14	38 / 17	44 / 19	29 / 14	32 / 14	39 / 17	52 / 22
Boeing 767-200	143789 / 78976	46.2	1.31	33 / 15	38 / 17	46 / 20	54 / 24	37 / 18	40 / 19	47 / 21	65 / 26
Boeing 767-300	159665 / 86070	47.5	1.21	38 / 17	45 / 19	54 / 23	63 / 27	43 / 20	48 / 21	58 / 24	78 / 32
Boeing 767-300ER	185520 / 88470	46.0	1.38	47 / 18	56 / 20	66 / 24	76 / 28	51 / 21	57 / 22	70 / 24	92 / 31

(Contd.)

Aircraft	All-up mass (kg)	Load on one main gear leg (%)	Tire pressure MPa	ACN for rigid pavement subgrade MN/m³				ACN for flexible pavement			CBR
				High	Medium	Low	Ultra low	High	Medium	Low	Very low
				150	80	40	20	15	10	6	3
Concorde	$\dfrac{185066}{78698}$	48.0	1.26	$\dfrac{61}{21}$	$\dfrac{71}{22}$	$\dfrac{82}{25}$	$\dfrac{91}{29}$	$\dfrac{65}{21}$	$\dfrac{72}{22}$	$\dfrac{81}{26}$	$\dfrac{98}{32}$
DC-3	$\dfrac{11430}{7767}$	46.8	0.31	$\dfrac{6}{4}$	$\dfrac{7}{5}$	$\dfrac{7}{5}$	$\dfrac{7}{5}$	$\dfrac{4}{3}$	$\dfrac{6}{4}$	$\dfrac{8}{5}$	$\dfrac{9}{6}$
DC8-63/73	$\dfrac{162386}{72002}$	47.6	1.34	$\dfrac{50}{17}$	$\dfrac{60}{19}$	$\dfrac{69}{23}$	$\dfrac{78}{26}$	$\dfrac{52}{18}$	$\dfrac{59}{19}$	$\dfrac{71}{22}$	$\dfrac{87}{29}$
DC9-51	$\dfrac{53338}{29336}$	47.0	1.17	$\dfrac{35}{17}$	$\dfrac{37}{17}$	$\dfrac{39}{18}$	$\dfrac{40}{19}$	$\dfrac{31}{15}$	$\dfrac{32}{15}$	$\dfrac{36}{16}$	$\dfrac{39}{19}$
MD-87	$\dfrac{68266}{33965}$	47.4	1.27	$\dfrac{45}{19}$	$\dfrac{47}{21}$	$\dfrac{49}{22}$	$\dfrac{50}{23}$	$\dfrac{39}{17}$	$\dfrac{42}{18}$	$\dfrac{46}{19}$	$\dfrac{50}{22}$
DC-10-30/40	$\dfrac{268816}{124058}$	37.6	1.21	$\dfrac{46}{20}$	$\dfrac{55}{21}$	$\dfrac{67}{25}$	$\dfrac{78}{29}$	$\dfrac{56}{23}$	$\dfrac{61}{23}$	$\dfrac{74}{26}$	$\dfrac{101}{33}$
Fokker-100	$\dfrac{44680}{24375}$	47.8	0.98	$\dfrac{28}{13}$	$\dfrac{29}{14}$	$\dfrac{31}{15}$	$\dfrac{32}{16}$	$\dfrac{25}{12}$	$\dfrac{27}{13}$	$\dfrac{30}{14}$	$\dfrac{32}{16}$
L1011-500	$\dfrac{225889}{108924}$	46.2	1.27	$\dfrac{50}{23}$	$\dfrac{59}{24}$	$\dfrac{72}{27}$	$\dfrac{84}{31}$	$\dfrac{60}{25}$	$\dfrac{65}{26}$	$\dfrac{79}{28}$	$\dfrac{107}{36}$

(ii) Contraction joints

The function of contraction joints is to provide controlled cracking of the pavement, when pavement contracts due to decrease in moisture content or a temperature drop. Contraction joints also decrease stresses caused by slap warping. Warping joints are thus not required to be provided separately. They relieve tensile stresses due to temperature, moisture and friction, thereby control cracking.

Typical contraction joints are shown in Fig. 10.30.

(iii) Construction joints

Construction joints are provided when two abutting slabs are placed at different times such as at the end of day's work. They facilitate construction of slab. The spacing between longitudinal joints is decided by the width and thickness of pavement. Load transfer is provided by using dowels and aggregate interlocking.

Typical construction joints are shown in Fig. 10.30.

16.8.2 Spacing of joints

The spacing between contraction joints depends upon the thickness of the slab, the type of aggregates and whether the slab is plain concrete or reinforced concrete.

The spacing between construction joint is decided by the width of paving machine and thickness of slab.

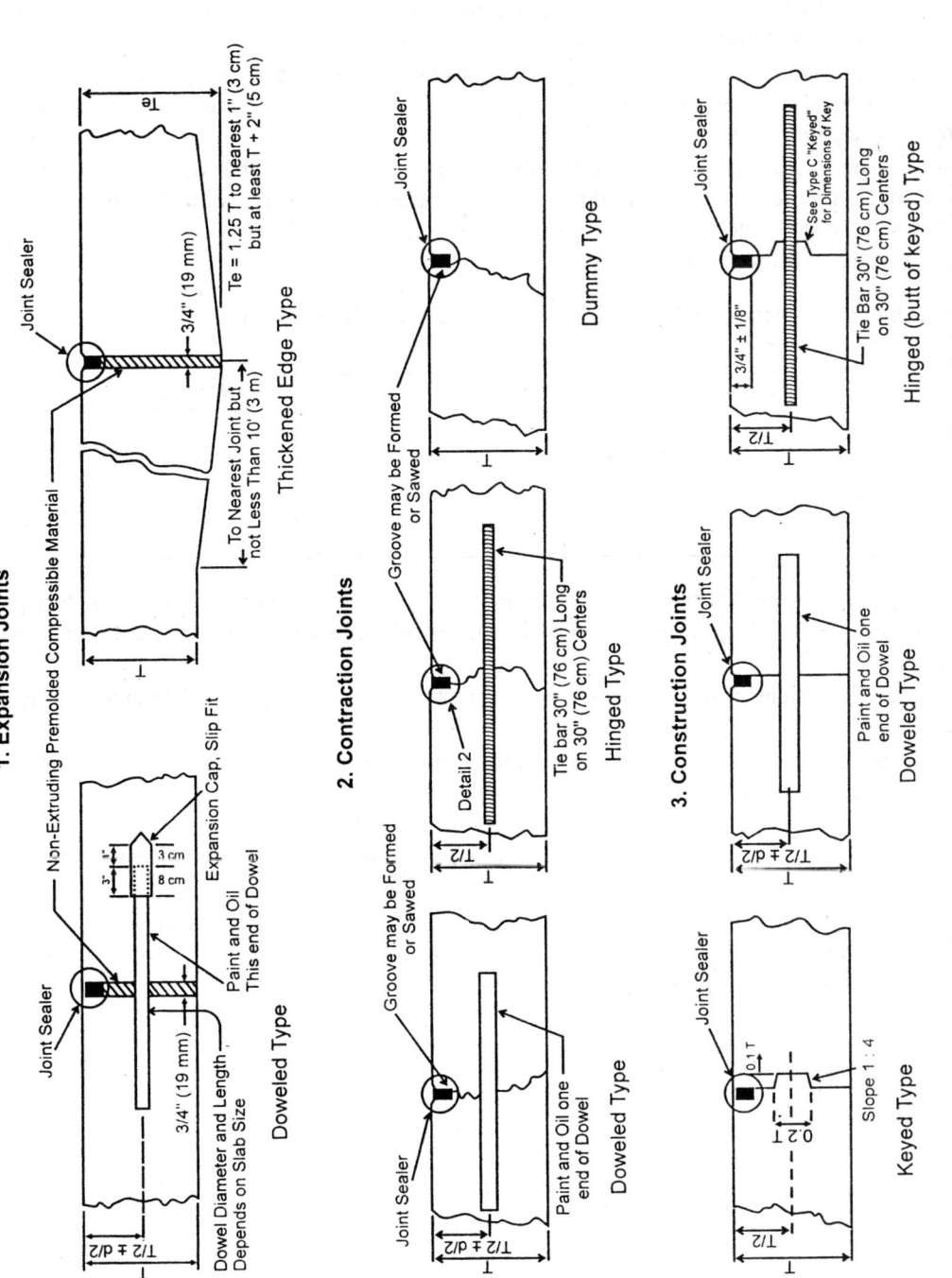

1. Expansion Joints

2. Contraction Joints

3. Construction Joints

Note: 1. Black shaded area is joint sealer

Fig. 10.30. Typical joints in concrete pavements (FAA).

The need and spacing for expansion joints is primarily depends upon slab thickness, thermal properties of concrete, seasonal temperatures and the temperature during construction.

Tabled 10.3 shows recommended maximum joint spacing for rigid pavements. Shorter spacings may be more convenient in some instances. The ratio of slab length to slab width should not exceed 1.25 in unreinforced pavements.

The rigid pavements supported on stabilised sub-base are subjected to higher warping and curling stresses than those supported on unstabilised base. In such cases joint spacing is function of radius of relative stiffness of the slab. The joint spacing should be selected such that the ratio of the joint spacing to the radius of relative stiffness is between 4 and 6.

Table 10.3. Recommended maximum joint spacing for rigid pavement without stabilised sub-base

Slab thickness (cm)	Transverse joint (m)	Longitudinal joint (m)
15	3.8	3.8
17.5–23	4.6	4.6
23–30.5	6.1	6.1
> 30.5	7.6	7.6

All joints should be sealed with a sealing compound to prevent infiltration of water or foreign matter into the joint space. The joint sealant must be capable of withstanding repeated extension and compression.

Dowel bars are load transfer devices which permit joints to open but prevent differential vertical displacement.

Tie bars are deformed steel bars which are used to hold certain joints in contact with each other. They do not transfer load from one slab to another.

CHAPTER 11

Airport Visual Aids-I (Markings)

11.0 INTRODUCTION

The visual aids support the air traffic control system for navigational help for approaching aircraft, and for control of aircraft and vehicles on the surface of the airport. The pilot is guided by visual aids during landing and taking off operations, in day and night in all weather conditions. The selection of visual aids to be used at an airport depends primarily on the visibility conditions under which operations are conducted and on the type of aircraft to be operated at the airport. They also depend upon the type of runways i.e. non-instrumental, instrumental approach or precision approach Category I, II or III. The type and system of visual aids to be used at an airport should be planned in the beginning during the initial planning stage so that requirement of additional land, clearance of obstacles, installations, ducts can be planned at the initial construction stage.

The different types of visual aids for navigation can be classified as :

(a) Indicator and signalling devices

(b) Markings

(c) Lights

(d) Signs

(e) Markers

11.1 INDICATOR AND SIGNALLING DEVICES

(a) Wind direction indicator

An airport must be equipped with at least one wind direction indicator, located at such a position so as to be visible from aircraft in flight or on movement area and in such a way as to be free from the effects of air disturbances caused by nearby objects.

The wind direction indicator should be in the form of a truncated cone made of fabric and should have a length of not less than 3.6 m and a diameter at the larger end of not less than 0.9 m. It should be constructed so that it gives a clear indication of the wind speed. The colour or colours should be so selected as to make the wind direction indicator clearly visible and understandable from a height of at least 300 m, having regard to the background. Generally a single colour preferably white or orange should be used. Where a combination of two colours is required to give adequate conspicuity against changing backgrounds, they should preferably be orange and white, red and white or black and white, and should be arranged in five alternate bands, the first and last bands being the darker colour.

The location of at least one wind direction indicator should be marked by a circular band 15 m in diameter and 1.2 m wide. The band should be centered about the wind direction indicator support and should be in a colour chosen to give adequate conspicuity, preferably white.

Provision should be made for illuminating at least one wind direction indicator at an airport intended for use at night. Fig. 11.1 shows a typical wind direction indicator. Fig. 11.2 shows wind director indicator supports. This cone is capable of illumination in the nights.

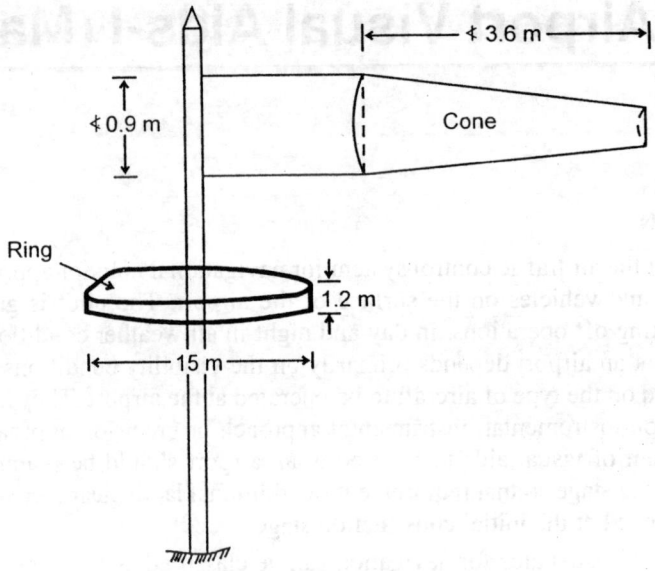

Fig. 11.1. Wind direction indicator.

(b) Landing direction indicator

The landing direction indicator should be located in a conspicuous place on the airport. It is in the form of a "T".

The shape and minimum dimensions of a landing "T" should be as shown in Fig. 11.3. The colour of the landing "T" shall be either white or orange, the choice being dependent on the colour that contrasts best with the background against which the indicator will be viewed.

Where required for use at night landing "T" shall either be illuminated or outlined by white lights.

(c) Signalling lamp

A signalling lamp shall be provided at a controlled airport in the airport control tower. The signalling lamp should be capable of producing red, green and white signals and capable of :

  (a) being aimed manually at any target as required.
  (b) giving a signal in any one colour followed by a signal in either of the two other colours.
  (c) transmitting a message in any one of the three colours.

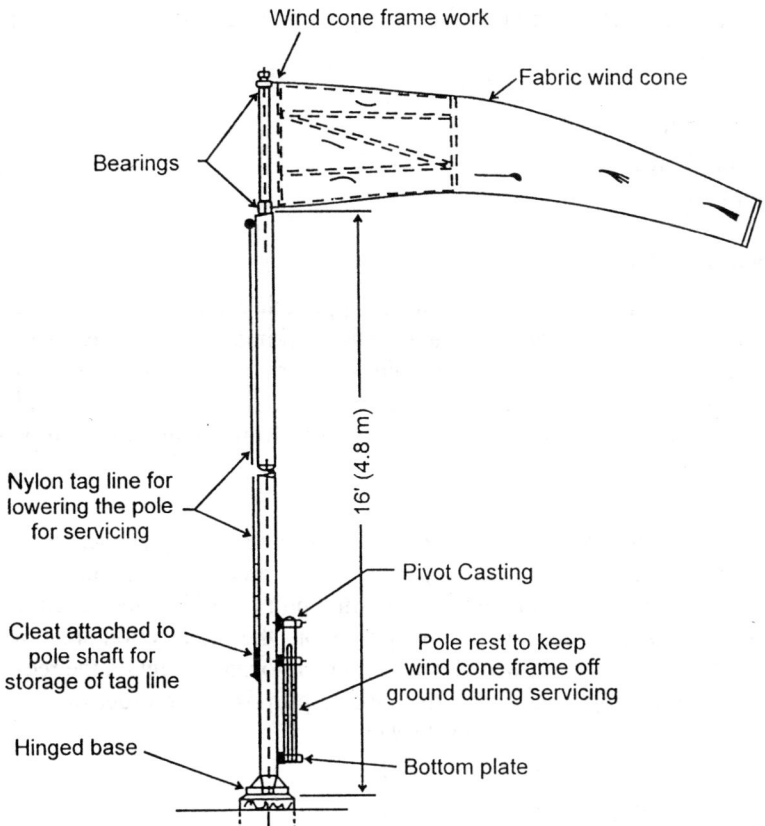

Fig. 11.2. Typical wind direction indicator support (FAA).

The beam spread should not be less than 1° nor greater than 3°, with negligible light beyond 3°. When the signalling lamp is intended for use in the day time the intensity of the coloured light should be not less than 6000 Cd.

11.2 AIRPORT MARKINGS

Airport markings help the pilot in proper, efficient and safe use of the airport facilities. To increase the conspicuity of markings on light coloured pavements, particularly in low visibility conditions, the contrast of the markings can be increased by outlining with a black border at least 15 cms in width. Outlining is also effective for highlighting holding position markings.

Striated markings may be used in areas subject to frost heave and may also be used for temporary markings in non-freeze areas. Striated markings consists of stripes

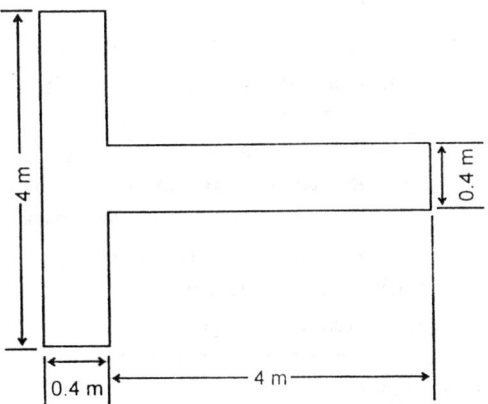

Fig. 11.3. Landing direction indicator (FAA).

10 cm to 20 cm in width separated by unpainted strips. The width of unpainted stripe may not exceed the width of painted stripe.

The airport markings can be grouped as :

(a) Runway markings

(b) Taxiway markings

(c) Other markings

11.2.1 Runway markings

Runway markings may differ depending upon the approach categories, i.e., visual vs non-precision instrument runway may have different markings. The markings on a runway may be upgraded to include elements that are not provided. For example, side stripes, touchdown zone markings etc. could be installed on a visual runway. If a runway has a displaced threshold, blast pad, stopway or wide shoulders, additional marking elements may be necessary. Runway markings are white in colour.

Runway marking precedence

Where runways intersect, the markings on the runway of the higher precedence continue through the intersection, while the markings of the runway of the lower precedence are interrupted except that the runway threshold marking, designation marking, aiming point marking and touchdown zone markings are relocated along the lower precedence runway to avoid the intersection area. Where aiming point markings are relocated, the threshold will also have to be relocated to retain the required distance from the threshold to the aiming point marking. For marking purposes, the order of preference and importance in descending order is as follows (Fig. 11.4) :

(a) Precision instrument runway, Category III

(b) Precision instrument runway, Category II

(c) Precision instrument runway, Category I

(d) Non-precision instrument runway

(e) Visual runway

11.3 RUNWAY MARKINGS

Following markings are used on runways :

(i) Runway designation marking

(ii) Runway centre line marking

(iii) Runway threshold marking

(iv) Runway aiming point marking

(v) Runway touchdown zone marking

(vi) Runway side stripe marking

(vii) Runway threshold bar

(viii) Demarcation bar

(ix) Arrows and arrow heads

(x) Chevrons

(xi) Holding position marking on runway

(xii) Runway shoulder marking

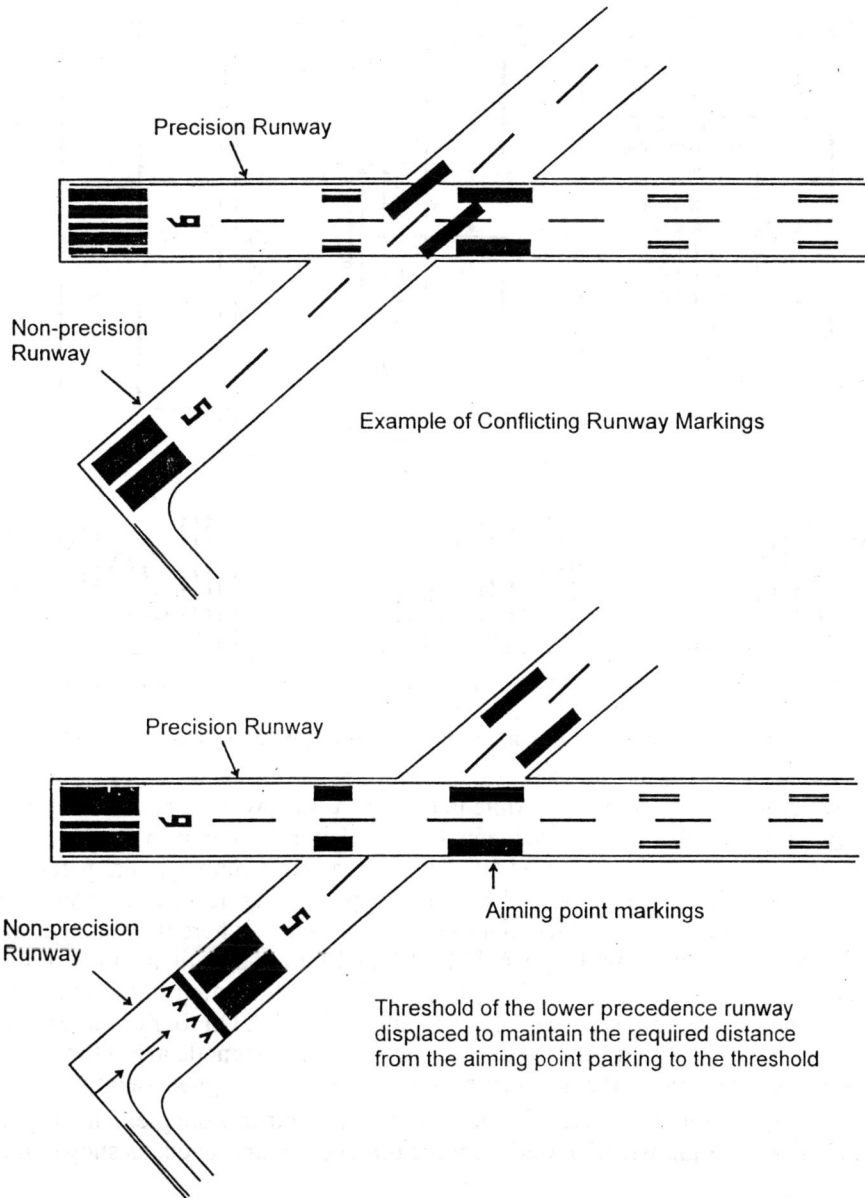

Precision Runway

Non-precision Runway

Example of Conflicting Runway Markings

Precision Runway

Non-precision Runway

Aiming point markings

Threshold of the lower precedence runway displaced to maintain the required distance from the aiming point parking to the threshold

Fig. 11.4. Examples of runway marking precedence (FAA).

11.3.1 Runway Designation Marking

A runway designation marking identifies a runway by its magnetic azimuth. They are located at the threshold of a paved runway, and also as far as practicable at the threshold of an unpaved runway, as shown in Fig. 11.5. If the runway threshold is displaced from the extremity of the runway, a sign showing the designation of the runway may be provided for aeroplanes taking off.

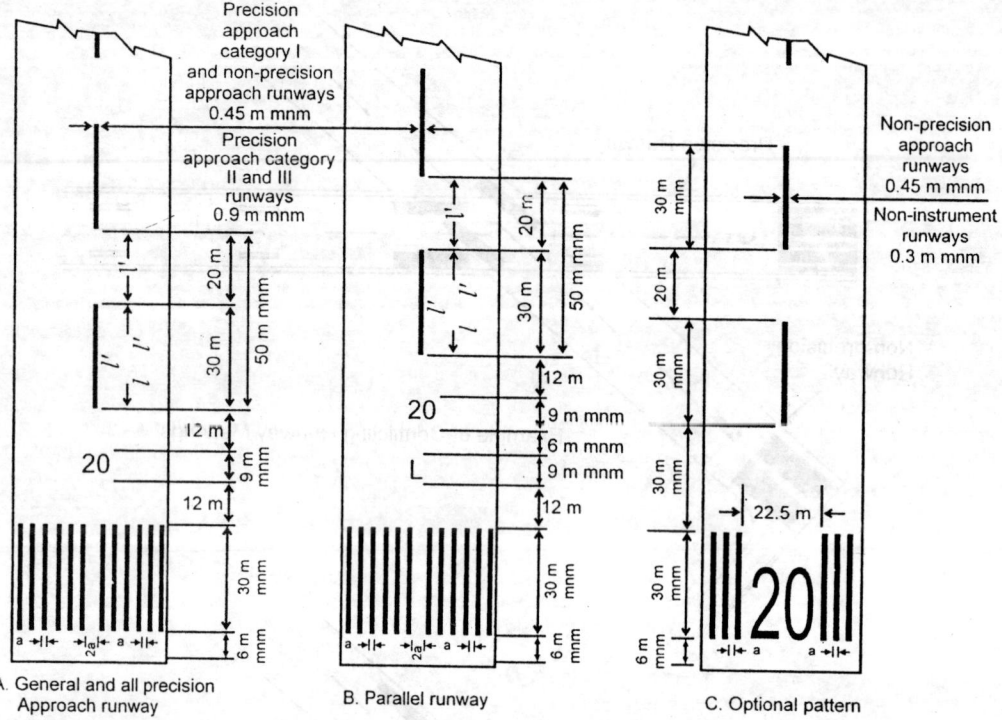

Fig. 11.5. Runway designation, centre line and threshold markings (FAA).

The runway designation markings are white in colour. A runway designation marking consists of a number and on parallel runways is supplemented with a letter as shown in Fig. 11.5. A single-digit runway designation number is not preceded by a zero. On a single runway, dual parallel runways and triple parallel runways, the designation number is the whole number nearest one-tenth of the magnetic azimuth when viewed from the direction of approach. For example, where the magnetic azimuth is 183 degrees, the runway designation marking would be 18; and for a magnetic azimuth of 87 degrees, the runway designation marking would be 9. For a magnetic azimuth ending in the number "5", such as 185 degrees, the runway designation marking can be either 18 or 19. On four or more parallel runways, one set of adjacent runways is numbered to the nearest one-tenth magnetic azimuth and the other set of adjacent runways is numbered to the next nearest one-tenth of the magnetic azimuth.

In the case of parallel runways, each runway designation number is supplemented by a letter, in the order shown from left to right when viewed from the direction of approach, as shown in the following examples :

(a) For two parallel runways having a magnetic azimuth of 182 degrees, the runways should be designated "18L", "18R".

(b) For three parallel runways having a magnetic azimuth of 87 degrees, the runways would be designated "9L", "9C", "9R".

(c) For four parallel runways having a magnetic azimuth of 324 degrees, the runways would be designated "32L", "32R", "33L", "33R".

(d) For five parallel runways having a magnetic azimuth of 138 degrees, the runways would be designated "13L", "13R", "14L", "14C", "14R".

(e) For six parallel runways having a magnetic azimuth of 83 degrees, the runways would be designated "8L", "8C", "8R", "9L", "9C", "9R".

The numbers and letters for runway designation marking are shown in Fig. 11.6. The dimensions should not be less than those shown, but where the numbers are incorporated in the threshold marking, larger dimensions shall be used in order to fill adequately the gap between the stripes of the threshold marking.

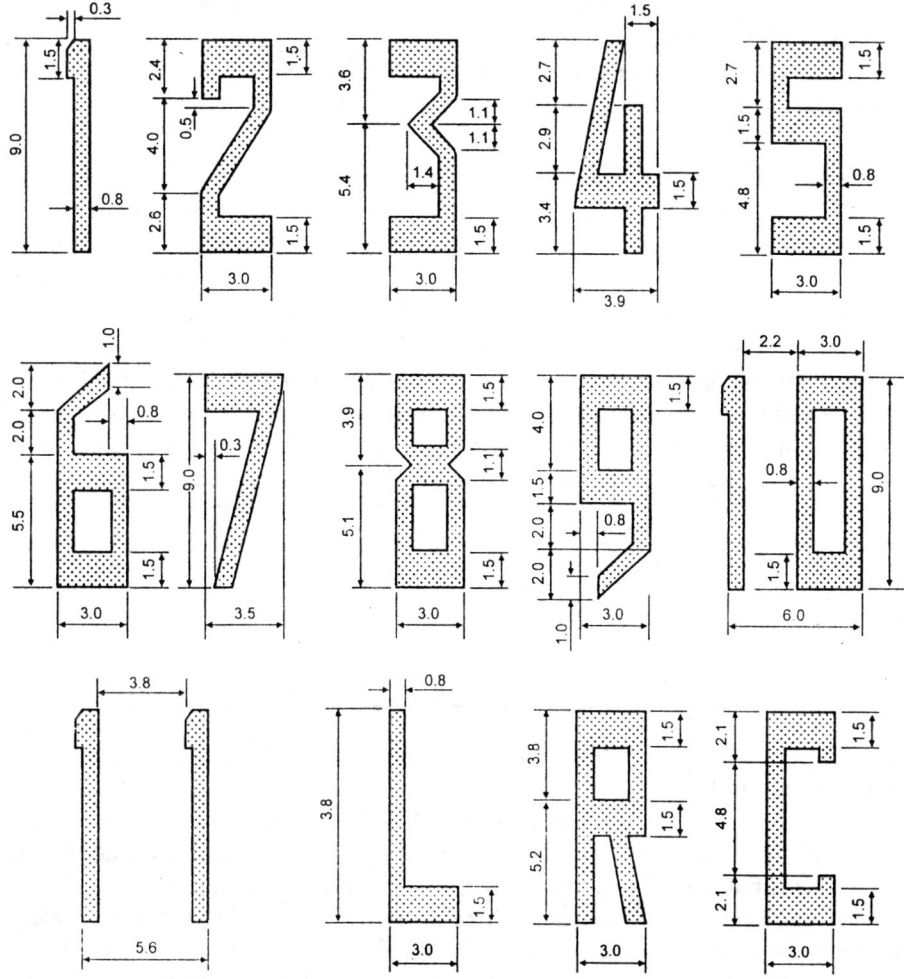

Note: All units are expressed in metres

Fig. 11.6. Form and proportions of numbers and letters for runway designation markings (FAA).

11.3.2 Runway Centre Line Marking

Runway centre line marking identifies the physical centre of the runway and provides alignment guidance during takeoff and landing operations. It is located along the centre line of the runway between the runway designation markings. Runway centre line markings are white in colour, as shown in Fig. 11.5.

A runway centre line marking shall consist of a line of uniformly spaced stripes and gaps. The length of stripes plus a gap shall be not less than 50 m or more than 75 m. The length of each stripe is 36 m and gaps are 24 m in length. Adjustments to accommodate the runway length are made near the runway midpoint. The minimum width of stripes is :
- 0.90 m for precision approach category II and III runways.
- 0.45 m for non-precision approach runway where code number is 3 or 4 and precision approach category I runways.
- 0.30 m on non-precision approach runways where the code number is 1 or 2 and on non-instrument runway (visual runways).

11.3.3 Runways Threshold Marking

A threshold marking identifies the beginning of the runway that is available for landing. A threshold marking is provided at the threshold of a paved instrument runway and of a paved non-instrument runway where the code number is 3 or 4 and the runway is intended for International Commercial air transport. It should be provided, as far as practicable, at the threshold of an unpaved runway as well.

The stripes of the threshold marking start 6 m from the runway threshold. They are white in colour.

A runway threshold marking shall consist of a pattern of longitudinal stripes of uniform dimensions disposed symmetrically about the centre line of a runway as shown in Fig. 11.5(A) and 11.5(B) for a runway width of 45 m. The number of stripes shall be in accordance with the runway width as follows:

Runway width	Number of stripes
18 m	4
23 m	6
30 m	/8
45 m	12
60 m	16

except that on non-precision approach and non-instrument runway 45 m or greater in width, they may be as shown in Fig. 11.5(C).

The stripes shall extend laterally to within 3 m of the edge of a runway or to a distance of 27 m on either side of a runway centre line, whichever results in the smaller lateral distance. Where a runway designation marking is placed within a threshold marking there shall be a minimum of three stripes on each side of the centre line of the runway. Where a runway designation marking is placed above a threshold marking, the stripes shall be continued across the runway. The stripes shall be at least 30 m long and approximately 1.80 m wide with spacings of approximately 1.80 m between them except that, where the stripes are continued across a runway, a double spacing shall be used to separate the two stripes nearest the centre line of the runway, and in the case where the designation marking is included within the threshold marking this spacing shall be 22.5 m.

11.3.4 Runway Aiming Point Marking

The aiming point marking serves as a visual aiming point for landing operations. The beginning of the aiming point marking commences 306 m from threshold. Aiming point markings are white in colour.

An aiming point marking consists of two **rectangular markings**, 45 m in length, located symmetrically on each side of the runway centre line as shown in Fig. 11.7. The width of each marking is 10 m for a runway width of 45 m or greater. The lateral spacing between the inner side of the marking is 21.6 m for a runway of width 45 m. For runway widths less than 45 m the width of markings and the lateral

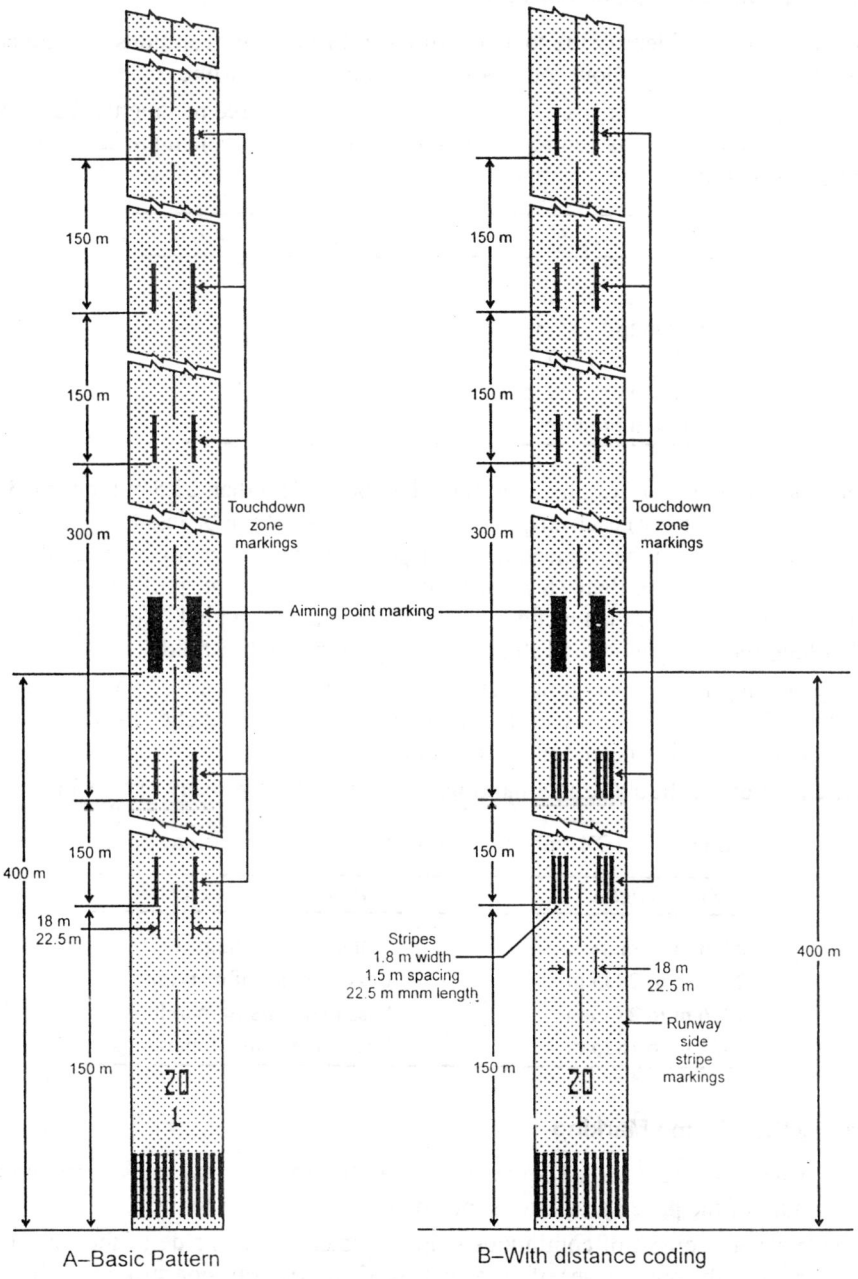

Fig. 11.7. Touchdown zone markings (illustrated for a runway with a length of 2100 m or more).

space between markings, is decreased in proportion to the decrease in runway width. Where touchdown zone markings are provided, the lateral spacing between the markings should be the same as that of the touchdown zone markings.

11.3.5 Runway Touchdown Zone Marking

Touchdown zone markings identify the touchdown zone for landing operations and are coded to provide distance information. The colour touchdown zone markings is white.

Touchdown zone markings consist of groups of one, two and three rectangular bars symmetrically arranged in pairs about the runway centre line as shown in Fig. 11.7. The number of such pairs related to runway length is as follows :

Runway length	Pair(s) of markings
Less than 900 m	1
900 m to 1200 m	2
1200 m to 1500 m	3
1500 m to 2100 m	4
2100 m or more	6

For pattern shown in Fig. 11.7(a), the markings shall be not less than 22.5 m long and 3 m wide. For the pattern shown in Fig. 11.7(b) each stripe of each marking shall not be less than 22.5 m long and 1.8 m wide with a spacing of 1.5 m between adjacent stripes. The lateral spacing between the inner sides of the rectangles shall not be less than 18 m nor more than 22.5 m, and preferably 18 m. The lateral spacing shall be equal to that of the touchdown zone light where provided. The pairs of markings shall be provided at longitudinal spacing of 150 m beginning from the threshold.

For runways having touchdown zone markings on both the ends, those pairs of markings which extend to within 270 m of the runway midpoint are eliminated. No touchdown zone markings eliminated where installed on only one end of the runway.

The requirement of touchdown zone markings on both ends of runway (Table 11.1).

Table 11.1. TDZ markings when installed on both runway ends.

Runway length	Markings on each end
2936 m or more	Full set of markings
2737 m to 2935 m	Less one pair of markings
1826 m to 2736 m	Less two pairs of markings
1527 m to 1825 m	Less three pairs of markings

11.3.6 Runway Side Stripe Marking

Runway side stripes provide a visual contrast between the runway and the surrounding terrain and delineates the width of the paved area that is intended to be used as a runway.

Runway side stripes consist of continuous stripes located on each side of the runway. The maximum distance between the outer edges of the stripes is 60 m. Runway side stripe markings are white in colour.

A runway side stripe marking should consist of two stripes one placed along each stripe approximately on the edge of each stripe approximately on the edge of runway, except that where the runway is greater than 60 m in width, the stripes should be located 30 m from the runway centre line.

A runway side stripe should have an overall width of at least 0.9 m on runway 30 m or more in width and at least 0.45 m on narrower runways. For precision instrument runways minimum width is 1 m, and on other runways width of stripes should be at least equal to the width of the runway centre line stripe. The stripes extend to the end of displaced threshold areas which are used for take offs.

11.3.7 Runway Threshold Bar

A threshold bar delineates the beginning of the runway that is available for landing when there is pavement aligned with the runway on the approach side of the threshold. A threshold bar is located on the landing runway at the threshold. The threshold bar is white in colour.

A threshold bar is 3 m in width and extends across the width of the runway.

11.3.8 Demarcation Bar

A demarcation bar delineates a runway with a displaced threshold from a blast pad, stopway or taxiway that precedes the runway. The demarcation bar is located on the blast pad, stopway or taxiway at the point of intersection with the runway as shown in Fig. 11.8.

The demarcation bar is yellow in colour. It is 1 m wide and extends across the width of the blast pad, stopway or taxiway.

11.3.9 Arrows and Arrow Heads

Arrows are used to identify a displaced threshold area and are useful for centre line guidance for take offs and/or rollouts. Arrow heads are used in conjunction with a threshold bar to further highlight the beginning of a runway where the use of chevrons is not appropriate.

Where a runway threshold is permanently displaced, arrows and arrowheads are provided in the portion of the runway before the displaced threshold, as shown in Fig. 11.9 where pavement area preceding a runway is used as a taxiway, arrowheads are provided prior to the threshold bar (Fig. 11.10).

Arrows and arrowheads used in a displaced threshold area are white in colour. Arrowheads used on a taxiway prior to a runway threshold are yellow.

When a runway threshold is temporarily displaced from the normal position, all markings prior to the displaced threshold shall be obscured except the runway centreline marking, which shall be converted to arrows.

Dimensions and spacing of arrows and arrowheads is given in Table 11.2.

Table 11.2. Dimensions of arrows and arrowheads

Runway width (W)	Number of arrows	Spacing between arrow heads	Spacing to runway edge
Equal to or more than 30 m	4	W/4	W/8
Less than 30 m	3	W/3	W/6
Less than 20 m	2	W/2	W/4

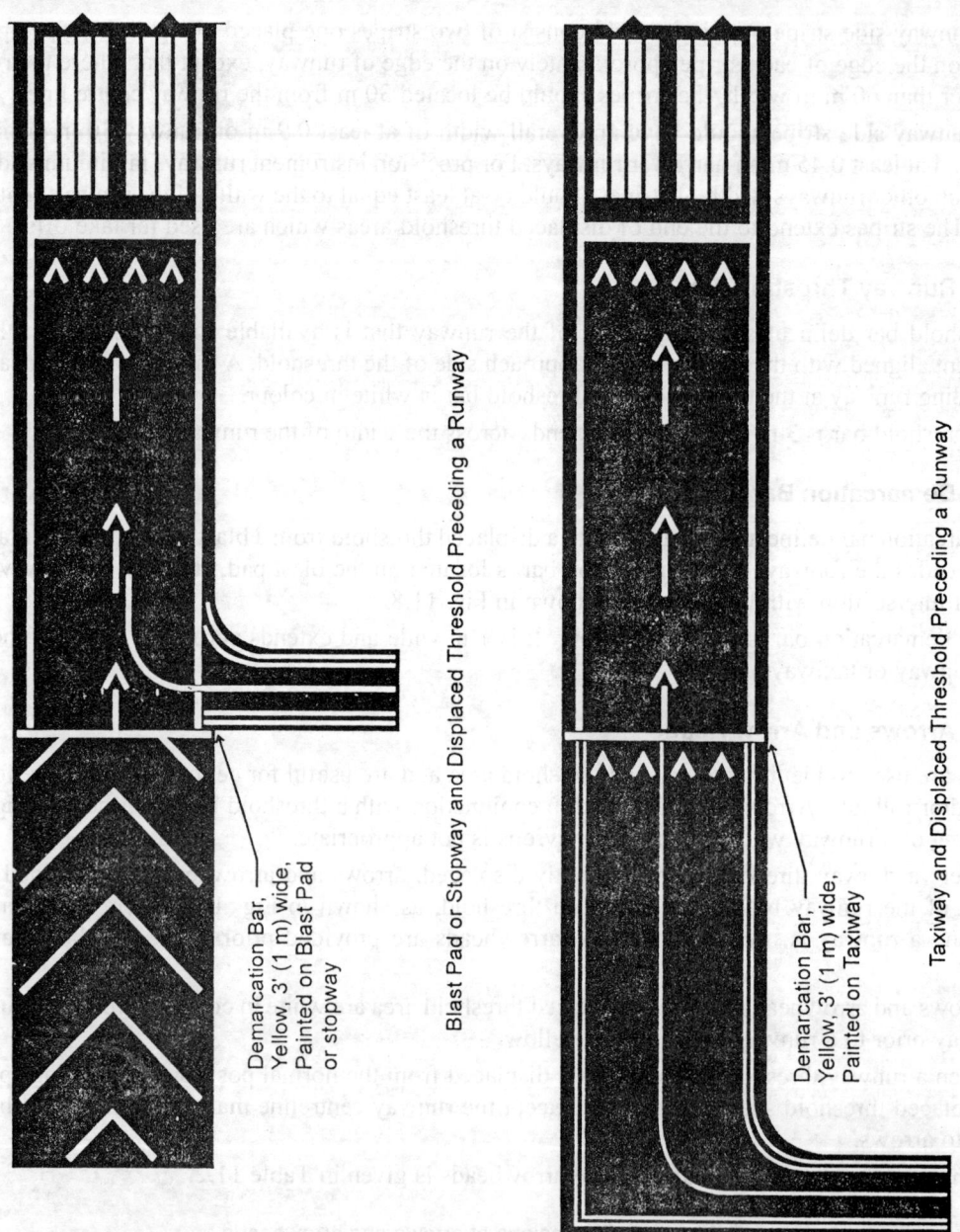

Fig. 11.8. Marking for blast pad or stopway or taxiway preceding a displaced threshold (FAA).

11.3.10 Chevrons

Chevrons are used to identify pavement areas unusable for landing, take off and taxing. Chevrons are located on pavement areas that are aligned with and contiguous to the runway as shown in Fig. 11.11. They are yellow in colour. Dimensions shown in Fig. 11.11 are in metres.

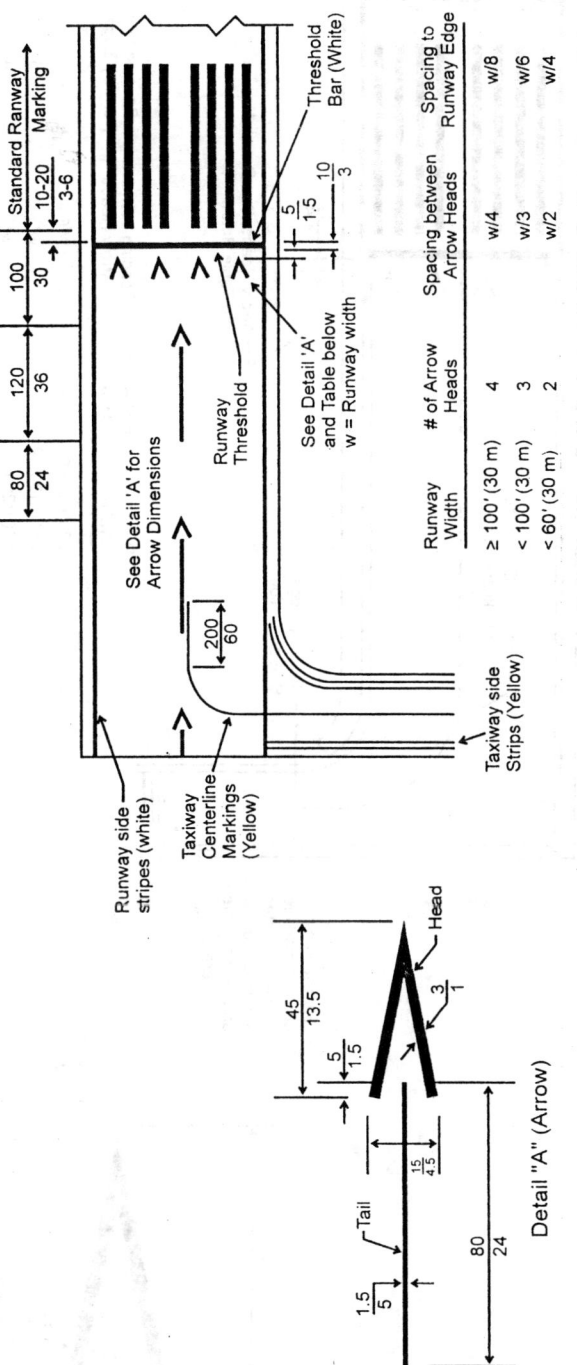

Fig. 11.9. Permanently displaced threshold markings (all dimensions in meters) (FAA).

Runway Width	# of Arrow Heads	Spacing between Arrow Heads	Spacing to Runway Edge
≥ 100' (30 m)	4	w/4	w/8
< 100' (30 m)	3	w/3	w/6
< 60' (30 m)	2	w/2	w/4

Detail "A" (Arrow)

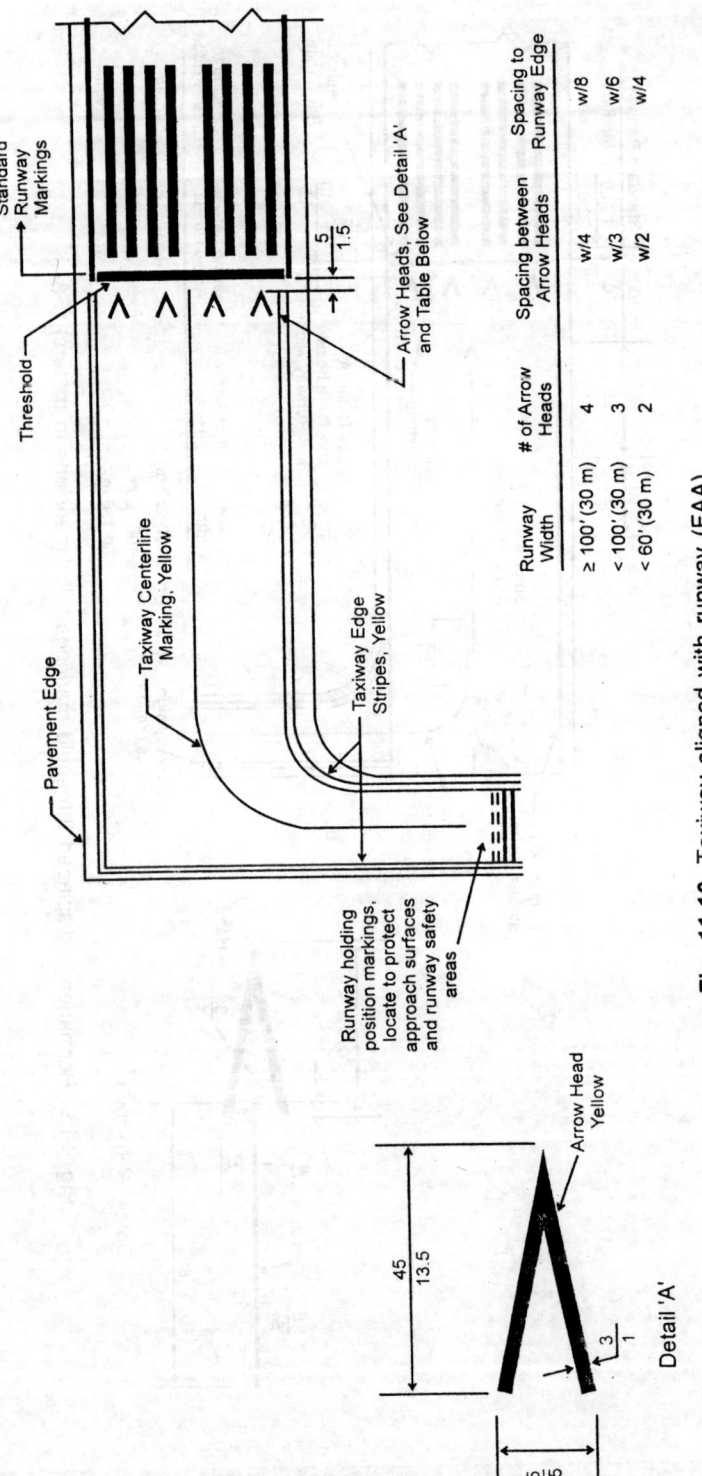

Fig. 11.10. Taxiway aligned with runway (FAA).

Runway Width	# of Arrow Heads	Spacing between Arrow Heads	Spacing to Runway Edge
≥ 100' (30 m)	4	w/4	w/8
< 100' (30 m)	3	w/3	w/6
< 60' (30 m)	2	w/2	w/4

Detail 'A'

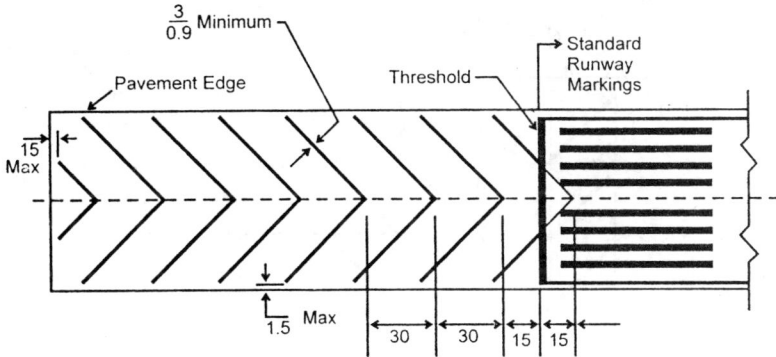

Fig. 11.11. Chevrons (FAA).

11.3.11 Holding Position Markings on Runway

These markings are installed on a runway where an aircraft is supposed to stop when the runway is normally used as a taxiway or used for "land and hold short operations".

Holding position markings for runway/runway intersection are determined on a case-by-case basis, as per standards of the concerned airport authority.

Holding position markings on runway are yellow in colour.

These markings are identical to taxiway/runway holding position markings. The solid lines of these markings are always on the side where the aircraft is to hold. The markings are installed perpendicular to the runway centre line and interrupt all runway markings except for the runway designation marking. The holding position markings should be moved so that they do not interrupt the designation marking.

11.3.12 Runway Shoulder Markings

Runway shoulder markings are used, when needed, as a supplement to runway side stripes to identify pavement areas contiguous to the runway sides that are not intended for use by aircraft. Runway side stripes are usually sufficient in defining the limits of usable pavements. Shoulder markings are generally needed where pilots have experienced problems identifying the runway from the shoulder thereby creating a need to delineate the shoulder as usable pavement.

Runway shoulder markings are located between the runway side stripes and the pavement edge as shown in Fig. 11.12.

The colour of runway shoulder markings should be yellow.

Runway shoulder markings consist of stripes 1 m in width and spaced 30 m apart. The stripes start at the runway midpoint, are slanted at an angle of 45 degrees to the runway centreline and are oriented as shown in Fig. 11.12.

11.4 TAXIWAY MARKINGS

All taxiways should have centreline markings and runway holding position markings wherever they intersect a runway. All the taxiway markings are yellow in colour. Following are the markings used on taxiways :

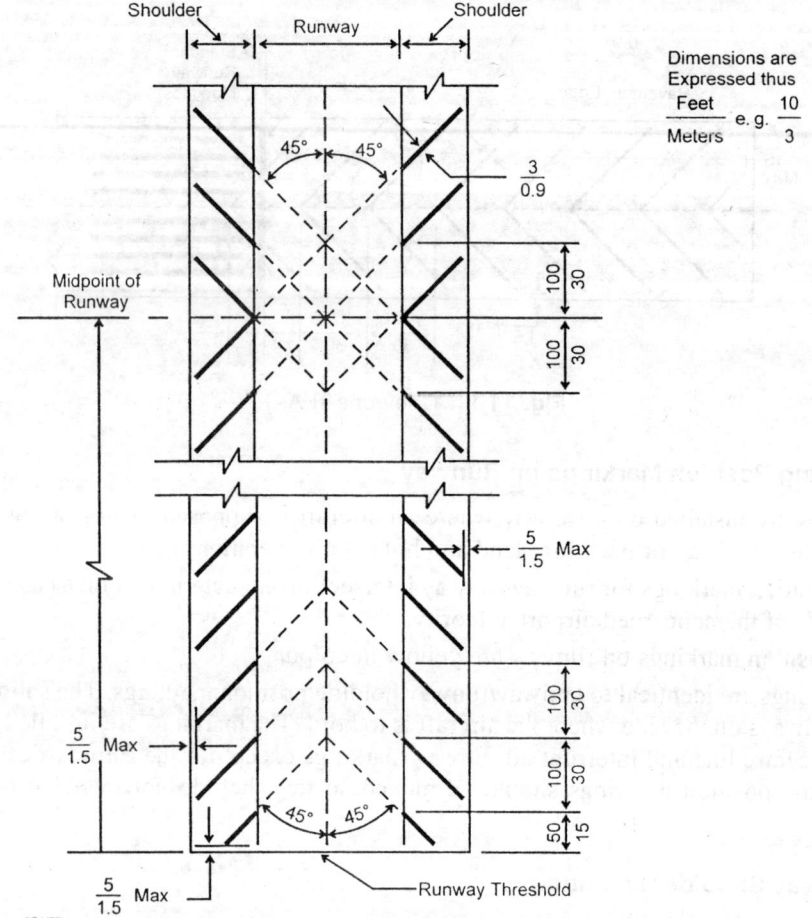

Fig. 11.12. Runway shoulder markings (dimensions in metres) (FAA).

(a) Taxiway centreline marking
(b) Taxiway edge marking
(c) Holding position markings on taxiways
(d) Taxiway shoulder markings
(e) Geographic position markings

11.4.1 Taxiway Centreline Marking

Taxiway centreline marking provides a visual cue to permit taxiing along a designated path. They are used on taxiways to provide guidance from the runway centreline to the point on the apron where aircraft stand markings commence.

On a straight section of a taxiway, the taxiway centreline marking should be located along the taxiway centreline. On a taxiway curve the marking should continue from the straight portion of the taxiway at a constant distance from the outside edge of the curve.

At an intersection of taxiway with a runway where the taxiway serves as an exit from the runway, the taxiway centreline marking should be curved into the runway centreline marking as shown in Fig. 11.13. The taxiway centreline marking should be extended parallel to the runway centreline marking for a distance of at least 60 m beyond the point of tangency where the code number is 3 or 4 and for a distance of at least 30 m where the code number is 1 or 2.

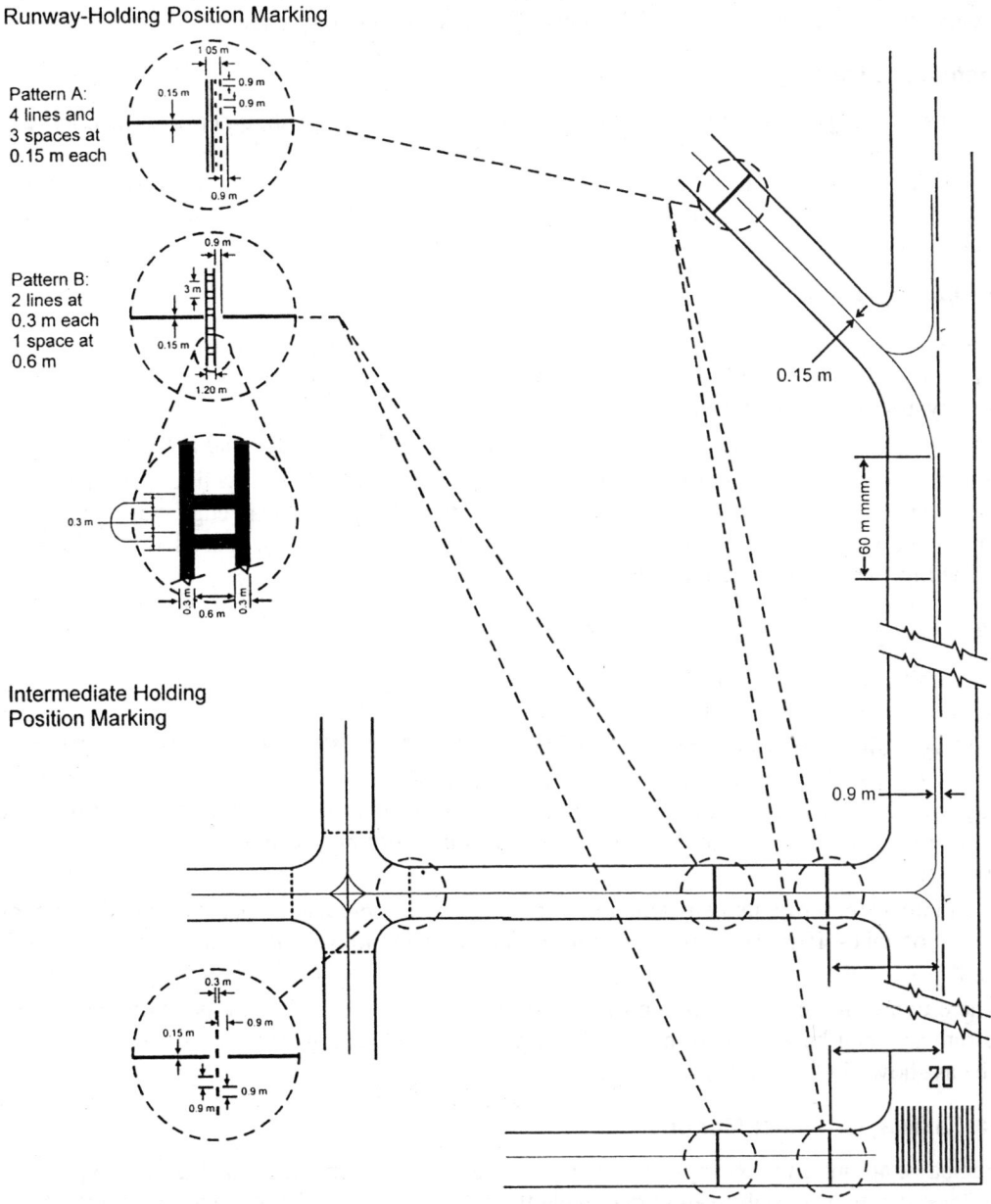

Fig. 11.13. Taxiway markings shown with basic runway markings (FAA).

A taxiway centreline marking shall be at least 15 cm in width and continuous in length except where it intersects a taxi-holding position marking as shown in Fig. 11.13.

11.4.2 Taxiway Edge Marking

Taxiway edge markings are used to delineate the edge of the taxiway. They are primarily used when the taxiway edge does not correspond with the edge of the pavement. Two types of markings are used depending upon whether the aircraft is supposed to cross the taxiway edge.

(i) Continuous Marking

Continuous markings are used to delineate the taxiway edge from the shoulder or some other contiguous paved surface not intended for use by aircraft. When an operational need exists, the continuous taxiway edge markings are to delineate the edge of the taxiway from a contiguous non-paved surface. Continuous taxiway edge markings are not to be used in situations where aircraft would be required to cross them.

(ii) Dashed Markings

Dashed taxiway edge markings are used when there is an operational need to define the edge of a taxiway or taxilane on a paved surface where the pavement contiguous to the taxiway edge is intended for used by aircraft e.g. an apron.

Taxiway edge markings are located on the taxiway at its defined edge and are yellow in colour.

Continuous taxiway edge markings consist of a continuous double yellow line with each line being at least 15 cm in width, spaced 15 cm apart edge to edge. Dashed taxiway edge markings consists of broken double yellow line, with each line being at least 15 cm in width, spaced 15 cm apart edge to edge. The lines are 4.5 m in length with 7.5 m gaps.

11.4.3 Holding Position Markings on Taxiways

Holding position markings on taxiways identify the location on a taxiway where an aircraft is supposed to stop when it does not have clearance to proceed onto the runway.

Holding position markings should be located on all taxiways that intersect runways. These markings are also located on taxiway crossing through the runway approach area so that an aircraft on the taxiway will not penetrate the surface used to locate the runway threshold, inner approach obstacle free zone, inner transitional obstacle free zone and clearway. Holding position markings on taxiways are yellow in colour.

At an intersection of a taxiway and a non-instrument, non-precision approach, precision approach category I or take-off runway, the taxi-holding position marking shall be as shown in Fig. 11.13, pattern A.

Where two or three taxi-holding positions are provided, the taxi-holding position marking closest to the runway shall be as shown in Fig. 11.13, pattern A, and the markings farther from the runway shall be as shown in Fig. 11.13, pattern B.

11.4.4 Taxiway Shoulder Markings

Holding bays, aprons and taxiways are sometimes provided with shoulder stabilisation to prevent blast and water erosion. This stabilisation may have the appearance of a full strength pavement, but is not intended for use by aircraft. Usually the taxiway edge marking defines this area, but sometimes conditions may exist such as stabilised islands or taxiway curves where confusion may occur as to which side of the edge stripe is full strength pavement. Where such conditions exist, taxiway shoulder markings should be used to indicate that the pavement is unusable.

The colour of taxiway shoulder markings is yellow. It is also acceptable to paint the stabilised area green.

The stabilised area is marked with 1 m yellow stripes perpendicular to the edge stripes as shown in Fig. 11.14.

On straight sections, the marks are placed at a maximum of 30 m spacing. On curves, the marks are placed at a maximum of 15 m apart between the curve tangents. The stripes are extended to 1.5 m from the edge of the stabilised area or to 7.5 m in length, whichever is less.

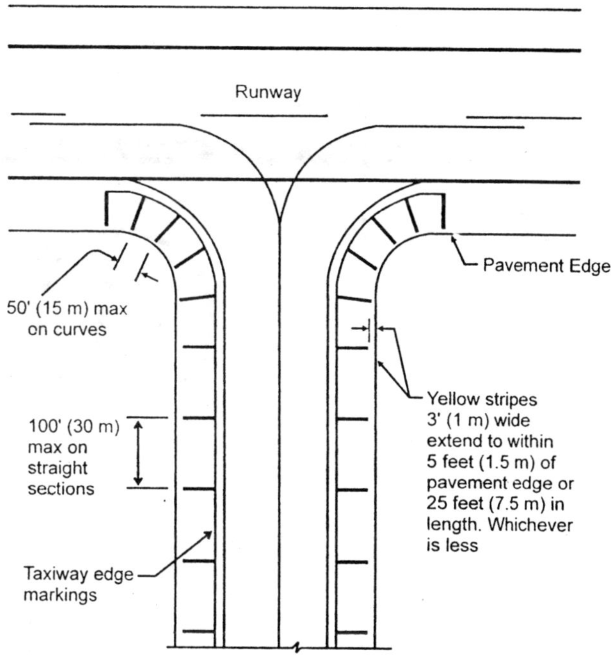

11.4.5 Geographic Position Markings

Fig. 11.14. Taxiway shoulder markings (FAA).

Geographic position markings are installed when points are necessary to identify the location of taxiing aircraft during low visibility operations. Low visibility operations are those that occur when the runway visible range (RVR) is below 360 m.

These markings are located along low visibility taxi routes designated in the airports plan. They are positioned to the left of the taxiway centreline in the direction of taxiing. On a particular airport, the airport operator in coordination with local airport traffic control tower, will determine where these markings are needed.

The geographic position marking is a circle with a diameter of 2.67 m. When installed on concrete or other light coloured pavement the circle is comprised of a 15 cm outer black ring contiguous to a 15 cm white ring with a pink circle with a diameter of 1.3 m in the middle as shown in Fig. 11.15.

When installed on asphalt or other dark coloured pavement, the white ring and block ring are reversed i.e., the white ring becomes the outer ring and the black ring becomes the inner ring.

Geographic position markings are designated with either a number or a number and letter. The number corresponds to the consecutive position of the marking on the route. When used the letter indicates the letter designation of the taxiway on which marking is located. The designation of the spot should be centred in the circle. The designation is black, has a height of 1.3 m.

11.5 OTHER MARKINGS

The other markings are used, as appropriate on airports. Some of the other markings used are :
  (a) Vehicle roadway markings
  (b) VOR receiver checkpoint markings
  (c) Aircraft stand markings
  (d) Nonmovement area boundary markings

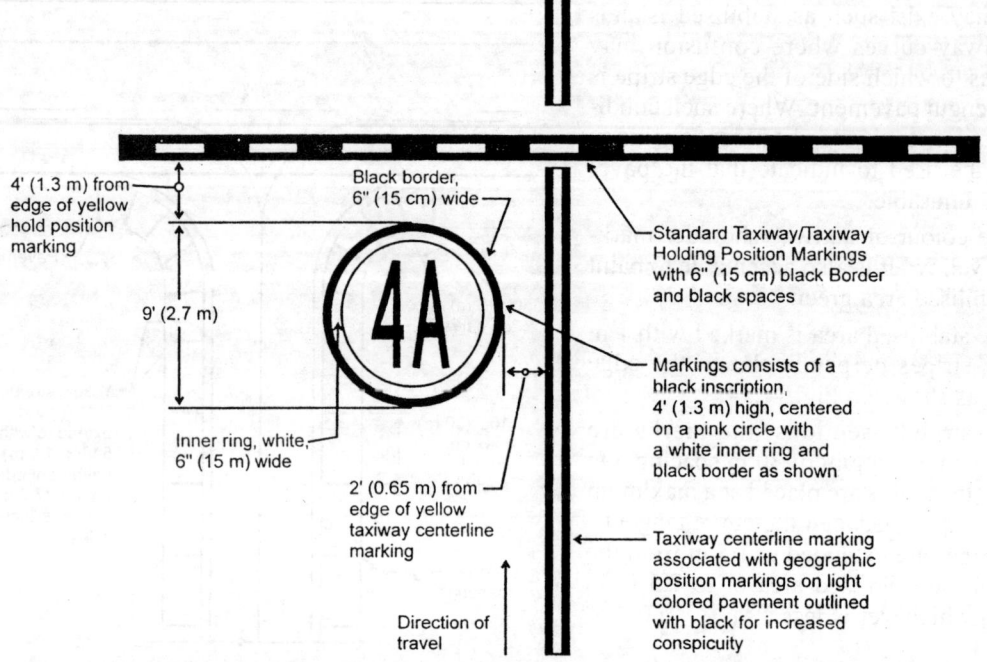

Fig. 11.15. Geographic position markings (FAA).

   (e) Apron safety line markings

   (f) Markings of permanently closed runways and taxiways

   (g) Temporarily closed runways and taxiways.

Markings for helicopters and vertiports are discussed in respective chapters on heliports and vertiports.

11.5.1 Vehicle Roadway Markings

The standard vehicle roadway markings are used to delineate roadways located on or crossing areas that are also intended for use by aircraft.

Vehicle roadway markings are white. Vehicle roadways are delineated on aircraft manoeuvering areas when there is a need to define a pathway for vehicle operations. A minimum spacing of 0.67 m must be maintained between the roadway edge marking and the non-movement area boundary marking.

Vehicle roadway markings consist of a solid line to delineate each edge of the roadway and a dashed line to separate lanes within the edges of the roadway. The edgelines and lane lines are both 15 m wide and the dashes for the lane lines are 4.5 m in length with a spaces of 7.5 m between dashes.

11.5.2 VOR receiver checkpoint markings

VOR receiver checkpoint markings allow the pilot to check aircraft instruments with navigational aid signals.

A VOR aerodrome checkpoint marking is located on the spot at which an aircraft is to be parked to receive the correct VOR signal. It is on airport apron or taxiway, but never on a runway.

VOR receiver checkpoints normally should not be established at distances less than 8 km from the facility nor should they be established on non-paved areas.

VOR receiver checkpoints are provided with painted markings and an associated sign. This consist of a circle 6 m in diameter and have a line width of 15 cm as shown in Fig. 11.16.

When it is preferable for an aircraft to be aligned in a specific direction, a line should be provided that passes through the centre of the circle on the desired azimuth. The line should extend 6 m outside the circle in the desired direction of heading and terminate in an arrowhead. The width of the line should be 15 cm as shown in Fig. 11.16.

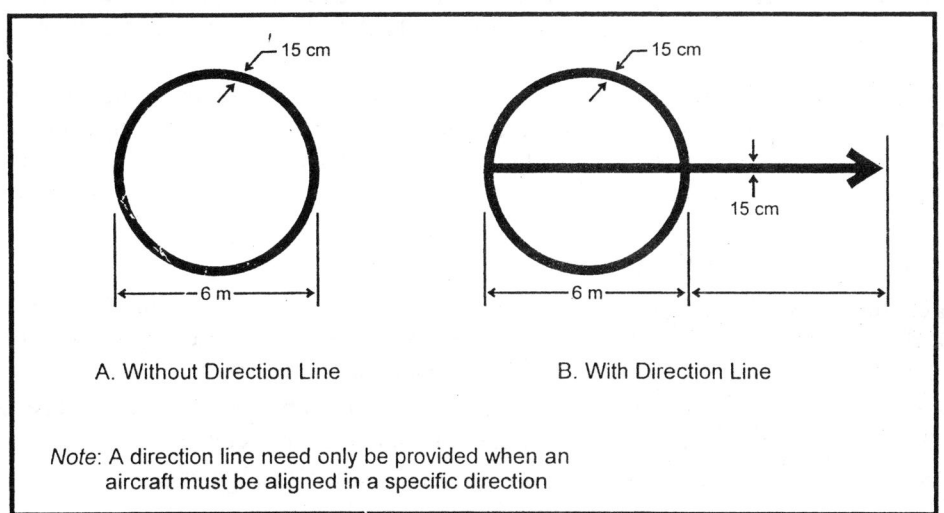

A. Without Direction Line   B. With Direction Line

Note: A direction line need only be provided when an aircraft must be aligned in a specific direction

Fig. 11.16. VOR checkpoint marking (FAA).

11.5.3 Aircraft Stand Markings

Aircraft stand markings should be provided for designated parking positions on paved aprons. These should be located so as to provide the clearances specified. There are several patterns for parking the aircraft as described in other chapter.

Aircraft stand markings should include such elements as stand identification, lead-in line, turn bar, turning line, alignment bar, stopline and load out line, as are required by parking configurations.

Lead-in, turning and lead-out lines should normally be continuous in length and have a width of not less than 15 cms. The curved portions of lead-in, turning and lead-out lines should have radii appropriate to the most demanding aircraft type for which the markings are intended.

Where it is intended that an aircraft proceed in one direction only, arrows pointing in direction to be followed should be added as part of the lead-in and lead-out lines.

A turn bar should be located at right angles to the lead-in line, a beam the left pilot position at the point of initiation of any intended turn. It should have a length and width of not less than 6 m and 15 m respectively and include an arrowhead to indicate the direction of turn.

An alignment bar should be placed so as to be coincident with the extended centre line of the airport in the specified parking position and visible to the pilot during the final part of the parking manoeuvre. It should have a width of not less than 15 cm.

A stepline should be located at right angles to the alignment bar, a beam the left pilot position at the intended point of stop. It should have a length and width of not less than 6 m and 15 cm respectively.

11.5.4 Non-Movement Area Boundary Marking

Non-movement area boundary markings are used when there is a need to delineate the movement area (i.e., area under air traffic control, from the non-movement area i.e. area not under air traffic control. This marking should be used only when it is considered necessary by airport operator and airport traffic control tower.

The movement area boundary marking is yellow. It is located on the boundary between the movement and non-movement area. In order to provide adequate clearance for the wings of taxiing aircraft, this marking should never coincide with edge of a taxiway.

The non-movement area boundary marking consists of two yellow lines, one solid and one dashed. The solid line is located on the non-movement areas side while the dashed yellow line is located on the movement area side. Each line is 15 cm in width with a 15 cm spacing between lines. The dashed are 1 m in length with a 1 m spacing between dashes.

11.5.5 Apron Safety Line Marking

Apron safety lines should be provided on paved aprons as required by the parking configurations and ground facilities. They are located so as to define the areas intended for use by ground vehicles and other aircraft serving equipment etc. to provide safe separation from aircrafts.

The apron safety lines should include such elements as wing tip clearance lines and service road boundary lines as required by the parking configurations and ground facilities.

An apron safe line should be continuous in length and at least 10 cm in width.

11.5.6 Marking of Permanently Closed Runway and Taxiways

For runways and taxiways which have been permanently closed, the runway threshold, runway designation and touchdown zone markings are obliterated and yellow crosses are placed at each end at a 300 m interval.

If the closed runway intersects an open runway, crosses should be placed on the closed runway on both sides of the open runway.

For taxiways, a yellow cross is placed on the closed taxiway at each entrance.

11.5.7 Markings of Temporarily Closed Runways and Taxiways

When it is necessary to provide a visual indication that a runway is temporarily closed, crosses are placed only at each end of the runway. The crosses are yellow in colour. Since the crosses are temporary, they are usually made of some easily removable material, such as plywood or fabric rather than painted on the pavement or fabric rather than painted on the pavement surface. Since these crosses are placed over white runway markings, their visibility can be enhanced by a 15 cm black border.

Sometimes a raised lighted cross may be placed on each runway end in lieu of the markings described above. The cross should be located with 75 m of the runway end.

On temporary closed taxiways, a yellow cross may be installed at each entrance to the taxiway.

Typical dimensions of crosses are shown in Fig. 11.17.

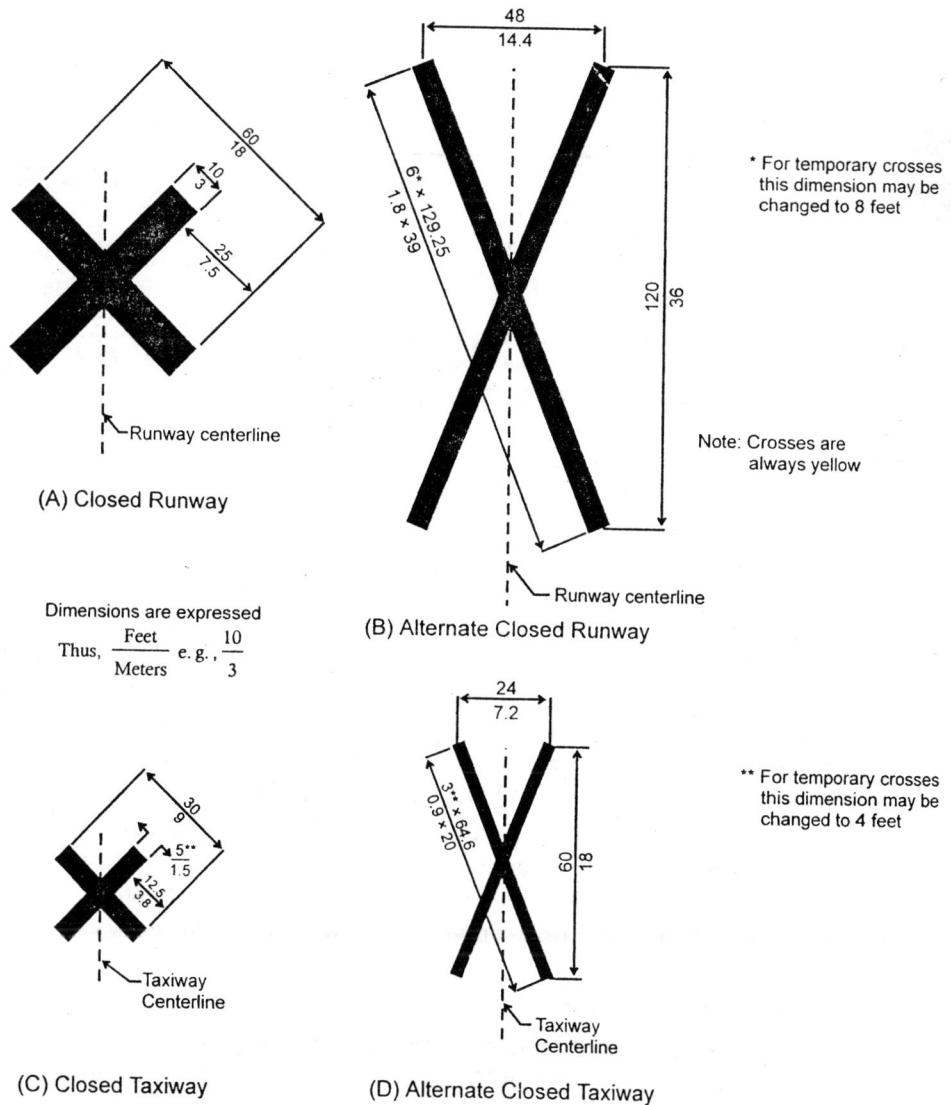

* For temporary crosses this dimension may be changed to 8 feet

Note: Crosses are always yellow

** For temporary crosses this dimension may be changed to 4 feet

(A) Closed Runway

Dimensions are expressed

Thus, $\dfrac{\text{Feet}}{\text{Meters}}$ e.g., $\dfrac{10}{3}$

(B) Alternate Closed Runway

(C) Closed Taxiway

(D) Alternate Closed Taxiway

Fig. 11.17. Closed runway and taxiway markings (FAA).

When all runways are closed temporarily, the runways are marked as described above the airport beacon is turned off.

Airport Visual Aids-II (Lightings)

12.0 INTRODUCTION

During the night time, the pilot is aided by the system of signal lights, which convey him certain information and help in safe landing and take-off. The pilot needs to know the alignment to bring the aircraft straight towards the runway, the height at which he is flying with reference to ground for landing, the distance away from the runway and roll guidance - the correct angle of the longitudinal axis of the aircraft in relation to ground surface. If the roll is not properly corrected the wing of the aircraft may strike the runway. The colours and pattern of airport lighting is standardised to provide the needed aeronautical information to the pilot, and prevent any confusion.

Any non-aeronautical ground light which by reason of intensity, configuration or colour near an airport, which might endanger the safety of aircraft should be extinguished, screened or otherwise modified so as to eliminate the source of danger or confusion. In particular attention should be directed to a non-aeronautical ground light visible from air within the areas described below :

(a) Instrument runway (code number 4)

Within the areas before the threshold and beyond the end of the runway extending at least 4500 m in length from the threshold and the runway end and 750 m either side of the extended runway centre line in width.

(b) Instrument runway (code number 2 or 3)

As in (a) above, except that the length should be at least 3000 m.

(c) Instrument runway (code number 1) non-instrument runway

Within the approach area.

12.1 LIGHT FIXTURES AND SUPPORTS

Two types of light fixtures and supports are used at the airports : Elevated lights and surface lights.

(1) Elevated lights

Elevated approach lights and their supporting structures within 300 m from the threshold (but not including the 300 m crossbar) or up to a distance from the runway end where the approach lights no longer constitute the major hazard to an aircraft overrunning the runway end or to an airborne aircraft inadvertently striking them, whichever distance is less, should be light weight and frangible. In cases where it is not possible to make such elevated approach lights and supporting structures lightweight and frangible, these characteristics should relate to at least the top 1.8 m of the structure.

Elevated lights on runway, stopway and taxiway should be light weight and frangibly mounted. Their height should be sufficiently low to preserve clearance for propellers and for engine pods of jet aircraft.

(2) Surface lights

Light fixtures inset in the surface of runways, stopways, taxiways and aprons should be so designed and fitted as to withstand being run over the wheels of an aircraft without damage either to the aircraft or to the lights themselves. The temperature produced by conduction or radiation at the interface between an installed inset light and an aircraft tire should not exceed 160°C during a 10 minute period of exposure.

12.2 LIGHT INTENSITY AND CONTROL

In dusk and poor visibility conditions during the day, lightings are more effective than marking. For lights to be effective during such conditions and in night, they must be of adequate intensity, for the minimum conditions of visibility.

The lights of an approach lighting system may be of higher intensity than the runway lighting, it is good practice to avoid abrupt changes in intensity as these could give a pilot a false impression that the visibility is changing during approach.

Where a high intensity lighting system is provided, a suitable intensity control should be incorporated to allow for adjustment of the light intensity to meet the prevailing conditions. Separate intensity controls or other suitable methods should be used to ensure that lighting systems for approach lights, runway edge lights, runway threshold lights, runway end lights, runway centre line lights, runway touchdown zone lights and taxiway centre line lights, when installed can be operated at compatible intensities. Standards of light intensities and beams are beyond the scope of this book.

12.3 EMERGENCY LIGHTING

At an airport provided with runway lighting and without secondary power supply, sufficient emergency lights should be installed, on at least the primary runway in the event of failure of the normal lighting system. Emergency lighting may also be useful to mark obstacles or delineate taxiways and apron areas.

When installed on a runway, the emergency lights should, as a minimum, conform to the configuration required for a non-instrument runway.

The colour of emergency lights should conform to the colour requirements for runway lighting, except that where the provision of coloured lights at the threshold and the runway end is not practicable, all lights may be variable white or as close to variable white as practicable.

12.4 LIGHTS AT AN AIRPORT

The lights used at an airport can be classified in the following categories :

  (a) Airport beacon

  (b) Identification beacon

  (c) Approach lighting system

     (i) For non-instrument and non-precision approach runway

     (ii) For precision approach runway category I

     (iii) For precision approach runway category II and II

  (d) Circling guidance lights

  (e) Runway leading lighting system

  (f) Runway threshold identification lights

  (g) Runway edge lights

  (h) Runway threshold and wing bar lights

  (i) Runway end lights

  (j) Runway centreline lights

  (k) Runway touchdown lights

  (l) Stopway lights

  (m) Taxiway centreline lights

  (n) Taxiway edge lights

  (o) Stop bars

  (p) Clearance bars

  (q) Taxi-holding position lights

  (r) Approach flood lighting

  (s) Visual docking guidance system

  (t) Airport stand manoeuvering guidance lights

  (u) Visual approach slope indicator system

12.4.1 Airport beacon

An airport beacon is provided at each airport intended for use in night, except in special circumstances when beacon is considered unnecessary due to airtraffic requirements, the conspicuity of the airport features in relation to its surroundings, and the installation of other visual aids useful in locating the airport.

The airport beacon is located on or adjacent to airport. The location should be such that the beacon is not shielded by objects in significant directions and does not dazzle a pilot approaching to land.

The airport beacon shows either green coloured flashes alternating with white flashes or white flashes only. The frequency of total flashes should be from 12 to 30 per minute, preferably not less than 20 per minute.

The coloured flashes emitted by beacons at water airports are yellow in colour unlike the green at the land airports.

The light from the beacon shows at all angles of azimuth. The vertical light distribution should extend upwards from an elevation of not more than 1° to an elevation sufficient to provide guidance at the maximum elevation at which the beacon is intended to be used.

12.4.2 Identification beacon

An identification beacon is provided at an airport which is intended for use at night and cannot be easily identified from the air by other visual means. It is located on the airport.

An identification beacon at land airport shows green light of peak intensity not less than 2000 cd. Identification beacons show yellow at water airports. The light should be emitted at all angles in azimuth and up to at least 45° above the horizontal.

The identification characters should be transmitted in the International Morse Code. The speed of transmission should be between 6 and 8 words per minute, the corresponding range of duration of the morse dots being from 0.15 to 0.2 seconds per dot.

12.4.3 Approach lighting system

The approach lighting system depends upon the type of approach used and can be grouped in following categories :
  (a) Non-instrument runway
  (b) Non-precision approach runway
  (c) Precision approach runway category I
  (d) Precision approach runway categories II and III

(a) Approach lighting system for non-instrument and non-precision approach runway

System of approach lights used for these types of approaches can be called simple approach lighting system. It consists of a row of lights on the extended centre line of the runway, extending, whenever possible, over a distance of not less than 420 m from the threshold with a row of lights forming a crossbar 18 m or 30 m in length at a distance of 300 m from the threshold. The lights forming the cross bar should be as nearly as practicable in a horizontal straight line at right angles to, and bisected by, the line of centreline lights.

The lights of the cross bar shall be spaced so as to produce a linear effect, except that when a cross bar of 30 m is used, gaps may be left on each side of centre line. These gaps should be kept to a minimum to meet local requirements and each should not exceed 6 m. 1 m and 4 m spacings for the cross bar lights are used.

The lights forming the centre line are placed at longitudinal intervals of 60 m, except that when it is desired to improve the guidance, an interval of 30 m may be used. The innermost light should be located either 60 m or 30 m from the threshold, depending on the longitudinal interval selected for centre line lights.

It is not physically possible to provide a centre line extending for a distance of 420 m from the threshold, it should be extended to 300 m so as to include the cross bar. If this is not possible, the centre line lights should be extended as far as practicable and each centre line light should then consist of a barrette at least 3 m in length. Subject to the approach system having a cross bar at 300 m from the threshold, an additional cross bar may be used at 150 m from the threshold.

The system should lie as nearly as practicable in the horizontal plane passing through the threshold, provided that :

(a) no light shall be screened from an approaching aircraft; and

(b) as far as possible, no object shall protrude through the plane of the approach lights within a distance of 60 m from the centre line of the system. Where this is unavoidable, as in the case of a single isolated object protruding through the plane of lights e.g. an ILS installation, the object should be treated as an obstacle and marked and lighted accordingly.

The lights of a simple approach lighting system shall be fixed lights and the colour of lights should be such as to ensure that the system is readily distinguishable from other aeronautical ground lights. Each centre line light shall consist of either :

(a) a single source; or

(b) a barrette at least 3 m in length.

When the barrette as in (b) is composed of lights approximately to point source, a spacing of 1.5 m between adjacent lights in the barrette is found satisfactory. It is advisable to use barrettes 4 m in length if it is anticipated that the simple approach lighting system will be developed into a precision approach lighting system.

When provided for a non-instrument runway the lights should show at all angles in azimuth necessary to a pilot on bas leg and final approach. The intensity of the lights should be adequate for all conditions of visibility and ambient light for which the system has been approved.

When provided for a non-instrument runway the lights should show at all angles in azimuth necessary to a pilot on bas leg and final approach. The intensity of the lights should be adequate for all conditions of visibility and ambient light for which the system has been approved.

When provided for a non-precision approach runway, the lights should show at all angles in azimuth necessary to the pilot of an aircraft which on final approach does not deviate by an abnormal amount from the path defined by the non-visual aid. The lights should be designed to provide guidance during both day and night in the most adverse conditions of visibility and ambient light for which it is intended that the system should remain usable.

(b) Approach lighting system for precision approach category I

A precision approach category I, lighting system consists of a row of lights on the extended centre line of the runway extending, wherever possible, over a distance of 900 m from the runway threshold with a row of lights forming a cross bar 30 m in length at a distance of 300 m from runway threshold. The lights forming the cross bar should be as nearly as practicable in a horizontal straight line at right angles to, and bisected by, the line of the centre line lights. The lights of cross bar shall be spaced so as to produce a linear effect, except that gap may be left on each side of the centre line. These gaps shall be kept to a minimum to meet local requirements and each shall not exceed 6 m.

The lights forming the centre line shall be placed at longitudinal intervals of 30 m with the innermost light located 30 m from threshold. The system should lie as nearly as practicable in the horizontal plane passing through the threshold, provided that :

(a) no light should be screened from an approaching aircraft; and

(b) as far as possible, no object protrude through the plane of the approach lights within a distance of 60 m from the centre line of the system. If it is unavoidable the object should be treated as an obstacle and marked and lighted accordingly.

The centre line and cross bar lights of a precision approach category I lighting system should be fixed lights showing variable white. Each centreline light is consist of either :

(a) a single light source in the innermost 300 m of the centre line, two light sources in the central 300 m of the centre line and three light sources in the outer 300 m of the centre line to provide distance information. Additional cross bar of lights to the cross bar at 300 m from threshold should be provided at 150 m, 450 m, 600 m and 750 m from threshold; or

(b) a barrete at least 4 m in length. The barrete is composed of light approximately to point source, a spacing of 1.5 m between adjacent lights in the barrete.

(c) Approach lighting system for precision approach category II and III

The approach lighting system consists of a row of lights on extended centre line of the runway, extending, wherever possible, over a distance of 900 m from the runway threshold. In addition, the system shall have two side rows of lights extending 270 m from threshold and two cross bars one at 150 m and one at 300 m from the threshold as shown in Fig. 12.1. The lights forming the centre line shall be placed at longitudinal intervals of 30 m with the innermost lights located 30 m from the threshold.

The lights forming the side rows shall be placed on each side of the centre line, at a longitudinal spacing equal to that of the centre line lights and with the fist light located 30 m from the threshold. The lateral spacing between the innermost lights of the side row shall be not less than 18 m nor more than 22.5 m, and preferably 18 m, but in any case shall be equal to that of the touch down zone lights.

The cross bar provided at 150 m from the threshold shall fill in the gaps between centre line and side row light. The cross bar provided at 300 m from the threshold shall extend on both sides of the centre line lights to a distance of 15 m from the centre line.

The system of lights should lie as nearly as practicable in the horizontal, plane passing through the threshold provided that :

(a) no light shall be screened from an approaching aeroplane; and

(b) as far as possible, no object shall protrude through the plane of the approach lights within a distance of 60 m from the centre line of the system. If unavoidable, the object shall be treated as an obstacle and marked and lighted accordingly.

The centre line of precision approach category II and III, lighting system for the first 300 m from the threshold shall consist of barrettes showing variable white, except that, where the threshold is displaced 300 m or more, the centre line may consist of single light sources showing variable white. The barrettes should be at least 4 m in length. When barrettes are composed of lights approximating to point source, the lights shall be uniformly spaced at intervals not more than 1.5 m.

Beyond 300 m from threshold each centre line light shall consist of either :

(a) a barrette as used on the inner 300 m; or

(b) two light sources in the central 300 m of the centre line and three light sources in the outer 300 m of the centre line. Additional cross bars of lights should be provided at 450 m, 600 m and 750 m from the threshold in this case. The outer ends of these crossbars shall lie on two straight lines that either are parallel to the centre line or converge to meet the runway centre line 300 m from threshold.

The side row shall consist of barrettes showing red. The length of a side row barrette and spacing of its lights shall be equal to those of the touchdown zone light barrettes.

The lights forming the cross bars shall be fixed lights showing variable white. The lights shall be uniformly spaced at intervals of not more than 2.7 m. The intensity of red lights shall be compatible with the intensity of the white lights.

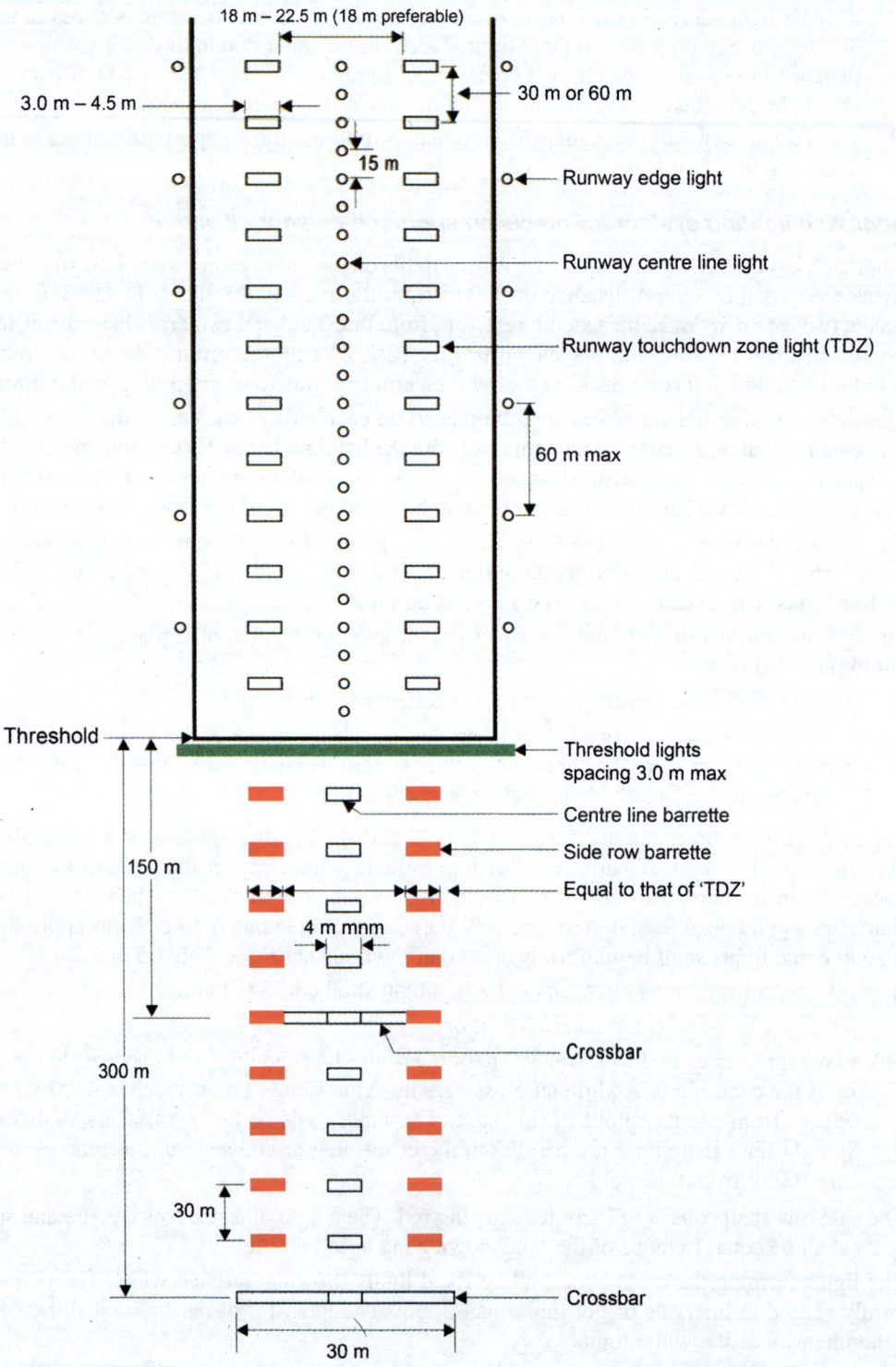

Fig. 12.1. Inner 300 m approach and runway lighting for precision approach runways II and III (ICAO).

12.4.4 Circling guidance lights

Circling guidance lights are provided when existing approach and runway lighting system do not satisfactorily permit identification of the runway and/or approach area to a circling aircraft.

The location and number of circling guidance lights to enable pilot to :

(a) join the downwind leg or align and adjust the aircraft's track to the runway at a required distance from it and to distinguish the threshold in passing; and

(b) keep in sight the runway threshold and/or other features which will make it possible to judge the turn on to base leg and final approach, taking into account the guidance provided by other visual aids.

Circling guidance lights should consist of :

(a) Lights indicating the extended centre line of the runway and/or parts of any approach lighting system; or

(b) lights indicating the position of the runway threshold; or

(c) lights indicating the direction or location of the runway; or

(d) a combination of such lights as is appropriate to the runway under consideration.

Circling guidance lights should be fixed or flashing lights of an intensity and beam spread, adequate for the conditions of visibility and ambient light in which it is intended to make visual circline approaches. The flashing light should be white and the steady light either white or gaseous discharge lights. Lights should be designed and installed in such a manner that they will not dazzle or confuse a pilot when approaching to land, taking off or taxiing.

12.4.5 Runway lead-in lighting systems

A runway lead-in lighting system should be provided where it is desired to provide visual guidance along a specific approach path, for reasons such as avoiding hazardous terrain or for purpose of noise abatement.

A runway lead-in lighting system should consist of a group of lights positioned so as to define the desired approach path so that one group may be sighted from the preceding group. The interval between adjacent groups should not exceed approximately 1600 m.

Runway lead-in lighting system may be curved, straight or a combination thereof. It should extend from a point as determined suitable up to a point where the approach lighting system (if provided) or the runway lighting system is in view.

Each group of lights of a runway lead-in lighting system should consists of at least three flashing lights in a linear or cluster configuration. The system may be augmented by steady burning lights where such lights would assist in identifying the system. The flashing lights should be white and the steady burning lights gaseous discharge lights. Where practicable, the flashing lights in each group should flash in sequence towards the runway.

12.4.6 Runway threshold identification lights

Runway threshold identification lights are installed at the threshold of a non-precision approach runway when additional threshold conspicuity is necessary or where it is not practicable to provide other approach lighting aids.

Runway threshold identification lights shall be located symmetrically about the runway centre line, in line with the threshold and approximately 10 m outside each line of runway edge lights. Runway threshold identification lights should be flashing white lights with a flash frequency between 60 and 120 per minute. The lights should be visible only in the direction of approach to runway.

12.4.7 Runway edge lights

Runway edge lights are provided for a runway intended for use at night or for a precision approach runway intended for use by day or night.

The runway edge lights are placed along the full length of the runway and shall be in two parallel rows equidistant from the centre line. They are placed along the edges of the area declared for use as the runway or outside the edges of the area at a distance of not more than 3 m.

Where the width of the runway exceeds 60 m, the distance between the rows of lights should be determined taking into account the nature of the operations, the light distribution characteristics of the runway edge lights and other visual aids serving the runway.

The lights shall be uniformly spaced in rows at intervals not more than 60 m for an instrument runway and at an interval of not more than 100 m for a non-instrument runway. The lights on opposite side of the runway axis shall be on lines at right angles to that axis. A intersections of runways, lights may be spaced irregularly or omitted, provided that adequate guidance remains available to the pilot.

Runway edge lights shall be fixed lights showing variable white except that :

(a) in the case of a displaced threshold, the lights between the beginning of the runway and the displaced threshold shall show red in the approach direction; and

(b) a section of the lights 600 m or one third of the runway length, whichever is the less, at the remote end of the runway from the end at which the take-off run is started, may show yellow.

The runway edge lights shall show at all angles in azimuth necessary to provide guidance to a pilot landing or taking off in either direction. When the runway edge lights are intended to provide circling guidance, they shall show at all angles in azimuth.

The intensity shall be at least 50 cd except that at an airport without extraneous lighting the intensity of the lights may be reduced to not less than 25 cd to avoid dazzling the pilot. For intensity of lights for runway edge lights on a precision approach runway higher intensities as prescribed by ICAO are needed (not included here).

12.4.8 Runway threshold and wing bar lights

Runway threshold lights are provided for a runway equipped with runway edge lights except on a non-instrument or non-precision approach runway where the threshold is displaced and wing bar lights are provided.

When a threshold is at the extremity of runway the threshold lights shall be placed in a row at right angles to the runway axis as near to the extremity of the runway as possible and in any case not more than 3 m outside the extremity.

When a threshold is displaced from the extremity of a runway, threshold lights shall be placed in a row at right angles to the runway axis at the displaced threshold.

Threshold lighting shall consist of :

(a) on a non-instrument or non-precision approach runway, at least six lights;

(b) on a precision approach runway category I, at least the number of lights that would be required if the lights were uniformly spaced at intervals of 3 m between the rows of runway edge lights; and

(c) on a precision approach runway category II or III lights uniformly spaced between the rows of runway edge lights at intervals of not more than 3 m.

The lights are equally spaced between the rows of runway edge lights or symmetrically disposed about the runway centre line in two groups, with the gap between the groups equal to the gauge of the touch down zone marking or lighting, where such is provided, or otherwise not more than half distance between the rows of runway edge lights.

Wing bar lights

Wing bar lights should be provided on a precision approach runway when additional conspicuity is considered desirable. Wing bar lights are provided on a non-instrument or non-precision approach runway where the threshold is displaced and runway threshold lights are required, but are not provided.

Wing bar lights are symmetrically disposed about the runway centre line at the threshold in two groups i.e., wing bars. Each wing bar shall be formed by at least five light extending at least 10 m outward from, and at right angles to, the line of the runway edge lights, with the inner-most light of each wing bar in the line of the runway edge lights.

Runway threshold and wing bar lights are fixed unidirectional lights showing green in the direction of approach to the runway. The intensity and beam spread of the lights shall be adequate for the conditions of visibility and ambient light in which the runway is used. Fig. 12.2 shows the arrangement of runway threshold and runway end lights.

12.4.9 Runway end lights

Runway end lights are provided for a runway equipped with runway edge lights. The runway end lights should be placed on a line at right angles to the axis as near to the end of the runway as possible and in any case, not more than 3 m outside the end.

Runway end lighting consists of at least six lights. The lights should be either :

(a) equally spaced between the rows of runway edge lights; or

(b) symmetrically disposed about runway centre line in two groups with the lights uniformly spaced in each group and with a gap between the groups of not more than half the distance between the rows of runway edge lights.

For a precision approach runway category III, the spacing between runway end lights, except between the two innermost lights if a gap is used, should not exceed 6 m.

The runway end lights are fixed unidirectional lights showing red in the direction of the runway. The intensity and beam spread of the lights should be adequate for the conditions of visibility and ambient light in which runway is used.

12.4.10 Runway centre line lights

The runway centre line lights are located along the centre line of the runway, except that the lights may be uniformly offset to the same side of the runway centre line by not more than 60 cm where it is not practicable to locate them along the centre line. The lights are located from the threshold to the end at a longitudinal spacing of approximately :

- 7.5 m or 15 m on a precision approach runway category III.
- 7.5 m, 15 m or 30 m on a precision approach runway category II or other runways.

Centre line guidance for take-off from the beginning of a runway to a displaced threshold should be provided by :

(a) an approach lighting system if its characteristics and intensity setting afford guidance required during take-off.

(b) runway centre line lights.

(c) barrettes of at least 3 m length and spaced at uniform intervals of 30 m as shown in Fig. 12.3, designed so that their photometric characteristics and intensity setting afford the guidance required during take-off without dazzling the pilot of an airport taking off.

At the landing operations centre line lights may be either switched off or reset the intensity.

The runway centre line lights shall be fixed lights showing variable white from the threshold to the point 900 m from the runway end. Alternate red and variable white from 900 m to 300 m from the runway end and red from 300 m to runway end, except that :

(a) where the runway centre line lights are spaced at 7.5 m intervals, alternate pairs of red and variable white lights shall be used on the sections from 900 m to 300 m from the runway end; and

(b) for runways less than 1800 m in length, the alternate red and variable white lights shall extend from the mid-point of the runway usable for landing to 300 m from the runway end.

12.4.11 Runway touchdown zone lights

Touchdown zone lights are used in the touchdown zone of a precision approach runway category II or III.

Touchdown zone lights extend from the tnreshold for a longitudinal distance of 900 m, except that on runways less than 1800 m in length, the system shall be shortened so that it does not extend beyond the mid-point of the runway. The pattern shall be formed by pairs of barrettes symmetrically located about the runway centre line. The lateral spacing between the innermost lights of a pair of barrettes should be equal to the lateral spacing selected for the touchdown zone marking. The longitudinal spacing between pairs of barrettes should be either 30 m or 60 m.

A barrette should be composed of at least three lights with a spacing between the lights of not more than 1.5 m. It should not be less than 3 m nor more than 4.5 m in length. Touchdown zone lights are fixed unidirectional lights showing variable white.

12.4.12 Stopway lights

Stopway lights are provided for a stopway intended to be used during night.

Stopway lights are placed along the full length of the stopway and shall be in two rows that are equidistant from the centreline and coincident with rows of the runway edge lights. Stopways lights shall also be provided across the end of a stopway on a line at right angles to the stopway axis as near to the end of the stopway as possible and in any case, not more than 3 m outside the end.

Stopway lights are fixed unidirectional lights showing red in the direction of the runway.

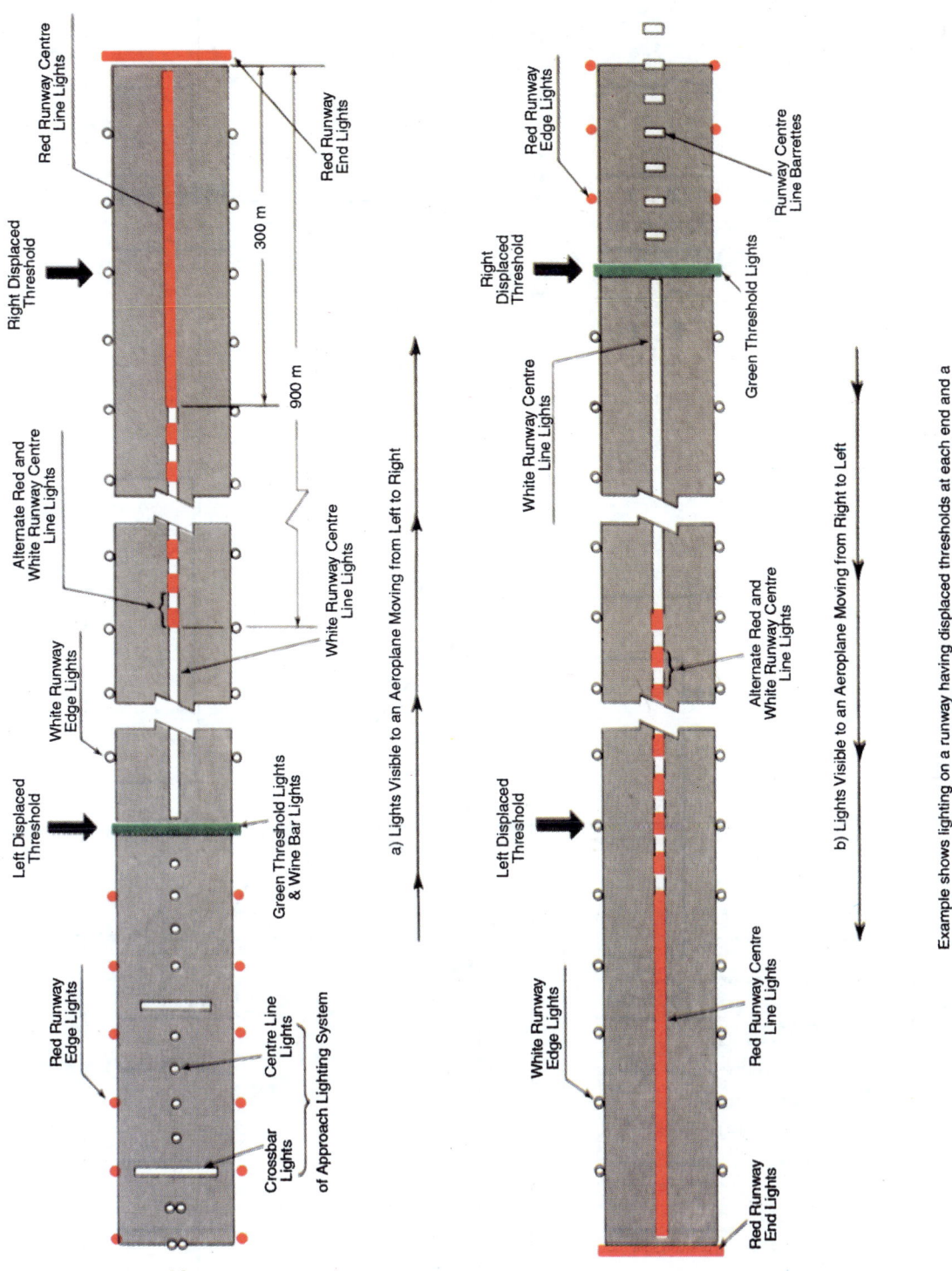

Fig. 12.3 Example of approach and runways lighting for runways with displaced thresholds (ICAO)

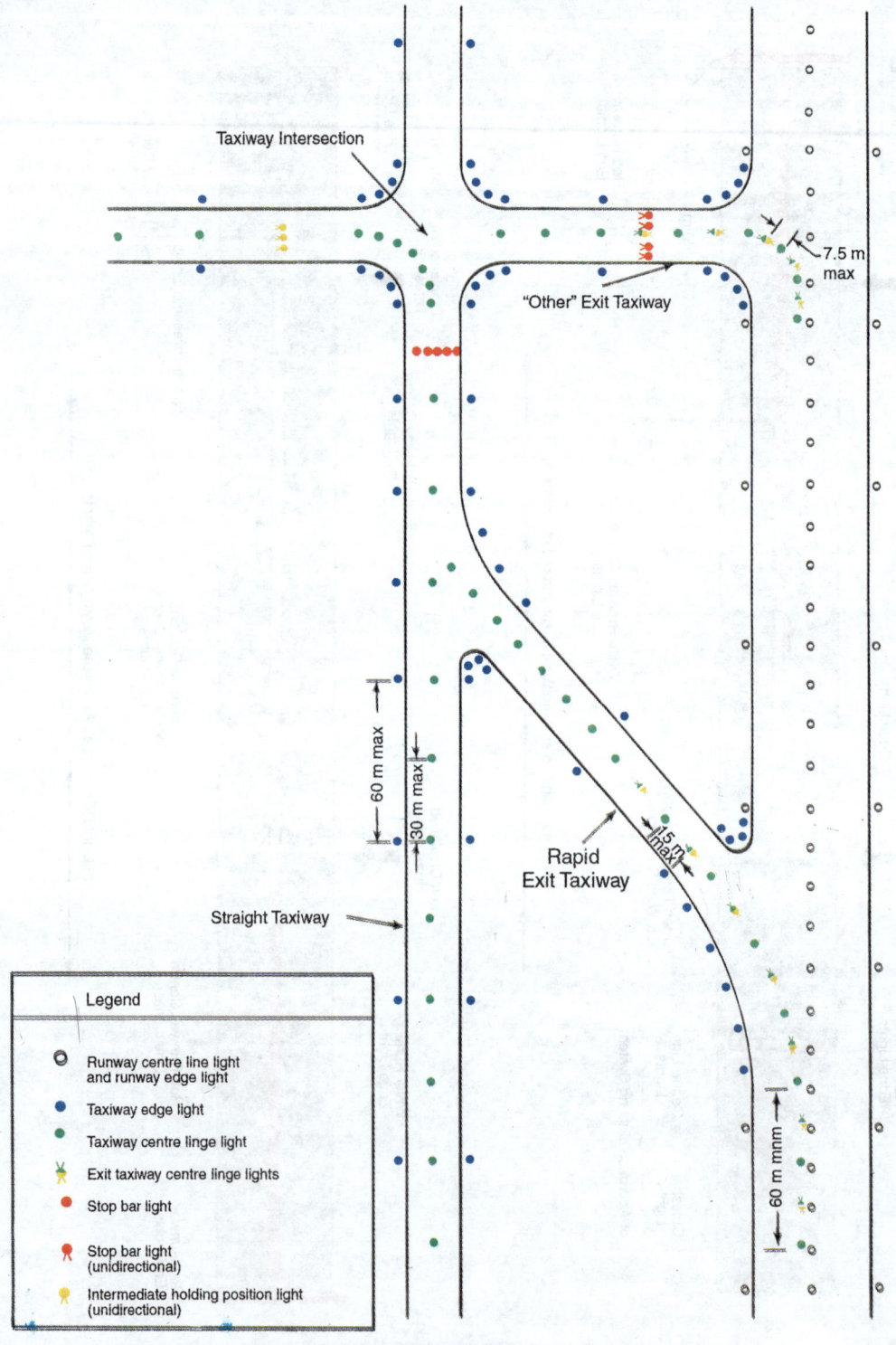

Taxiway Intersection

"Other" Exit Taxiway

7.5 m max

60 m max

30 m max

15 m max

Rapid Exit Taxiway

Straight Taxiway

60 m mnm

Legend

○ Runway centre line light and runway edge light

● Taxiway edge light

● Taxiway centre linge light

✖ Exit taxiway centre linge lights

● Stop bar light

🔴 Stop bar light (unidirectional)

🟡 Intermediate holding position light (unidirectional)

Fig.12.4. Taxiway lighting (ICAO)

12.4.13 Taxiway centre line lights

Taxiway centre line lights are provided on an exit taxiway, taxiway and apron intended for use in runway visual range conditions, so as to provide continuous guidance from the runway centre line to the point on the apron where aircraft commence manoeuvring for parking, except that these lights need not be provided where there is low volume of traffic and taxiway edge lights and centre line marking provide adequate guidance. Where there may be a need to delineate the edge of a taxiway e.g., on a rapid exit taxiway, narrow taxiway or in snow conditions this may be done with taxiway edge lights or markers.

Taxiway centre line lights on a taxiway other than an exit taxiway shall be fixed lights showing green with beam dimensions such that the light is visible only from aeroplanes on or in the vicinity of the taxiway.

Taxiway centre line lights on an exit taxiway shall be fixed lights. Alternate taxiway centre line lights shall show green and yellow from their beginning near the runway centre line to the perimeter of the ILS critical/sensitive area or the lower edge of the inner transitional surface, whichever is farthest from the runway; and thereafter all lights shall show green as in Fig. 12.4. The light nearest to the perimeter shall always show yellow. The lights shall be bi-directional where the taxiway is used in both directions.

Care is necessary to limit the light distribution of green lights on or near a runway so as to avoid possible confusion with threshold lights.

Taxiway centre line lights should normally be located on the taxiway centre line marking, except that they may be offset by not more than 30 cm where it is not practicable to locate them on the marking.

Taxiway centre line lights on a straight section of a taxiway should be spaced at longitudinal intervals of not more than 30 m, except that :

(a) larger intervals not exceeding 60 m may be used where, because of the prevailing meteorological conditions, adequate guidance is provided by such spacing;

(b) intervals less than 30 m should be provided on short straight sections; and

(c) on a taxiway intended for use in RVR conditions of less than a value of the order of 400 m, the longitudinal spacing should not exceed 15 m.

Taxiway centre line lights on a taxiway curve should continue from the straight portion of the taxiway at a constant distance from the outside edge of the taxiway curve. The lights should be spaced at intervals such that a clear indication of the curve is provided.

On a taxiway intended for use in RVR conditions of less than a value of the order of 400 m, the lights on a curve should not exceed a spacing of 15 m and on a curve of less than 400 m radius the lights should be spaced at intervals of not greater than 7.5 m. This spacing should extend for 60 m before and after the curve.

Spacing on curves that have been found suitable for a taxiway intended for use in RVR conditions of the order of 400 m or greater are :

Curve radius	Light spacing
upto 400 m	7.5 m
401 m to 899 m	15 m
900 m or greater	30 m

12.4.13.1 *Taxiway centreline lights on rapid exit taxiways*

Taxiway centre line lights on rapid exit taxiway should commence at a point at least 60 m before the beginning of the taxiway centre line curve and continue beyond the end of the curve to a point on the centre line of the taxiway where an aeroplane can be expected to reach normal taxiing speed. The lights on that portion parallel to the runway centreline should always be at least 60 m from any row of runway centre line lights, as shown in Fig. 12.5.

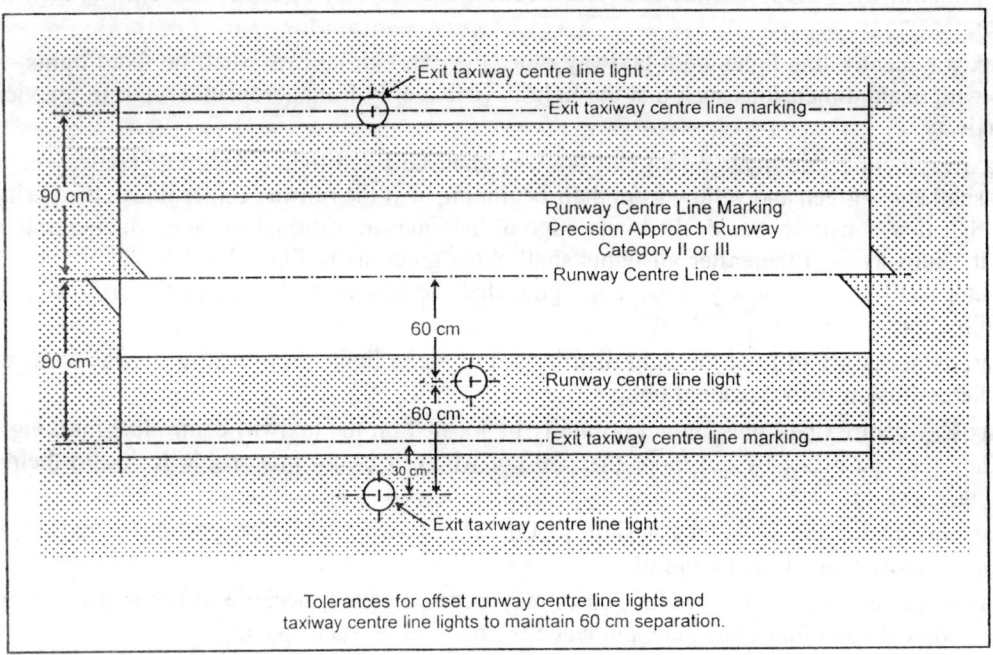

Fig. 12.5. Offset runway and taxiway centre line lights (ICAO).

These lights should be spaced at longitudinal intervals of not more than 15 m, except that, where runway centre line lights are not provided a greater interval not exceeding 30 m may be used.

12.4.13.2 *Taxiway centre line lights on other exit taxiways*

Taxiway centre line lights on exitways other than rapid exit taxiways should commence at the point where the taxiway centre line marking begins to curve from the runway centre line and follow the curved taxiway centreline marking at least to the point where the marking leaves the runway. The first light should be at least 60 cm from any row of runway centre line lights as shown in Fig. 12.5. The lights should be spaced at longitudinal intervals of not more than 7.5 m.

12.4.14 Taxiway edge lights

Taxiway edge lights provided on a holding bar, apron, etc. intended for use at night and on a taxiway not provided with taxiway centre line lights and intended for use at night, except that taxiway edge lights need not be provided where, considering the nature of the operations, adequate guidance can be achieved by surface illumination or other means.

Taxiway edge lights on a straight section of a taxiway should be spaced at uniform longitudinal intervals of not more than 60 m. The lights on a curve should be spaced at intervals less than 60 m so that a clear indication of the curve is provided.

The lights should be located as near as practicable to the edges of the taxiway, holding bay or apron, etc., or outside the edges at a distance of not more than 3 m.

Taxiway edge lights shall be fixed lights showing blue. The lights shall show up to at least 30° above the horizontal and at all angles in azimuth necessary to provide guidance to a pilot taxiing in either direction. At an intersection, exit or curve the lights shall be shielded as far as practicable so that they cannot be seen in angles of azimuth in which they may be confused with other lights.

12.4.15 Stop bars lights

A stop bar should be provided at each taxi-holding position associated with a runway intended for use in runway visual range conditions of values between the order of 400 m and 800 m.

One or more stop bars, as appropriate, should be provided at a taxiway intersection or taxi-holding position when it is desired to supplement markings with lights and to provide traffic control by visual means.

Where the normal stop bar lights might be obscured (from a pilot's view), for example, by snow or rain, or where a pilot may be required to stop the aircraft in a position so close to the lights that they are blocked from view by the structure of the aircraft, then a pair of elevated lights should be added to each end of the stop bar.

Stop bars shall be located across the taxiway at the point where it is desired that traffic stop. Where the additional lights are provided, these lights shall be located not less than 3 m from the taxiway edge.

Stop bars shall consist of lights spaced at intervals of 3 m across the taxiway, showing red in the intended direction(s) of approach to the intersection or taxi-holding position.

Stop bars installed at a taxi-holding position shall be unidirectional and shall show red in the direction of approach to the runway.

The lighting circuit shall be designed so that :

(a) stop bars located across entrance taxiways are selectively switchable; and

(b) stop bars located across taxiways intended to be used only as exit taxiways are switchable selectively or in groups.

A stop bar is switched on to indicate that traffic stop and switched off to indicate the traffic proceed.

12.4.16 Clearance bars

A clearance bar is provided at a taxiway intersection where it is desirable to define a specific aeroplane holding limit and there is no need for stop-and-go signals as provided by a stop bar.

Clearance bars shall be located at a point between 30 m to 60 m from the near edge of the intersecting taxiway.

Clearance bars shall consist of at least three fixed unidirectional lights showing yellow in the direction of approach to the intersection with a light distribution similar to taxiway centre line lights if provided. The lights shall be disposed symmetrically about, and at 90° to, the taxiway centre line, with individual lights spaced 1.5 m apart.

12.4.17 Taxi-holding position lights

Taxi-holding position lights should be provided at a taxi-holding position intended for use in runway visual range conditions less than a value of the order of 800 m and at other taxi-holding positions where enhanced conspicuity of the location of the holding position is necessary.

Where provided, taxi-holding position lights shall be located at each side of a taxi-holding position as close as possible to the taxiway edge.

Taxi-holding position lights shall consist of two alternatively illuminated yellow lights.

The light beam shall be unidirectional and aligned so as to be visible to the pilot of an aeroplane taxiing to the holding position.

The intensity of the light should be adequate for the conditions of visibility and ambient light in which the use of the holding position is intended but should not dazzle the pilot.

The lights shall be illuminated alternately between 30 and 60 cycles per minute. The light suspension and the illumination periods shall be equal and opposite in each light.

12.4.18 Apron flood lighting

Apron floodlighting should be provided on an apron and on a designated isolated aircraft parking position, intended to be used at night.

Apron floodlights should be located so as to provide adequate illumination on all apron service areas, with a minimum of glare to pilots of aircraft in flight and on the ground, aerodrome and apron controllers, and personnel on the apron. The arrangement and aiming of floodlights should be such that an aircraft stand receives light from two or more directions to minimize shadows.

The spectral distribution of apron floodlights shall be such that the colours used for aircraft marking connected with routine servicing, and for surface and obstacle marking, can be correctly identified. The average illuminance should be at least the following :

Aircraft stand

- *Horizontal illuminance* : 20 lux with a uniformity ratio (average to minimum) of not more than 4 to 1; and
- *Vertical illuminance* : 20 lux at a height of 2 m above the apron in relevant directions.

Other apron areas

- *Horizontal illuminance* : 50 per cent of the average illuminance on the aircraft stands with a uniformity ratio (average to minimum) of not more than 4 to 1.

12.4.19 Visual docking guidance system

A visual docking guidance system should be provided when it is intended to indicate by a visual aid, the precise positioning of an aircraft on an aircraft stand.

The factors to be considered in evaluating the need for a visual docking guidance system are in particular : the number of aircraft using the aircraft stand, weather conditions, space available on the apron and the precision required at the parking position. Other means such as marshallers may be considered.

The system should provide both azimuth and stopping guidance. The azimuth guidance unit and the stopping position indicator should be adequate for use in all weather, visibility and pavement conditions for which the system is intended both by day and night, but should not dazzle the pilot.

The azimuth guidance unit and the stopping position indicator should be of a design such that :

(a) a clear indication of failure is available to the pilot if either or both fail to give the required information; and

(b) they can be turned off when the aircraft stand is not to be used.

The azimuth guidance unit and stopping position indicator should be located in such a way that there is continuity of guidance between the aircraft stand markings, the aircraft stand manoeuvring guidance lights, if present, and the visual docking guidance system. The accuracy of the system should be adequate for the type of loading bridge and fixed aircraft servicing installations with which it is to be used.

The system should preferably be usable by all types of aircraft for which the aircraft stand is intended, without selective operation. If selective operation is required to prepare the system for use by a particular type of aircraft, then the system should provide an identification of the selected aircraft type to both the pilot and the system operator as a means of ensuring that the system has been set properly.

Azimuth guidance unit

The azumuth guidance unit should be located on the extension of the stand centre line ahead of the aircraft so that its signals are visible from the cockpit of an aircraft throughout the docking manoeuvre and aligned for use by the pilot occupying the left seat. The azimuth guidance unit should provide self-evident left/right guidance which enables the pilot to acquire and maintain the lead-in line without overcontrolling.

When azimuth guidance is indicated by colour change, green should be used to identify the centre line and red for deviations from the centre line.

The stopping position indicator should be located in conjunction with, or sufficiently close to, the azimuth guidance unit so that a pilot can observe both the azimuth and stop signals without turning the head. It should preferably be usable by the pilots occupying both the left and right seats, but should in any case be usable by the pilot occupying the left seat. The stopping position information provided by the indicator for a particular aircraft type should not be significantly affected by possible variations in pilot eye height and/or viewing angle.

The stopping position indicator should show, preferably without selective operation by ground personnel, the stopping position for the aircraft for which guidance is being provided, and should provide closing rate information to enable the pilot to gradually decelerate the aircraft to a full stop at the intended stopping position. Provision of closing rate information over a distance of approximately 10 m is considered satisfactory.

When stopping guidance is indicated by colour change, green should be used to show that the aircraft can proceed and red to show that the stop point has been reached.

12.4.20 Aircraft stand manoeuvring guidance lights

Aircraft stand manoeuvring guidance lights should be provided to facilitate the position of an aircraft on an aircraft stand intended for use in poor visibility conditions, unless adequate guidance is provided by other means.

Aircraft stand manoeuvring lights shall be collocated with the aircraft stand markings. These lights shall be fixed yellow lights, visible throughout the segments within which they are intended to provide guidance.

The lights used to delineate lead in, turning and lead out lines should be spaced at intervals of not more than 7.5 m on curves and 15 m on straight sections. The lights indicating a stop position are fixed, unidirectional, showing red.

The intensity of the lights should be adequate for the condition of visibility and ambient light in which aircraft stand is used. The lighting circuit should be so designed that the lights may be switched on to indicate that stand is to be used and switched off to indicate that it is not to be used.

12.4.21 Visual approach slope indicator systems

A visual approach slope indicator system is provided to serve the approach to a runway whether or not the runway is served by other visual approach aids or by non-visual aids, where one or more of the following conditions exist :

(a) the runway is used by turbojet or other aeroplanes with similar approach guidance requirements;

(b) the pilot of any type of aeroplane may have difficulty in judging the approach due to :

  (i) inadequate visual guidance such as is experienced during an approach over water or featureless terrain by day or in the absence of sufficient extraneous lights in the approach area by night, or

  (ii) misleading information such as is produced by deceptive surrounding terrain or runway slopes;

(c) the presence of objects in the approach area may involve serious hazard if an aeroplane descends below the normal approach path, particularly if there are no non-visual or other visual aids to give warning of such objects;

(d) physical conditions at either end of the runway present a serious hazard in the event of an aeroplane undershooting or over-running the runway; and

(e) terrain or prevalent meteorological conditions are such that the aeroplane may be subjected to unusual turbulence during approach.

For deciding which runway of an airport should receive first the visual approach slope indicator system, following factors should be considered :

(a) frequency of use;

(b) seriousness of the hazard;

(c) presence of other visual and non-visual aid;

(d) type of aeroplanes using the runway; and

(e) frequency and type of adverse weather conditions under which the runway will be used.

For the standards and specifications of the visual approach slope indicator systems, special publications of ICAO should be referred.

CHAPTER 13

Airport Visual Aids-III (Signs and Markers)

13.0 INTRODUCTION

Signs are installed to provide pilots with information, necessary for the safe and efficient operation of an airport. A properly designed and standardised sign system is an essential component of surface movement.

As the functional layout of each airport is different, the number of signs needed to provide the pilot with the necessary guidance information may differ. Following points, however, should be kept in view for planning a system of signs.

- A sign should be placed as near to the edge of the pavement as possible for easier visibility by the pilot.
- Signs should be of light weight and frangibly mounted.
- Those located near a runway or a taxiway should be sufficiently low to preserve clearance for propellers and for the engine pods of jet aircrafts.
- Signs should be rectangular with the longer side horizontal.
- The only signs on the movement area having red colour should be mandatory signs.
- The inscription on a sign, dimensions, width and height etc. should be adequate to be legible from the cockpit of an aircraft.
- Unless otherwise specified, signs should always be placed on the left side. If signs are installed on both sides at the same location, the sign faces should be identical.
- Information signs should not be collocated with mandatory, location, direction or destination signs.

For general specifications on size, location and installation of signs, readers should refer to specific latest editions of ICAO or FAA.

13.1 TAXIWAY DESIGNATION SYSTEM

Before developing a taxiway guidance sign system, a simple and rational method of designating taxiways should be adopted. The following general guideline may be followed in establishing taxiway designation system.

- It should be simple and logical.
- Letters of the alphabet should be used for designation of taxiways.
- Designation of taxiway should start at one end of the airport and continue to the opposite end e.g. north to south or east to west.
- Where there are more taxiways than letters of the alphabet, then double letters such as "AA" should be used. An exception is permitted for a major taxiway in having numbers, such as a taxiway parallel to a runway or a taxiway adjacent to a ramp area. In such cases the short taxiway could be designated as "A1", "A2", "A3" etc. Numbers alone and the letter "I", "X", and "O" are never used as they could be mistaken for a runway number.
- All separate, distinct taxiway segments should be designated.
- No separate, distinct taxiway should have the same designation as any other taxiway.
- Taxiway designation should not be changed when there is no significant change in direction of the taxiing route.
- Designating taxiways by reference to a direction of travel or to a physical object should be avoided. For example 'inner', 'outer', 'parallel' should not be used.

Fig. 13.1 shows an example of designating taxiways.

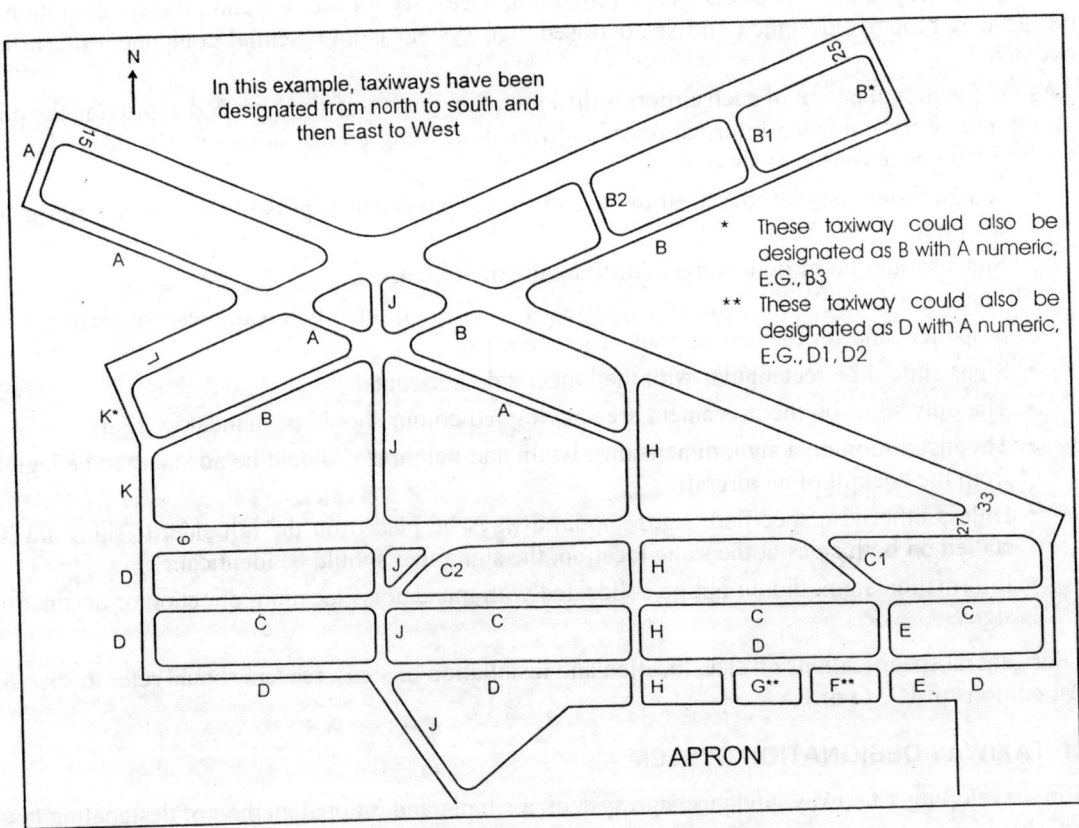

Fig. 13.1. Designation for taxiways (ICAO).

13.2 TAXIWAY SIGN SYSTEM

The following guidelines should be followed in taxiway guidance sign system at a particular airport :

- A holding position sign and taxiway location sign should be provided at the holding position of any taxiway that provides access to a runway.
- A holding position sign should be provided on any taxiway at the boundary of the ILS critical area or the runway approach area when it is necessary to protect the navigational signal, airspace, or safety area for a runway.
- A holding position sign should be provided on any taxiway at the boundary of the ILS critical area or the runway approach area when it is necessary to protect the navigational signal, airspace, or safety area for a runway.
- A holding position sign should be provided on any runway where that runway intersects another taxiway.
- A sign array consisting of taxiway direction sign should be installed prior to each taxiway/taxiway intersection if an aircraft would normally be expected to turn at or hold short of the intersection. The direction sign in the array should include the taxiway designation an arrow for each taxiway that an aircraft would be expected to turn onto or hold short of.
- A runway exit sign should be provided along each runway for each normally used runway exit.
- Designation signs should be provided at suitable point.
- Standard highway stop signs should be provided on vehicle roadways at the intersection of each roadway with a runway or taxiway.
- Addition signs should be installed on the airport where they are necessary to eliminate confusion e.g., it may be necessary to provide a taxiway location sign at the entrance to a taxiway from an apron area where there are several such entrances.

13.3 TYPE OF SIGNS

The signs of the following types are used on the airport :

1. Mandatory instruction signs
2. Information signs
3. Direction signs
4. Destination signs
5. Other signs

13.3.1 Mandatory signs

A mandatory sign is provided when it is intended to identify, by a sign, a location beyond which an aircraft or vehicle should not proceed, unless authorised by the aerodrome control tower.

Mandatory signs include taxiway/runway intersection signs, holding position signs, 'no entry' signs.

A taxiway/runway intersection sign should be located at least on one side of the taxi-holding positions marking facing the direction of approach to the runway. Wherever possible a taxiway/runway intersection sign should be located on each side of a taxiway. Where a taxiway/runway intersection sign is provided only on one side it shall be on the left-hand side as viewed by the pilot.

A 'no entry' sign should be located at the beginning of an area to which entrance is prohibited wherever physically possible a 'no entry' sign should be located on each side of a taxiway. Where a 'no entry' sign is provided only on one side it shall be on the left hand side of the pilot.

A mandatory sign consists of an inscription in white on a red background. When intended for use in night or during poor visibility a mandatory instruction sign should be illuminated either internally or externally.

The inscription on a taxiway/runway intersection sign consists of the runway designations of both extremities of the intersection runway. Where appropriate following inscriptions are used :

Inscription	Use
No entry	To indicate that entrance to an area is prohibited
The runway designation of a particular runway extremity	To indicate a taxi-holding position located at a runway extremity
The runway designations of both extremities of a runway	To indicate a taxi-holding position located at other runway intersections
Cat II	To indicate a category II taxi-holding position
Cat II/III	To indicate a joint category II/III taxi-holding position

Mandatory instruction signs include the following :

(a) Holding position signs for taxiway/runway intersections

The inscription on a holding position sign at taxiway/runway intersection is the runway number(s) such as "15-33" as shown in Fig. 13.2. The runway numbers are separated by a dash and their arrangement indicates the direction to the corresponding runway threshold "15-33" indicates that the threshold for runway is "15" is to the left and the threshold for runway "33" is to the right.

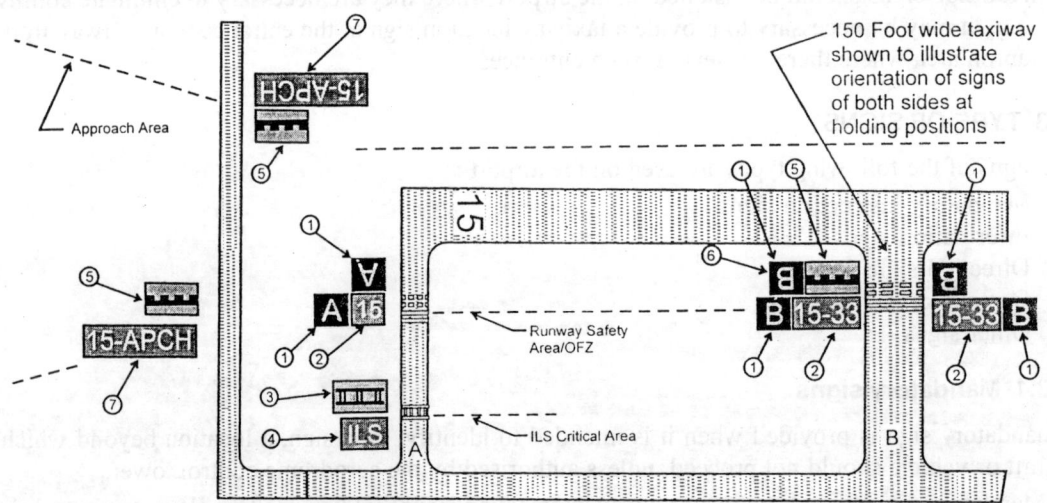

1. Taxiway Location Sign
2. Holding Position Sign
3. ILS Critical Area Boundary Sign
4. ILS Holding Position sign
5. Runway Safety Area/OFZ and Runway Approach Area Boundary Sign
6. Taxiway Location Sign – Optional, Depending on Operational Need
7. Holding Position Sign for Approach Areas

Fig. 13.2. Example of holding position signs (ICAO).

The sign at each take off end contains the inscription only for the take-off runway while other signs contain both runway designation numbers.

In some geometrical configurations of runways and taxiways, it is necessary to install hold position signs on both sides of the taxiway. These configurations include :

- taxiways which are 50 m or greater in width.
- taxiways where the painted hold position markings extend across an adjacent holding bay etc.
- taxiways where the painted holding position markings do not extend straight across the taxiway.
- taxiway where painted hold position markings are located a short distance from an intersection with another taxiway, as in this situation the pilot would have difficulty seeing the hold position sign on left.

(b) Holding position signs for runway/runway intersections

Signs used to identify runway/runway intersection are identical to signs used for taxiway/runway intersection. For runways 50 m or less in width, only one sign is needed. For runways more than 50 m in width or the runways of any width which are used for land and hold short operations or normally used for taxiing, signs on both sides of the runways are needed. Signs should be located at a distance from the intersecting runway to meet the clearance requirements of the intersecting runway.

(c) Holding position signs for ILS critical areas

The inscription on a sign for an ILS critical areas is shown in Fig. 13.3 where the distance between the runway hold line and the holdline for an ILS critical area is 50 m or less, one holdline may be installed, provided it will not affect capacity.

Fig. 13.3. ILS holding position sign (ICAO).

(d) Holding position signs for runway approach areas

The inscription on a sign for runway approach area is the associated runway designation followed by a dash and the abbreviation as shown in Fig. 13.4.

The sign is installed on taxiways located in approach areas where an aircraft on a taxiway would either cross through the runway safety area or penetrate the airspace required for the approach or departure runway. This sign should not be installed on runways.

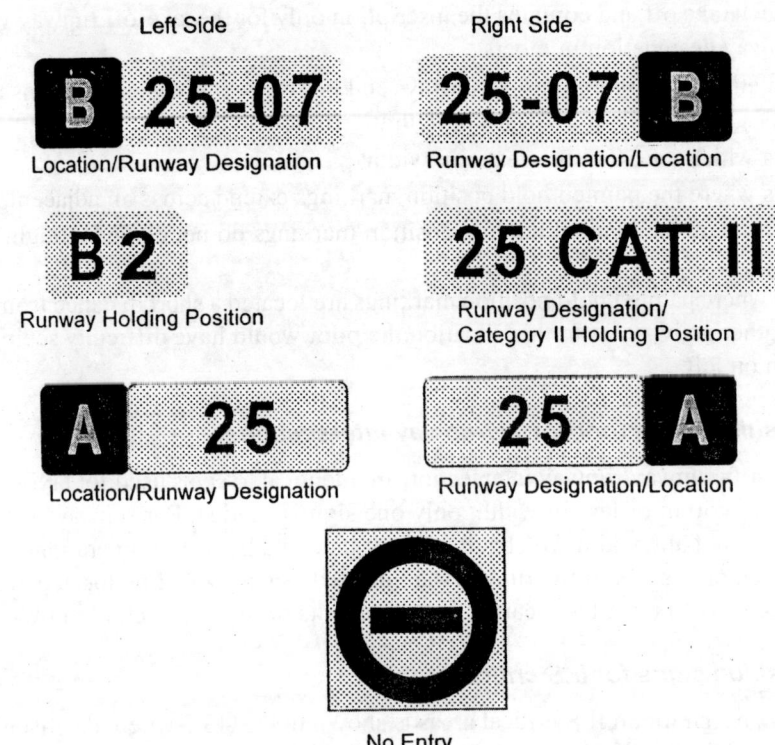

Fig. 13.4. Mandatory instruction signs (ICAO).

(e) No entry sign

No entry sign indicates that entry into a particular area is prohibited to aircraft. The sign inscription is as shown in Fig. 13.4.

13.3.2 Information signs

An information sign is provided when it is intended to indicate, by a sign, a specific location or destination on the movement area or to provide other information. For example when a VOR airport checkpoint is established, it shall be indicated by a VOR airport checkpoint marking and sign. A VOR airport checkpoint sign is located as near as possible to the checkpoint so that the inscriptions are visible from the cockpit.

A sign on a taxiway indicating location should be located on the left hand side of the taxiway. At an intersection of a taxiway with another taxiway, the sign should be located prior to the intersection.

A sign indicating a destination should be located on the same side of a taxiway (left or right) as the direction to the location to be indicated. If the destination is straight ahead, the sign should be located either on the left or on the right. At intersection location the destination sign should be placed before the intersection.

An information sign should consists of either yellow inscriptions on a black background or black inscriptions on a yellow background. These are illuminated internally or externally when intended for night use or coated with retroreflecting materials.

The inscriptions on a sign indicating a destination should include an arrow showing the direction to be followed with a number, word or abbreviation that indicates the destination.

Information signs include the following :

(a) Taxiway location sign

These signs identify the taxiway on which an aircraft is located. The signs have yellow inscription on a black background with a yellow border and do not contain arrows.

(b) Runway location signs

These signs are installed on runway when two runway ends are in proximity which could create confusion. A typical sign is shown in Fig. 13.5.

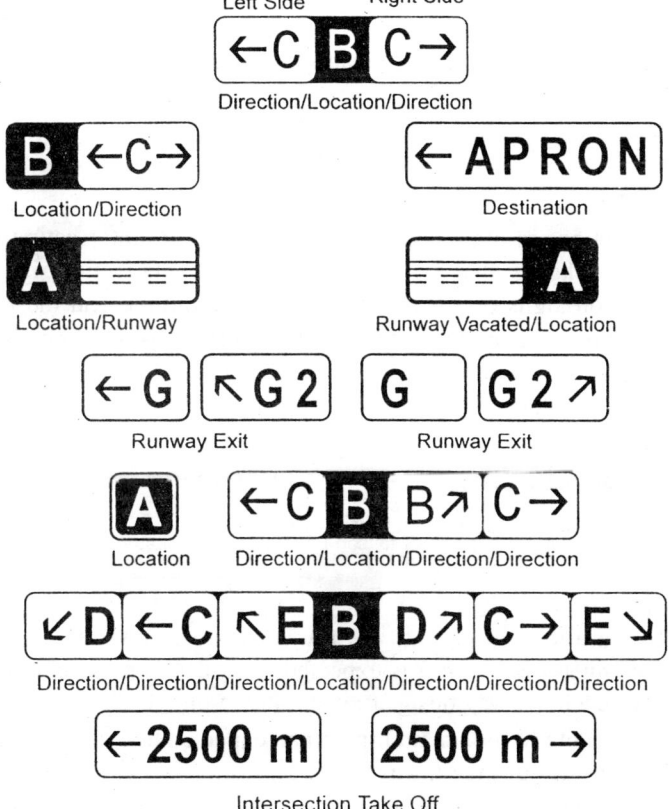

Fig. 13.5. Information sign.

These signs should be located to identify clearly the runways for pilot and contain the runway designation only for the one runway end. The signs have yellow inscription on a black background with a yellow border and do not carry arrows. Fig. 13.5 shows other information signs also.

(c) Runway safety area/OFZ and runway approach boundary signs

These signs identify the boundary of the runway safety area/OFZ or the runway approach area to pilot who are exiting these areas. They have a black inscription that depicts the holdline marking on a yellow background.

These signs are only used at controlled airports on taxiways where the controller commonly ask the pilot to report "clear of the runway". The pilot can use the sign as a guide in deciding when to report back to the controller.

(d) ILS critical area boundary sign

These signs identify the boundary of the ILS critical area to pilot who are existing this area. They have a black inscription that depicts the ILS holdline marking on a yellow background.

These signs are used at controlled airports on taxiways where the controller commonly asks pilots to report "clear of the ILS critical area". The pilot can use the sign as a guide in deciding upon the report back to the controller.

This would not normally be required on taxiways having colour coded centreline lights, but may be desirable in areas where the centreline lights are obscured by snow or ice. These signs are installed only on the back side of ILS holding position signs.

13.3.3 Direction signs

These signs indicate directions of other taxiways leading out of an intersection. The signs have black inscriptions on a yellow background and always contain arrows. Direction signs should not be collocated with holding position signs or installed between the holdline and the runway. Signs used to indicate the direction of taxiways on the opposite side of a runway should be located on the opposite side of the runway.

Examples of typical direction signs are :

(a) Taxiway direction sign

A typical taxiway direction sign is shown in Fig. 13.5.

(b) Runway exit sign

Signs for runway exists are located prior to the runway/taxiway intersection on the side and in the direction of the runway where the aircraft is expected to exit. A runway exit sign should never have more than one arrow for each taxiway designation as shown on the sign.

If a taxiway crosses a runway and an aircraft can be expected to exit on either side, then exit signs should be located on both sides of the runway.

For taxiways that are intended only to be used as exits from the runway in one direction, such as taxiways located near the end of the runway or intersecting the runway at an acute angle, the signs should be installed only for the runway direction in which they are intended to be used.

When two acute-angle taxiways (i.e., high speed exits), intended to be used in opposite directions, intersect the runway at a common point, the exit signs should be located prior to the common point intersection rather than in the area between the two exits.

13.3.4 Destination signs

Destination signs have block inscriptions on a yellow background and always contain an arrow. These signs indicate the general direction to a remote location. These signs are used when location signs and direction signs alone would not adequately guide a pilot to the desired destination. Destination signs are more beneficial at uncontrolled airports.

(a) Outbound destination signs

Outbound destination signs are used to identify directions to the take off runways. These routes usually begin at the entrance to a taxiway from an apron area. The inscription is the runway number plus an arrow indicating the direction as shown in Fig. 13.5.

More than one runway number, separated by a dot, may be shown where the taxiing route is common to both runways.

(b) Inbound destination signs

Major destination areas are usually shown on inbound destination signs. For example, at many airports, sign indicating the way to the apron may be adequate, whereas, at other airports, it may be necessary to make a distinction between passenger aprons, cargo aprons, military aprons or between aprons in different locations on the airport such as north apron, east apron etc.

At appropriate points closer to the major destination areas, destination signs should be provided to indicate specific areas which are designated for parking services, passenger handling, military aircrafts etc.

The inscription on destination signs should contain a minimum of three letters which should be selected so that no confusion could exist with other taxiway guidance signs. Common abbreviations used for inbound destinations are :

Apron : General parking, servicing, and loading areas

Fuel : Areas where aircraft are fueled or serviced

Term : Gate positions at which aircraft are loaded or unloaded

Civil : Areas set aside for civil aircraft

Mil : Areas set aside for military aircraft

Pax : Areas set aside for passenger handling

Cargo : Areas set aside for cargo handling

INTL : Areas set aside for handling international flights

FBO : Fixed-base operator

13.3.5 Other signs

Other type of signs that may be used on an airport are :

(a) Airport identification signs

This sign is provided at an airport where there is insufficient alternative means of visual identification.

The airport identification sign should be placed on the airport so as to be legible, at all angles above the horizontal. It consists of the name of the airport.

The colour selected for the sign should give adequate conspicuity when viewed against its background. The characters should have a height of not less than 3 m.

(b) Aircraft stand identification signs

An aircraft stand identification marking should be supplemented with an aircraft stand identification sign where feasible.

This sign should be located so as to be clearly visible from the cockpit of an aircraft prior to entering the aircraft stand.

An aircraft stand identification sign should consist of either yellow inscriptions on a black background or black inscriptions on a yellow black ground. A sign intended for use at night or during conditions of poor visibility should be illuminated either internally or extermally.

(c) Roadway signs

Vehicle roadways that intersect runways or taxiways should have a standard retroreflective highway stop sign or yellow sign on them prior to the intersection.

These signs should be located at the edge of the applicable runway safety area or taxiway safety area. They should be on frangible mounts and restricted to a height that does not interfere with airport using the runway or taxiway.

(d) Runway distance remaining signs

Runway distance remaining signs are used to provide distance remaining information to pilot during takeoff and landing operations. These signs are located along the side(s) of the runway and the inscription consists of a white numeral on a black background as shown in Fig. 13.6 to indicate the runway distance remaining in increments of 300 m (1000′).

The signs may be configured by either of the three different methods, described below :

(i) Preferred method

This method consists of double faced signs located only on one side of the runway. In this method, sign should be placed on the left side of the runway as viewed from the most often used direction. Signs may also be placed on the right side of the runway where necessary due to runway/taxiway separation distances or because of conflicts between intersecting runways or taxiways.

(ii) Alternate method number 1

This method uses single-faced signs installed on both sides of the runway. The advantage of this method is that the runway distance remaining can be more accurately reflected in cases where the runway length is not an exact multiple of 300 m (1000′).

(iii) Alternate method number 2

This method uses double faced signs installed on both sides of the runway. The advantage of this method is that runway distance is displayed on both sides of the runway which is particularly advantageous when a sign on one side has to be omitted because of a clearance conflict.

The method choosen should be based on cost considerations and adaptability to the specific airport configuration.

Displaced threshold areas which are used for take off and/or roll out are treated as part of the runway for purpose of locating the signs.

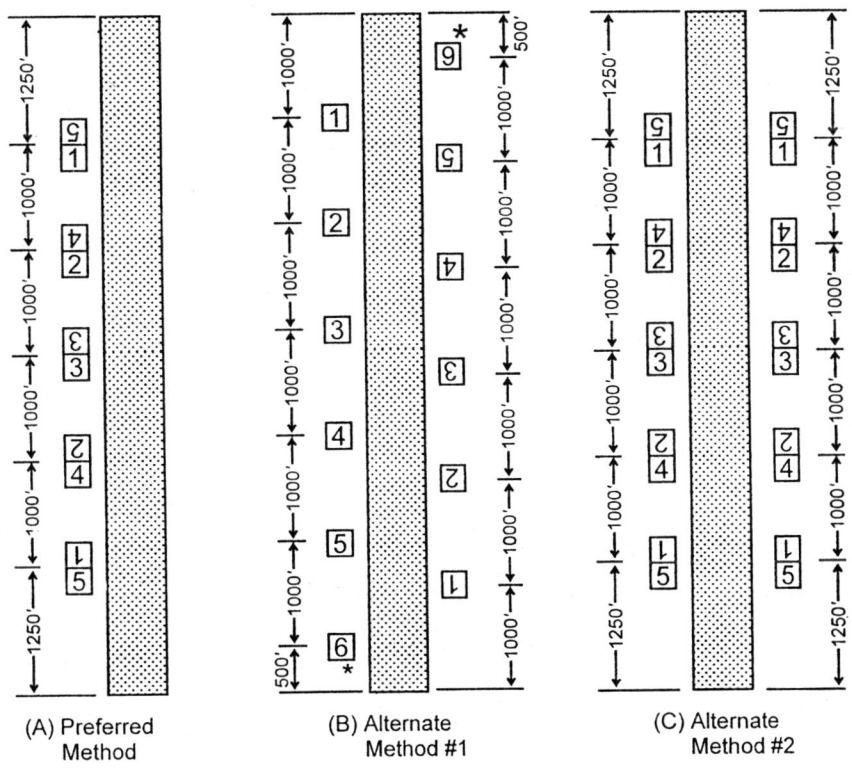

| (A) Preferred Method | (B) Alternate Method #1 | (C) Alternate Method #2 |

Notes:
1. Examples Based on a 6500 Foot Runway
2. Signs Less than 1000 Ft. from Takeoff end, as Indicated by Asterisks in Alternate Method #1, May be Omitted

Fig. 13.6. Runway distance remaining sign configurations (ICAO).

The sign system should be designed so that they are illuminated at all times, the runway edge lights are illuminated.

The sign should be of the same size on any particular runway. The choice of size involves several factors such as effectiveness, aircraft clearance, jet blast and snow removal operations. Normally a larger sign, located closer to edge is more effective.

13.4 MARKERS

Markers used on the airports should be light in weight and frangibly mounted. These located near a runway or taxiway should be sufficiently low to preserve clearance for propellers and for the engine pods of jet aircraft.

The following markers are commonly used :

(a) Unpaved runway edge markers

(b) Stopway edge markers

(c) Edge markers for snow covered runways

(d) Taxiway edge markers

(e) Taxiway centre line markers

(f) Unpaved taxiway edge markers

(g) Taxiway ending markers

(h) Boundary markers

13.4.1 Unpaved runway edge markers

Markers should be provided when the extent of an unpaved runway is not clearly indicated by the appearance of its surface compared with that of the surrounding ground.

When runway lights are provided, the markers should be incorporated in the light fixtures where there are no lights, markers of flat rectangular or conical shape should be placed so as to delimit the runway clearly.

The flat rectangular markers should have a minimum size of 1 m by 3 m and should be placed with their long dimension parallel to the runway centre line. The conical markers should have a height not exceeding 50 cm.

13.4.2 Stopway edge markers

Stopway edge markers should be provided when the extent of a stopway is not clearly indicated by its appearance, compared with that of the surrounding ground.

The stopway edge markers are sufficiently different from any runway edge marker, used to ensure that the two types of markers can not be confused.

These markers consist of small vertical boards, comouflaged on the reverse side, as viewed from the runway.

13.4.3 Edge markers for snow covered runways

The edge markers for snow covered runways should be used to indicate the usable limits of a snow-covered runway when the limits are not otherwise indicated. Runway lights are used to indicate the limits.

Edge markers for snow-covered runways should be placed the sides of the runway at intervals of not more than 100 m and should be located symmetrically along the runway centre line at such a distance from the centre line that there is adequate clearance for wing tips. Sufficient markers should be placed across the threshold and end of the runway.

Edge markers for snow covered runways should consist of conspicuous objects such as evergreen trees about 1.5 m high or light-weight markers.

13.4.4 Taxiway edge markers

Taxiway edge markers should be provided on a taxiway where the code number is 1 or 2 and taxiway centre line or edge lights or taxiway centre line markers are not provided.

They are installed at least at the same locations as the taxiway edge lights would have been used. A taxiway edge marker is retroreflection blue in colour. The marked surface as viewed by the pilot should be a rectangle and should have a minimum viewing area of 150 cm^2. They should be light in weight and frangible.

13.4.5 Taxiway centre line markers

Taxiway centre line markers should be provided on a taxiway where code number is 1 or 2 and taxiway centre line or edge lights or taxiway edge markers are not provided. Also they should be provided on a taxiway where code number is 3 or 4 and taxiway centre line lights are not provided if there is a need to improve the guidance provided by the taxiway centre line markings.

Taxiway centre line markers should be installed at least at the same location as would taxiway centre line lights had they been used. Taxiway centre line markers should normally be located on the taxiway centre line marking except that they may be offset by not more than 30 cm where it is not practicable to locate them on the marking.

A taxiway centre line marker shall be retroreflective green. The marked surface as viewed by the pilot should be a rectangle and have a minimum viewing area of 20 cm². They should be so designed and fitted as to withstand being run over by the wheels of an aircraft without damage either to the aircraft or to the markers themselves.

13.4.6 Unpaved taxiway edge markers

Where the extent of an unpaved taxiway is not clearly indicated by its appearance, compared with that of surrounding ground markers should be provided.

Where taxiway lights are provided, markers should be incorporated in the light fixtures. Where there are no lights, markers of conical shape should be placed so as to delimit the taxiway clearly.

13.4.7 Taxiway ending markers

The sign system does not provide a sign to indicate that a taxiway does not continue beyond an intersection. A frangible, retroreflective barrier should be installed on the far side of the intersection if the normal visual cues such as marking and lighting are inadequate.

13.4.8 Boundary markers

Boundary markers are provided at an airport where landing area has no runway. They are spaced along the boundary of the landing area at intervals not more than 90 m, if the conical type is used and not more than 200 m if the type shown in Fig. 13.7 is used.

The boundary markers should be of a form similar to that shown in Fig. 13.7 or in the form of a cone not less than 50 cm high and not less than 75 cm in diameter at the base.

The markers should be coloured at contrast with the background against which they will be seen. A single colour, orange or red or two contrasting colours, orange and white or alternatively red and white, should be used, except where such colours merge with the background.

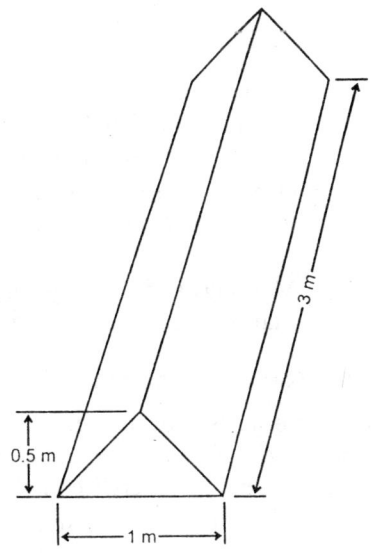

Fig. 13.7. Boundary markers (ICAO).

13.5 OTHER VISUAL AIDS

13.5.1 Visual aids for denoting obstacles

The marking and/or lighting of obstacles is done to reduce hazards to aircrafts by indicating the presence of the obstacles.

Some overall guidance for the objects to be marked and/or lighted are induced in the following paragraphs :

- A fixed obstacle that extends above a take-off climb surface within 3000 m of the inner edge should be marked and if the runway is used at night should be lighted.
- A fixed object, other than an obstacle, adjacent to take-off climb surface should be marked and lighted for night use.
- A fixed object that extends above an approach or transitional surface within 3000 m of the inner edge of the approach surface should be marked and lighted for night use.
- A fixed obstacle above a horizontal surface should be marked and lighted for night use.
- A fixed object that extends, above an obstacle protection surface shall be marked and lighted for night use.
- All elevated objects within following specified distances from the centre line of a taxiway, an apron taxiway or aircraft stand taxilane should be marked and if the taxiway, apron taxiway or aircraft stand taxilane is used at night, lighted.

Code letter	Taxiway (other than aircraft stand taxilane) centre line to object (m)	Aircraft stand taxilane centre line to object (m)
A	16.25	12
B	21.25	16.5
C	26.0	24.5
D	40.5	36
E	47.5	42.5

- In areas beyond the limits of the obstacle limitation surfaces, at least those objects which extend to the height of 150 m or more, above the ground elevation should be regarded as obstacles and should be marked and lighted for night use.
- Overhead wires, cablt etc. should be marked and their supporting towers also marked unless they are lighted by high intensity obstacle lights by the day, provided on the supporting towers.

Markings and lightings may be omitted when obstacle is shielded by another fixed obstacle. Markings may be avoided when obstacle is lighted by high-intensity obstacle light by day.

13.5.2 Marking of objects

All fixed objects to be marked are generally coloured, but if this is not practicable, markers or flags are displayed on or above them, except those that are sufficiently conspicuous by their shape, size or colour need not be otherwise marked.

Colouring of the objects should show chequered pattern. The pattern may consist of rectangles of not less than 1.5 m and not more than 3 m on a side, the corners being of darker colour (Fig. 13.8). The colour of the pattern should contrast each with the other and with the background against which they

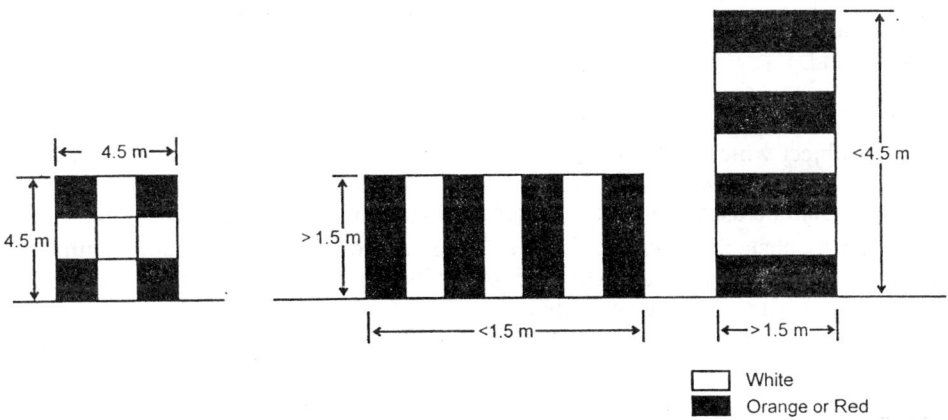

Fig. 13.8. Basic marking patterns (ICAO).

will be seen. Orange and white or alternatively red and white should be used, except such colours merge with the background.

An object should be coloured to show alternating contrasting bands. Bands should be perpendicular to the longest dimension. An object should be coloured in a single conspicuous colour if its projection on any vertical plane has both dimensions less than 1.5 m. Orange or red may be used, except where these colours merge with the background.

When mobile objects are marked by colour, a single conspicuous colour, preferably red or yellowish green for emergency vehicles and yellow for service vehicles should be used.

(a) Use of markers

Markers displayed on or adjacent to objects shall be located in conspicuous positions so as to retain the general definition of the object and shall be recognizable in clear weather from a distance of at least 1000 m for an object to be viewed from the ground in all directions in which an aircraft is likely to approach the object.

The shape of markers should be distinctive to ensure that they are not mistaken for markers employed to convey other information. A marker displayed on an overhead wire, cable etc. should be spherical and have a diameter of not less than 60 cm.

A marker should be of one colour. When installed, white and red or white and orange markers should be displayed alternately.

(b) Use of flags

Flags used to mark objects should be displayed around, on top of, or around the highest edge of the object. When flags are used to mark extensive objects or groups of closely spaced objects, they shall be displayed at least every 15 m.

Flags used to mark fixed object shall not be less than 0.6 m square and flags used to mark mobile objects, not less than 0.9 m square. Flags should be orange in colour or a combination of two triangular sections, one orange and the other white, or one red and the other white, except that where such colours merge with the background, other conspicuous colours should be used.

Flags used to mark mobile objects shall consist of a chequered pattern, each square having sides of not less than 0.3 m. Orange and white or alternatively red and white shall be used if they donot merge with the background.

(c) Use of lights

The presence of object which must be lighted should be indicated by low, medium or high obstacle lights or a combination of such lights. High-intensity obstacle lights are used for day use as well as night use. They should be used to indicate presence of an object if its height is 150 m or more.

Medium-intensity obstacle lights should be used, either alone or in combination with low-intensity obstacle lights, where the object is an extensive one or its height is greater than 45 m. A group of trees or buildings are regarded as an extensive object.

Where the use of low-intensity obstacle lights is inadequate or an early special warning is required, then medium or high intensity obstacle lights should be used.

The lights can be located on the top of the object, top of a tower etc. depending upon the local conditions. The number and arrangement of lights should be such that the object is indicated from every angle in azimuth.

Low intensity obstacle lights on fixed object should be red lights having an intensity sufficient to ensure conspicuity, but not less than 10 cd of red light. Low intensity obstacle lights on mobile objects shall be flashing lights either red or preferably yellow. The flash sequence should be between 60 and 90 per minute. The effective intensity of the flash shall be not less than 40 cd of red or yellow light.

Medium-intensity obstacle lights should be flashing red lights, except that when used in conjunction with high intensity obstacle lights they should be flashing white lights. The flash frequency should be between 20 and 60 per minute. The effective intensity of the flash shall be not less than 1600 cd of red light.

High-intensity obstacle lights should be flashing white light. The effective intensity of high intensity obstacle light located on an object other than a tower supporting overhead wires or cables should be variable and dependent on the background luminance as follows :

Background luminance	Effective intensity
Above 500 cd/m^2	200,000 cd minimum
50 to 500 cd/m^2	20,000 cd $\pm$ 25% cd
Less than 50 cd/m^2	4000 $\pm$ 25% cd

High intensity obstacle lights located on an object other than a tower supporting overhead wires and cables etc. should flash simultaneously at a rate between 40 and 60 per minute.

The effective intensity of a high-intensity obstacle light located on a tower supporting over head wires, cables etc. should be variable and dependent on the background luminance as follows :

Background luminance	Effective intensity
Above 500 cd/m^2	100,000 cd minimum
50 to 500 cd/m^2	20,000 cd $\pm$ 25% cd
Less than 50 cd/m^2	4,000 $\pm$ 25% cd

High intensity obstacle lights located on a tower supporting overhead wires, cablet etc. should flash sequentially, first the middle light, second the top light and last, the bottom light. The interval between flashes of the lights should approximate the following ratios.

Flash interval between	Ratio of cycle time
Middle and top light	1/13
Top and bottom light	2/13
Bottom and middle light	10/13

The cycle frequency should be 60 per minute.

13.6 VISUAL AIDS FOR DENOTING RESTRICTED USE AREAS

Some markings should be placed denoting restricted use of areas such as :

    (i) Closed runway and taxiway or part thereof
    (ii) Non-load bearing surfaces
    (iii) Pre-threshold area
    (iv) Un-serviceable areas

13.6.1 Closed runways and taxiways

A closed marking should be displaced on a temporarily closed runway or taxiway or portion thereof, except that such marking may be omitted when the closing is of a short duration and adequate warning by air traffic service is provided.

On a runway a "closed" marking should be placed at each end of the runway or portion thereof, declared closed, and additional markings should be so placed that the maximum interval between markings does not exceed 300 m on a taxiway 'closed' marking shall be placed at least at each end of the taxiway or portion thereof closed.

The closed marking shall be of the form of a cross with minimum dimensions as shown in Fig. 13.9 and should be of a single contrasting colour yellow or white.

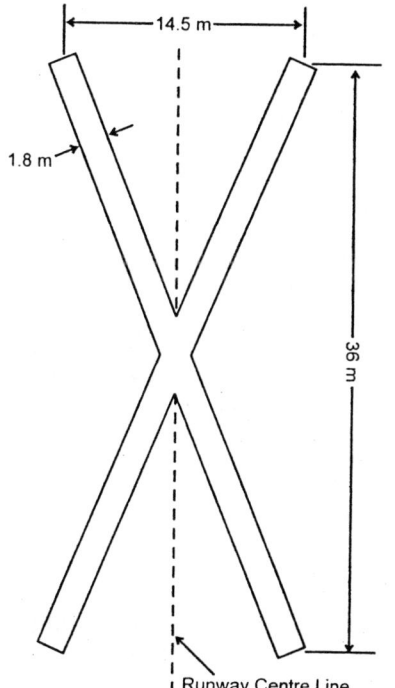

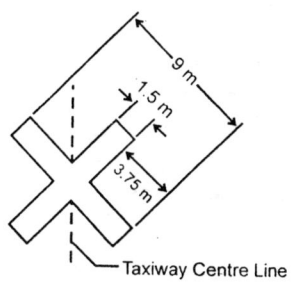

Illustration a) Closed Runway Marking   Illustration b) Closed Taxiway Marking

Fig. 13.9. Marking for a closed runway, taxiway (ICAO).

When a runway or taxiway or portion thereof is permanently closed, all normal runway and taxiway markings shall be obliterated. Lighting on a closed runway or taxiway should not be operated, except as required by maintenance.

13.6.2 Non-load bearing surfaces

Shoulders of taxiways, holding bays and aprons and other non-load bearing surfaces which cannot readily be distinguished from load-bearing surfaces and which if used by aircraft, might result in damage to aircraft should have the boundary between such areas and the load bearing surface marked by a taxi side stripe marking as described in chapter on markings. A taxi side stripe marking should be placed along the edge of the load bearing pavement, with the outer edge of the marking approximately on the edge of the load-bearing pavement.

The taxi side stripe marking consists of a pair of solid lines, each 15 cm wide and spaced 15 cm apart and the same colour as the taxiway centre line marking.

13.6.3 Pre-threshold area

When the surface before a threshold is paved and exceeds 60 m in length and is not suitable for normal use by aircraft, the entire length before the threshold should be marked with a chevron marking.

A chevron marking should be of conspicuous colour and contrast with the colour used for the runway markings. It should preferably be yellow in colour. It should have an overall width of at least 0.9 m.

A chevron marking should point in the direction of the runway and placed as shown in Fig. 13.10.

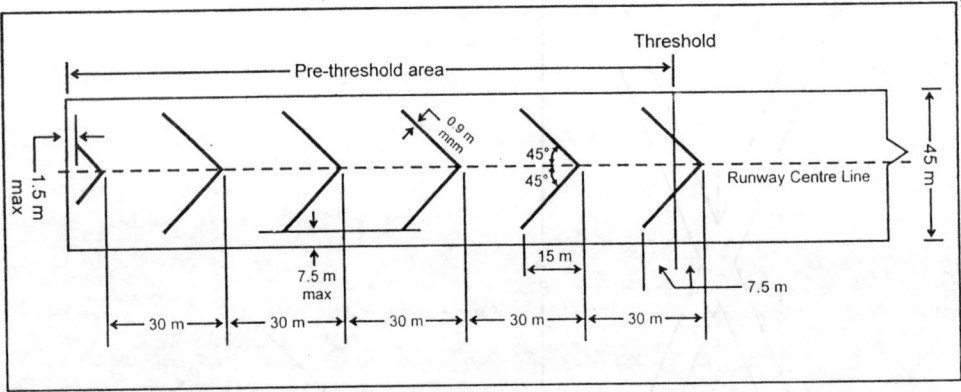

Fig. 13.10. Pre-threshold marking (ICAO).

13.6.4 Unserviceable areas

Unserviceability markers should be displayed wherever any portion of a taxiway, apron or holding bay is unfit for the movement of aircrafts, but it is still possible for aircraft to bypass area safety on a movement area used at night. Unserviceability lights should be used.

These markers and lights warn pilots of a hole in a taxiway or apron pavement or outlining a portion of pavement, such as on an apron, that is under repair. During major repairs or unserviceability for long time runways and taxiways are normally closed.

Unserviceability markers should be placed at intervals sufficiently close as to delineate the unserviceability area. They consist of conspicuous upstanding devices such as flags, cones, lights or marker boards.

An unserviceability light shall consist of a red fixed light or a red or yellow flashing light, of sufficient intensity to ensure conspicuity considering the intensity of the adjacent lights and general level of illumination against which it would normally be viewed. However, in no case the intensity of less than 10 cd of red light should be used. The red or yellow flashing light should have an effective intensity of not less than 5 cd.

An unserviceability cone should be at least 0.5 m in height and red, orange or yellow or any one of these colours in combination with white.

An unserviceability flag should be at least 0.5 m square and red, orange or yellow or any one of these colours in combination with white.

An unserviceability marker board should be at least 0.5 m in height and 0.9 m in length, with alternate red and white or orange and white vertical stripes.

CHAPTER 14

Airport Traffic Control

14.0 INTRODUCTION

The first rules for the prevention of collision between aircrafts were prescribed by the International Commission for Air Navigation (ICAN) in 1919, which came into force on July 11, 1922. Radio direction-finder devices were in operation in USA in 1920.

Between 1928 and 1932 intermediate frequency directive radio range stations of the crossed-loop type were installed along the key airways in USA. These ranges were later redesigned to use vertical tower radiators instead of loop antennas, joining the broadcast stations to form the simultaneous radio range and communication stations which transmit radio range signals and communication to aircraft.

Extensive modernization and expansion of the radio facilities took place during 1937-39. By the end of 1944, Federal Airways system of USA included over 400 radio range and airway communication stations, more than 200 very high frequency (VHF) fan markers and VHF radio ranges and instrument landing systems was in process of commissioning.

As the amount of flying increased, to the point of traffic congestion, it was found necessary to regulate the movement of aircraft both at the terminal and in the vicinity of the terminal, to prevent possibility of collision and to permit the movement of aircrafts on schedule. As the first attempt for airport traffic control flagmen were stationed on the airport to signal take-off and landing clearance to pilot. In 1931 radio equipped airport traffic control tower by which necessary instructions could be transmitted by radio directly to the pilot was established at Cleveland airport. The tower kept the times of arrival and departure of planes along the airways. When the plane was estimated to be within 5 to 8 kms of the airport, the control tower would transmit to the pilot height, visibility, the direction and velocity of wind, the number of planes expected to arrive at the airport within the next few minutes and the number in the air around the airport. If the plane was equipped with a transmitter, the pilot would acknowledge receipt of the information and might request additional information. The course of the aircraft was then followed carefully from the time it entered the traffic circling the airport until it landed. During this time the control tower kept the pilot informed by the position of all planes in the air in his vicinity and the movement of all planes on the ground which might have a bearing on his landing.

It was in 1938 federal regulation of airport traffic control came into effect in USA. In 1941 Civil

Aeronautics Administration (CAA), entered the field of airport traffic control to operate control towers with increasing army, navy and civil air traffic activity.

14.1 AIRWAY TRAFFIC CONTROL

With the establishment of radio aids to air navigation, the techniques of flying by instruments were also developed. In 1933 standards for proficiency in instrument flying were prescribed for pilots of airlines engaging in instrument flight operation. In 1935, the first airway traffic control centre came into being, in USA at New York. Othere were established at Chicago (April 1936) and Cleveland (June 1936). Airlines then established "Interline Safety Agreements" providing for the regulation of airline traffic by specifying procedures for the exchange of positon reports, altitudes to be flown and methods of making approaches. The control areas of above traffic control centres was about 80 kms from the airport at which the centre was located. Their sole communication system was party-line telephone circuit connecting the centre with local airline radio stations, the control tower. Flight data were posted with chalk on a large black board and the positions of aircraft were plotted with markers on a map table. In 1939 new type flight progress board was designed, consisting of a series of removable metal slats in which removable paper strips could be inserted. These flight progress boards eliminated the old blackboard method of posting flight data and the plotting of aircraft positions on a map table.

During the world war in 1942, there was a tremendous increase in operation of army and navy aircraft. This resulted in rapid expansion of airway traffic control system in USA and other countries.

In 1936, regulations were issued governing instrument flying on the civil airways. These regulations required that pilots must be properly qualified and aircraft to be properly equipped for instrument flight. It was also required that the pilot submit a flight plan and fly at a prescribed height above sea level, depending upon the direction of his flight. "Air Traffic Rules" were issued in 1937, as a result of experience with the original air traffic rules and experience in the operation of airway traffic control centres. Under these rules all arriving aircrafts were handled strictly in accordance with their sequence of arrival.

14.2 PURPOSE OF AIR TRAFFIC CONTROL

The basic purpose of air traffic control is protection of life and property by avoidance of collisions, through protective rules and regulations for air traffic. Aircrafts must be properly constructed, provided with necessary equipment and maintained in good condition. The pilot must be well trained and licensed or certified to undertake flight.

Laws, rules and regulations are not sufficient to ensure safety. Air traffic control covers two distinct aspects : airway traffic control and airport traffic control. Airway traffic control extends out along all the airways crossing the country and is provided by a number of air route traffic control centres. Its planning is done in advance. Airport traffic control is localised to a small zone around an airport. It functions by caring for immediate situations, through control towers located in the terminal area. Both have important purposes and transition from one to another must be effected properly, smoothly and efficiently with proper coordination.

Various purposes of air traffic control can be summarised as :

(a) **Safety :** Fundamental purpose of air traffic control is safety through :
   - providing guidance to the pilot whose vision may be obstructed, during landing and take off.
   - providing separation and right of way guidance to arriving and departing aircrafts when weather conditions prevent seeing.

ICAO has classified visibility into three categories :

Category I : weather conditions which provide a forward visibility of at least 800 m.

Category II : forward visibility of 400 m.

Category III : zero visibility conditions.

(b) **Efficiency :** Proper air traffic control helps in :
- expeditious movement of traffic.
- time tables and schedules of flights are maintained within reasonable limits, avoiding delays, disappointments and mis connections.
- avoiding waste of time so that airports are not waiting for business and space in the airways is not idle facilities are thus effectively utilised.

(c) **Economy :** Delays cause loss. Every excess minute spent in air means so much monetary loss for fuel, wear and tear and other expenses. Air traffic control is designed to avoid delays as far as possible, and thus protects waste. Idle planes, airports not used to capacity and airways not carrying the traffic of which they are capable represent wasted investment.

14.3 COMPARISON OF AIRWAY AND AIRPORT TRAFFIC CONTROL

Table 14.1. Comparison of airway and airport traffic control systems

S. No.	Item of comparison	Airway traffic control	Airport traffic control
1.	Management	Managed by national authority on international norms provided by ICAO	Managed by local personnel at civil airports and military aerodromes
2.	Location	Control centres are located on or near large airports, for greater convenience and to eliminate the express by additional communication facilities	Control towers are located at airports so that operator will have a clear view of landing area and surrounding airspace
3.	Jurisdiction	Jurisdiction is over all aircrafts flying in its control area, including the airspace over all airports within its area	It is exercised over aircraft and vehicular traffic operating on the landing area and over aircraft in flight within the vicinity of the landing area
4.	Extent of control	It is or may be continuous while the aircraft is in flight on the civil airways. All time during flight pilots have contact with one controlling station enroute	It is more immediate. Short lived in relation to each flight
5.	Method of communication	Communications are transmitted by interphone, radio facilities	It is exercised directly between tower personnel and aircraft pilot by radio, light signals etc.
6.	Traffic clearances	Traffic clearances and instructions are given as per approved flight plan and necessary amendments in view of safety when necessary	Traffic clearances and instructions are given on the authority of control personnel as pertain to a particular flight, in the landing area
7.	Information to pilots	Information regarding other flights that will operate within certain time and altitude limits of aircraft concerned is provided to pilot	Information concerning time checks, altimeter settings, surface winds and other weather conditions, runway to be used etc. are provided to pilot using the airport

14.4 AIR TRAFFIC CONTROL TYPES

There are two types of flight rules for air traffic, determined by weather conditions. They are called 'Visual Flight Rules (VFR) and Instrument Flight Rules (IFR). Flight under visual flight rules (VFR) may be described as 'clear weather' flying, when the weather conditions are such that aircraft can maintain safe separation by visual means by the pilots themselves. The other type of flight in accordance with instrument flight rules (IFR), may be described as "bad weather' or 'blind flying', when visibility is poor or the height of the clouds falls below the visual meteorological conditions. In such situations (IFR conditions) air traffic control personnel guide the safe separation between aircraft.

Under VFR conditions there is very little of air traffic control. Air control monitor intervenes only when apparent conflicts develop between aircrafts, while under IFR conditions the positive air traffic control is exercised by control personnel strictly. VFR and IFR basically require the assignment of aircraft to specific routes and altitudes and maintenance of minimum separation between aircrafts. To avoid possibility of mid-air collision at high speeds and density of traffic IFR are prescribed regardless of the weather conditions. This is called **"positive control airspace"** which is used where high speed jets operate at airport radar service area (ARSA) as well as the airspace at or above 6000 m above mean sea level, in which jet fly from one airport to another enroute.

14.5 AIRPORT RADAR SERVICE AREAS

For all VFR and IFR aircrafts an airspace around the airports where in Air Traffic Control (ATC), provides continuous guidance is used. These are called "Airport Radar Service Area" (ARSA).

An ARSA is usually established at busy airports, for safety of aircraft departing from or arriving at the airport. An ARSA consists of airspace from the airport surface up to an elevation of 1330 m (4000′) above ground level from airport surface within a radius of 9.25 kms (5 nmi) of the airport and from an elevation of 400 m (1200′) above ground surface to an elevation of 1330 m (4000′) above the ground surface between 9.25 km (5 nmi) and 18.5 km (10 nmi) radially from the airport, as shown in Fig. 14.1. Pilots are required to establish two-way radio communication with ATC before entering an ARSA. Except in an emergency, all operations within an ARSA must be in compliance with ATC clearance and instructions.

ARSA services include the following :

(a) Within the ARSA

   (i) Radar sequencing of all aircrafts arriving at the primary airport.

   (ii) Standard IFR separation between IFR flights.

   (iii) Between IFR and VFR aircrafts – traffic advisories and conflict resolution so that radar targets do not touch or 170 m (500′) vertical separation.

   (iv) Between VFR aircraft – traffic advisories and as appropriate "safety alerts".

(b) Within the outer area

The same services are provided for aircrafts operating within the outer area, as within ARSA, when two-way communication and radar contact are established.

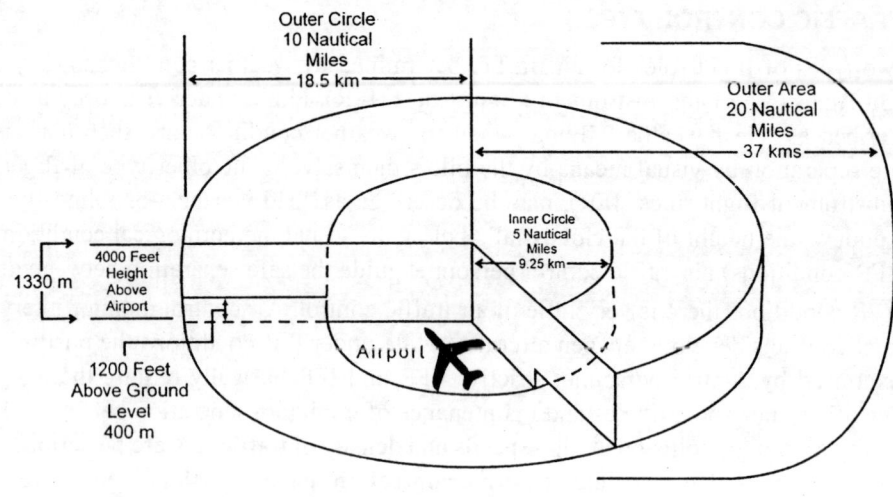

Note: The normal radius of the Outer Area, will be
20 nm, with some site specific vairatins.
37 km

Fig. 14.1. Airport Radar Service Area ARSA (FAA).

(c) Beyond the outer area

   (i) Standard IFR separation.
   (ii) Basic radar service.
  (iii) Safety alerts, as appropriate.

(d) Satellite and secondary airport operations

In some locations ARSA may overlap the airport traffic area of a secondary airport. In order to allow that control tower to provide service to aircraft portions of the overlapping ARSA may be procedurally excluded when the secondary airport tower is in operation. ARSA service to aircraft inbound to these airports will be discontinued when the aircraft is instructed to contact the tower.

Aircrafts departing secondary controlled airports will not receive ARSA service until two-way communication and radar contact have been established.

Service beyond the outer area, depends upon workload and may be terminated if workload dictates. ARSA services to aircraft proceeding to a satellite airport are terminated at a sufficient distance to allow time to change to the appropriate tower or advisory frequency.

14.6 AIRSPACE CLASSIFICATION

Airspace is classified in 6 categories by Federal Aviation Agency, USA.

14.6.1 Class A airspace (positive control areas)

Airspace in which only IFR operations are permitted. Pilots must be instrument-rated and have ATC clearance to enter this airspace. The airspace is designated from 6000 m (18000 ft) above mean sea level to flight level 20,000 m (60,000 ft).

14.6.2 Class B airspace (terminal control areas)

Airspace in which both VFR and IFR operations are permitted, but an ATC clearance is required to enter the airspace. Aircraft must be equipped with appropriate frequencies and a very high-frequency omnirange radio receiver for IFR operations. All aircrafts are subject to ATC separation and VFR aircrafts must remain clear of clouds.

14.6.3 Class C airspace (airport radar service areas)

Airspace in which entry requires prior two-way radio-communications with ATC. ATC provides traffic advisories and conflict resolution between aircrafts. VFR aircrafts must maintain visual flight rules. Special VFR operations are approved by ATC when basic VFR minima cannot be maintained.

14.6.4 Class D airspace (airport traffic areas and control zones at airports with control towers)

Airspace established around airports with operating air traffic control towers. When it is consistent with safety, nearby uncontrolled airports may be included. Pilots must establish and maintain radio contact with ATC prior to entering and while operating in this airspace. The ceiling upper limit is normally 830 m (2500 ft) above ground level. Weather minima are basic VFR.

14.6.5 Class E airspace (general controlled airspace and control zones at airports without control tower)

The airspace above 4830 m (14500 ft) above mean sea level. In addition airspace above 400 m (1200 ft) above ground level or 230 m (700 ft) above ground level for the transition areas for instrument approaches. Basic VFR minima are 1 km visibility, 150 m (500´) below clouds, 300 m (1000 ft) above clouds, and 600 m (2000 ft) horizontally from clouds. Above 3000 m (10000 ft) mean sea level, the VFR minima are 1.6 km (5 mi), 300 m (1000 ft) below clouds, 300 m (1000 ft) above the clouds, and 1.6 km (1 mi) horizontally from clouds.

14.6.6 Class E airspace (uncontrolled airspace)

The airspace not under control of any ATC facility. Instrument flight rules require that, prior to departure or en route, the pilots file a flight plan with the air route traffic control tower/centre. Each flight plan indicates the aircraft's destination, desired route and desired altitudes.

14.7 AIRWAYS AND AREA NAVIGATION

Aircrafts flying from one point to another follow the designated routes. These are called airways. With the use of jet aircrafts in 1958, aircraft flew at increased altitudes and jet routes were established. The jet routes are delineated by the aids to navigation on the ground. While for other aircrafts airways extend from 400 m (1200 ft) above the terrain to 6000 (18000 ft) above mean sea level, jet routes extend from 6000 m (18000 ft) to 15000 m (45000 ft) above MSL. Above 15000 m (45000 ft) there are no designated routes and aircrafts are handled on an individual basis.

For many years all aircrafts were required to fly on designated routes or jet routes. VOR stations used to delineate the routes. This resulted in congestion on some routes and were not the shortest so resulting in additional fuel consumption, flight time and cost. Area navigation technique called "RNAV" is a method of air navigation which permits aircraft movement on any desired course within the limits of a self-contained system capability.

Area navigation provides a more flexible routing capability that allows for better utilisation of air space, reducing delay and costs. Area navigation is accomplished by the installation of special computers in the aircraft which are tuned to VOR stations. Each station provides information on the distance to the station and the azimuth of the aircraft relative to the station.

Area navigation can also be used in the vertical plane. It is than called "VNAV". It can also include time reference capacity. A properly equipped aircraft can arrive at a specified point in space with no need for ground vectoring or directions and could also be at that point at a specified altitude and time. Thus it has four-dimensional capability giving altitude, latitude, longitude and time. Advantages of area navigation are increasing air space capacity, enhancing safety, reducing workload of the pilot and the air traffic controller.

14.8 JURISDICTION AND CONTROL SYSTEMS

To properly manage the air traffic in the system, on national airways, there are three types of facilities used as described below :

    (a) Air route traffic control facility,

    (b) Terminal control facility, and

    (c) Oceanic facilities.

14.8.1 Air route traffic control facilities

Air Route Traffic Control Centres (ARTCC) control the movement of en route aircrafts along the airway, jet routes and in outer parts of the air space. They have specified geographical area. At the boundary point of the one control area centre, control of the aircraft may be transferred to an adjacent centre, or an approach control facility.

ARTCC are generally not located at the airports and are concerned primarily with control of aircraft operating under instrument flight rules. ARTCC approves the flight plan, along the route of the flight. Each ARTCC area is divided into sectors in both horizontal and vertical plane. Each sector is suitably staffed, depending on the traffic, degree of automation provided and complexity of movement.

Each sector is normally provided with one or more air route surveillance radar (ARSR) units to cover entire sector, and very high and ultra high radio communication frequencies for communication. The controller in turn assigns a specific frequency to the pilot.

14.8.2 Terminal approach control facility

Terminal approach control facility monitors the air traffic in the airspace surrounding airports. Its jurisdiction is from the boundary area of the air traffic control tower at an airport to a distance of up to 80 kms from the airport and to an altitude ranging up to 5600 m. This is commonly called as terminal area.

This facility receives aircraft from the ARTCC and guides them to one of several airports for arriving planes and vice versa for departing planes. Various types and degree of automation in an approach control facility are provided depending on the volume of traffic to be handled.

14.9 AIRPORT TRAFFIC CONTROL TOWER

The airport traffic control tower is the facility which supervises, directs and monitors the arrival and departure of aircrafts at the airport and in the airspace within about 8 kms from the airport. It provides following basic services :

- Issue traffic control instructions and information to aircraft taking off, landing and approaching for landing.
- Issue instructions to aircraft taxiing for the purpose of preventing collision between aircraft on landing area and between aircraft and vehicular traffic on the landing area.
- Issue appropriate advice to pilots of aircraft when meteorological conditions which may not be known to the pilot.
- Relay message concerning operation, dispatch, and control of aircraft between pilots of the aircrafts and operation offices, weather bureau offices, airway traffic control centres and other appropriate agencies.
- Issue special instructions to pilots, when necessary regarding safety, concerning aeronautical facilities, emergency landing areas, obstructions, landmarks, restrictions and regulations etc.
- Start and plan crash and emergency procedures when an emergency occurs on or in the vicinity of the landing area.

The airport control tower must be so located that it will provide an unobstructed view of all runways, taxiways, and approaches to the airport. The usual practice at civil airports is to place the control tower structure on top of the highest building, usually the administration building.

The design of control tower structure should be functional, to provide for :
- Sufficient height to obtain an unobstructed view of the airport from the selected location.
- Minimum floor space consistent with requirements of accommodating equipment and personnel.
- Unobstructed visibility for 360° is a horizontal plane and as much vertical visibility as is practicable, or in any case not less than 70° visibility in vertical plane measured upward from the horizontal plane.
- Window glass so arranged as to obtain the minimum of reflections both day and night.
- Effective interior night lighting so that instruments may be observed without detracting from exterior vision.
- Air conditioning and heating apparatus for the comfort of personnel and protection of equipment.

In an isolated structure, supporting the tower cabin is a shaft structure, which generally contains only stairs, elevators and necessary utility spaces. Thus shaft is purely a structure to hold up the cab at the proper height. With the expansion of many airport facilities the heights and size of the cab of modern control towers have been increasing. New towers have heights of 60 to 70 m and cab floor area of 90-100 sq.m. or more. Fig. 14.2 shows a typical modern control tower.

14.10 AIDS FOR AIR NAVIGATION

Various aids to air traffic control can be classified in following categories :
(i) *External aids* : Those that are located on the ground.
(ii) *Internal aids* : Those installed in the cockpit of the aircraft.
(iii) *Special aids* : Used primarily for flying over the oceans.

Other methods of classifying could be :
(i) Aids used during the en route portion of flight.
(ii) Aids used in terminal area or near airport.

Fig. 14.2. A typical control tower.

External enroute aids (over land)

   (i) Very high frequency ominrange radio stations (VOR)

  (ii) Distance measuring equipment (DME)

 (iii) Air route surveillance radar (ARSR)

External terminal aids (overland)

   (iv) Instrumental landing system (ILS)

    (v) Microwave landing system (MLS)

   (vi) Precision-approach radar (PAR)

  (vii) Airport surveillance radar (ASR)

 (viii) Approach lighting system (ALS)

   (ix) Airport surface detection (ASD)

    (x) Visual-approach slope indicators (VASI)

   (xi) Runway end identifier lights (REIL)

External enroute aids (over water)

(xii) Long-range aerial navigation (LORAN)

Internal enroute aids (over water)

(xiii) Doppler navigation system

(xiv) Inertial navigation system

Internal enroute aids (over land)

(xv) Global positioning system (GPS)

Description of AIDS for air traffic control

A brief description of these aids is covered in following paragraphs.

OVER LAND AIDS

14.10.1 Very high frequency ominrange radio : Stations (VOR)

These stations are located on the ground and they send radio signals in all directions, which can be followed by an aircraft. There are 360 radial courses or routes at $1°$ interval from a VOR station, from $0°$ pointing towards magnetic north increasing to $359°$ in a clockwise direction.

VOR station is a small square building. Its broadcast frequency is just above FM radio stations. A system of VOR stations establishes the network of airways and jet routes.

Aircraft is equipped with VOR receiver in the cockpit which can be tuned to the desired VOR frequency. A pilot can choose the route to follow to the VOR station. A position deviation indicator in the cockpit specifies the position of the aircraft relative to the direction of the desired radial and whether aircraft is to the left or right of the radial route.

14.10.2 Distance measuring equipment (DME) and TACAN

Distance measuring equipment (DME) is installed at VOR stations. These stations equipped with this facility are called "VORTAC" facility. The distance measuring equipment shows to the pilot the slant distance between the aircraft and a particular VOR station.

There is another facility for distance measuring introduced in USA. It is called "TACAN" which stands for "Tactical Air Navigation" and was developed for tactical needs of the military. This aid combines azimuth and distance measuring into one unit and is operated in the ultra high frequency band. A station where this facility is used is called 'VORTAC'.

14.10.3 Air route surveillance radar (ARSR)

Long range radar for tracking an enroute aircraft is used. These radars provide air traffic controller with a visual display of the position of each aircraft so they can monitor the spacing and intervene if necessary in some situations. The radar is used for guiding the aircrafts and thus provides an aid to air navigation.

10.10.4 Instrument landing system (ILS)

Instrumental landing system (ILS) is most commonly used system. It consists of two radio transmitters located at the airport. One radio beam is called the 'localiser' and the other is the 'glide slope'. The localiser indicates to the pilot whether he is towards left or right of the correct position for approach to the runway.

To further help the pilot on his ILS approach two low power fan markers called 'ILS markers' are installed so that pilot will know how far along the approach to the runway they have progressed. The first is called the 'outer marker' (LOM) and is located about 7 kms from the end of the runway. The other marker is called 'middle marker' (MM) and is located about 1 km from the end of the runway.

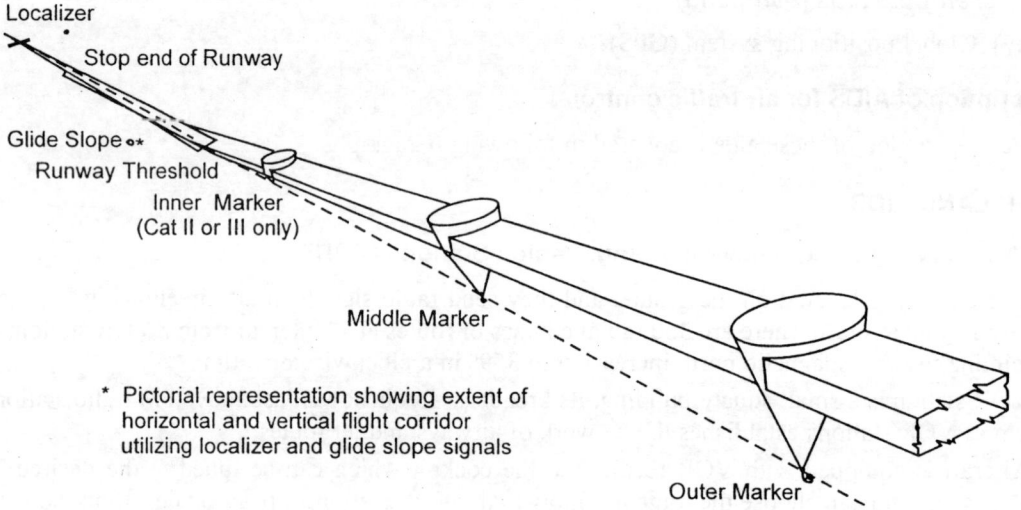

Fig. 14.3. Typical instrumental landing system (FAA).

For category II, operations, when visibility is quite poor, an additional marker called "inner marker (IM) is located 300 m from the end of the runway. The inner marker is placed so as to alert pilot that they must have visual reference with the ground at that point and if they do not, abandon the approach.

When the plane passes over a marker, a light is lit in the cockpit with a high pitched tone sound. At some airports, the fan markers are replaced with DME using the VORTAC or TACAN at the airport.

ILS thus provides landing facility even when weather and visibility is poor, however, it has following limitations.

- The functioning of localiser and glide slope transmitters is affected by the close proxomity of moving objects such as vehicular and aircraft traffic.
- Stationary objects nearby can also cause a deterioration of the signals.
- Abrupt changes of slope in the proximity to the antennas are not permitted, or else signals will not be transmitted properly.
- The glide slope beam is not reliable below a height of about 65 m above the runway.
- The ILS provides only one path in space, which all aircraft must follow if they are using the system.
- The area adjacent to the antennas must be smooth and kept clear of any obstruction such as buildings and taxiing aircraft, otherwise beam will be distorted.

14.10.5 Microwave landing system (MLS)

It is a more sophisticated landing system than ILS. ILS is based on signals reflecting from the surface of the ground. Some aircraft, particularly short takeoff and landing (STOL) type, can use a steeper approach angle of 7°, than conventional aircraft, which use 2.5° to 3° approaches. ILS is unable to provide for these differing needs. Only limited number of frequency channels are available for the ILS.

To overcome the limitations of ILS, the 'Microwave Landing System' (MLS) is developed. Instead of providing only one glide slope as in ILS, the MLS provides for a number of slopes. In the horizontal plane, the MLS provides for any desired route as long as they are within an area that is from 20° to 60° on each side of the runway centreline, whereas ILS provides only one route to the runway. Distance measuring capability can be incorporated into the MLS, providing the pilot with continuous information on the aircraft distance from the end of the runway and avoiding the need for establishing markers as in ILS. The MLS is far less susceptible to interference from surrounding objects than ILS.

With the MLS, pilot can choose any desired route to runway at any glide slope within the vertical coverage of the system.

The other big advantage of the MLS is the reduction of noise since aircraft can be kept at higher altitudes before they make the descent to the airport. Another advantage is the elimination of the requirement that all aircraft, large or small, follow a common approach route to the runway.

14.10.6 Precision approach radar (PAR)

The precision approach radar (PAR) was developed by the military people during world war II, to provide a mobile unit that is not dependent on airborne navigation equipment. PAR system gives controller a picture of the descending aircraft both in plan and elevation i.e., one half the radarscope is in plan and one-half is in elevation. The controller can determine whether an aircraft is on the glide path or whether it is on the correct alignment.

Instructions from the controller to pilot are given by voice communication, and therefore, no airborne navigation equipment is necessary. Commercial aircrafts generally use ILS, because PAR places too much dependence on the controller. At airports where both ILS and PAR are provided, commercial pilots use ILS but often request that they be monitored by PAR.

14.10.7 Airport surveillance radar (ASR)

Airport surveillance radar (ASR) is installed at major airport, to provide the control tower operator with an overall picture of what is going on within the airspace surrounding the terminal. ASR rotates through 360° and the picture is received on the radar scope in the control tower. The range of ASR varies from 50 kms to 100 kms. It shows the aircraft in their relative horizontal position on the radarscope as bips. ASR does not indicate the altitude of aircraft.

14.10.8 Approach lighting system (ALS)

While landing, most critical situation comes when aircraft breaks through the overcast and the pilot changes from instrument to visual conditions. Only few seconds are available for the pilot to make this transition and complete landing.

As an aid in making this transition approach lighting system (ALS) is provided. Details on number and type of lights used in ALS, has been described in the chapter on lighting.

14.10.9 Airport surface detection

A specially designed radar called 'airport surface detection equipment' or ground radar is provided at busy airports to help the controllers in regulating taxiing aircraft, when they can not see the aircraft in poor visibility conditions. The system provides the air traffic controller, in control tower a pictorial display of the runways, taxiways and terminal area, with radar indicating the positions of aircraft and other vehicles moving on the surface of the airport.

14.10.10 Visual-approach slope indicators (VASI) and precision approach path indicator (PAPI)

Through a system of of lights, the proper approach slope to the runway (like the glide slope in ILS) is provided through the VASI and PAPI. 'Precision Approach Path Indicator' system is a recently developed being used in USA. PAPI system gives a more definitive indication of approach slope to the pilot. It uses only a single set of electronic devices at one point down the runway.

VASI systems are used for day or night operations during good (VFR) weather conditions, while PAPI can be used in bad weather as well.

PAPI system consists of a unit with four lights on either side of the approach runway. By utilising a colour scheme, the pilot is able to ascertain five approach angles relative to the proper glide slope, compared to three in VASI system. PAPI system provides an instant transition from one colour indication to another as a reaction to the descent path of the aircraft. It is a one bar system, while VASI uses two bars and thus it is economical to operate and maintain and the pilot does not have to look at two bars for glide slope indications. Fig. 14.4 shows a typical PAPI system.

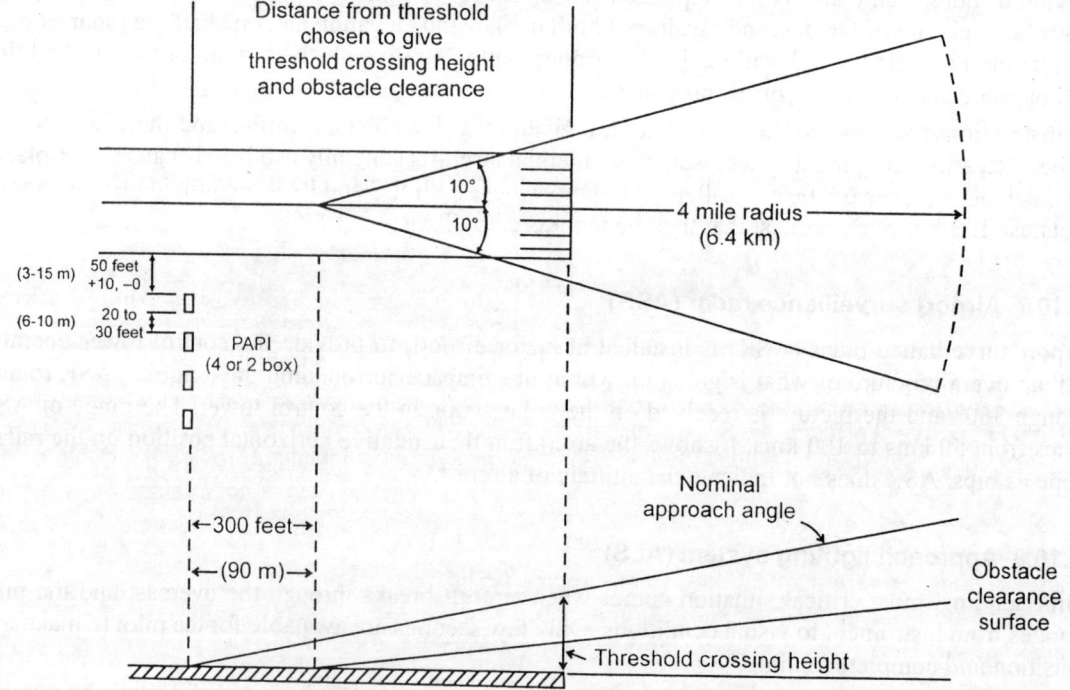

Fig. 14.4. Typical PAPI system (FAA).

14.10.11 Runway end identifier lights (REIL)

Runway end identifier lights (REIL) are used to provide the pilot a positive visual identification of the approach end of the runway, when there are no approach light.

These lights and their configuration is described in chapter on lighting.

14.10.12 Over water aids (external) : Long-range aerial navigation (LORAN)

The principal overwater aid to navigation is LORAN which consists of stations located on the ground. The system consists of a master station and a slave station located some distance from the master. The master station sends radio signals into space and at the same time one signal goes to slave station, where it is delayed a specified amount of time and then sent into space. Thus a contour of equal time differences can be drawn in space. The same can be done at another slave and master station resulting in another contour. The intersection of the two contours establishes a position in space.

In the aircraft the LORAN receiver tunes in on two master and slave stations, establishing an intersection of two time-difference contours in space. The range of LORAN is affected by the time of day, being greater at night than during day light. LORAN requires the use of a navigator in the cockpit.

14.10.13 Over water aids (internal) Doppler navigation system

This system is a long-range-radar type aid that provides the pilot with :
   (i) the ground speed,
  (ii) angle of the aircraft axis relative to the desire course,
 (iii) distance of the aircraft right or left of the desired courses, and
 (iv) distance to the destination or way point.

The basis of doppler system is that aircraft sends to the ground four beams of continuous wave energy (8800 mc), two forward and two backwards. The change in frequency of the energy return from the ground is measured. This change in frequency is known as the "doppler shift" and is proportional to the aircraft speed in the direction the beams are pointing.

14.10.14 Inertial navigation system

This system provides all the information that the doppler system provides as well as the wind speed and direction, latitude and longitude of the aircraft at any instant, and time to reach the next way point. This system is development of the space programme and is quite accurate and reliable.

14.10.15 Global positioning system (GPS)

This system GPS is a recent space-based satellite radio system, designed to provide highly accurate position and speed information on the continuous global basis. The system is not affected by weather. It is expected that GPS will have a horizontal accuracy of 100 m, and is likely to become an external navigational aid in near future.

The global positioning system can provide external overland and overwater precise navigational assistance to aircrafts and also in conducting terminal area precision approaches to runways.

CHAPTER 15

Airport Terminal Building

15.0 INTRODUCTION

The airport terminal is the facility that provides the connection between the aircraft and the vehicles providing ground transport. The different number , size and length of stay of the aeroplanes in the air and ground vehicles imply quite dissimilar amount of space on the airside and landside of the terminal. Typically, on airside, the stands for stationing the aircrafts require a much bigger space than curb needed to provide for cars, buses and other landside vehicle. How to balance these conflicting requirements on either side, is the essential question of terminal planning and design. The basic purpose of the landside of an airport is to effect the transfer of the air traveller and their baggage from ground transportation to air transportation in minimum time and with minimum amount of confusion and discomfort and vice versa. It is interface between ground travel and air travel.

15.1 FUNCTIONS OF TERMINAL BUILDING

The landside includes following basic elements :

1. The terminal building :
   (a) Passenger building
   (b) Cargo building (if required separately)
2. Access roadways, and
3. Parking for vehicles used for the ground transportation of passengers.

A simple diagram of the basic passenger-handling system is shown in Fig. 15.1

   (a) *The primary functions* include airport access and egress, including road system, parking sideworks, lobbies etc., ticketing and checkin, international clearance, assembly at the aircraft, enplanning and deplanning the aircraft and baggage chain, baggage and cargo moving equipment, security etc.

   (b) *The secondary functions* include services (restrooms, restaurants, public telephone) and meeting and greeting space, lavatories etc.

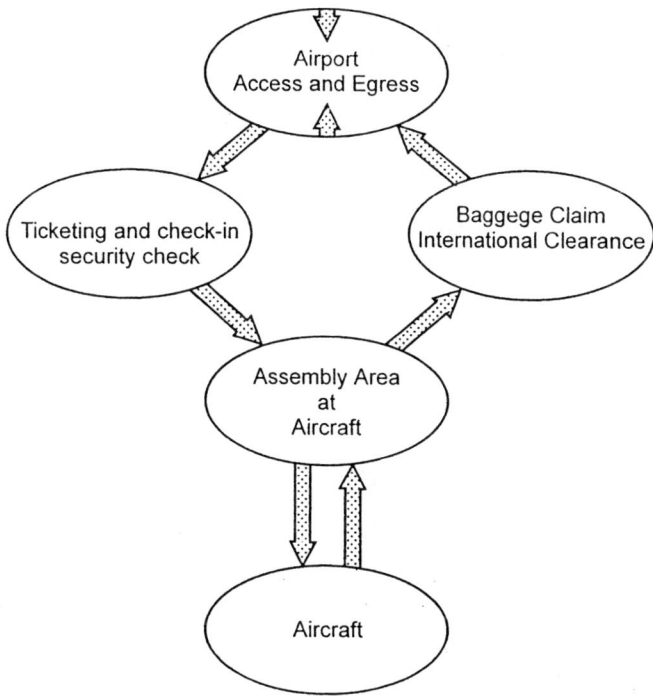

Fig. 15.1. Airport landside passenger-handling system.

Each of these primary functions is associated with one or more physical facilities as shown in the Table 15.1. Airport landside is more commonly called passenger terminal building area or passenger terminal complex.

Table 15.1. Passenger-handling system

Primary function	Physical facilities	Airport landside
Access and egress to and from airport	Private automobile	Curbside
		Parking space
		Curbside
		Open lot
		Structural parking
		Valet
		Short term
		Long term
	Rental car	Curbside
		Ready and return areas
	Taxicab, limousine, bus	Curbside
		Dispatch facility
		Staging area
	Train	Station
	Private aircraft	Aircraft parking position
	Public air carrier	Aircraft parking position

(Contd.)

Primary function	Physical facilities	Airport landside
Ticketing	Home or office (mail)	
	Ticket counter	
	Remote	
	At terminal	Counter
Check-in	At rail station	
	At downtown terminal	
	At airport terminal	Curbside
		Ticket counter
		Walk-through check-in counters
		Aircraft gate
		Parking lot or garage
		Ground transportation center
International clearance	At airport terminal	
	Passport control	Counter
	Immigration	Counter
	Public health	Counter
	Customs	Claim and inspection facility
		Direct from cart
		Linear tray
		Diverter
		Carousel
		Conveyor belt
Assembly area for aircraft	At airport terminal	Queeing area (concourse)
		Departure lounge
		Transporter
		Main waiting room
		VIP rooms
		In-transit lounge
Enplaning and deplaning	At airport terminal	Above apron level loading
		Loading bridges
		Transporter
		Apron level loading
		Open boarding stairs
		Covered boarding stairs
		Escalators
Baggage claim	At airport terminal	Baggage claim area
		Direct from cart
		Linear tray
		Diverter
		Carousel
		Conveyor belt
Circulation	At airport terminal	Concourses
		Moving sidewalks
		Escalators
		Elevators
		Shuttle trains
		Buses
		Carts

Radical changes in aircraft technology and explosive growth in passenger demand, created the need for greater emphasis on development of a terminal complex to meet these growing demands as effectively as possible. It used to be that terminals and related facilities for servicing passengers accounted for only one fourth or less of the cost of an airport and most of the money had to be spent on making runway and taxiways, longer, wider and stronger, to accommodate new generation of aircraft, especially jets. This has to be changed now, as these new aircrafts require much larger terminal facilities that were needed before, and much higher costs.

The landside traffic is of 3 categories: people, baggage and cargo (including mail). The people include air passengers, employees of the airport complex and airport visitors and greeters. Air passengers essentially represent the independent variable that governs all activities of the terminal building.

Air passenger demand significantly fluctuates as a function of hour of the day, day of the week or month of the year.

15.2 AIRPORT LANDSIDE FUNCTIONAL FLOW REPRESENTATION

Fig. 15.2 shows the airport landside functional flow of the overall traffic movements and activity within the airport. Each facility may represent a network of subfacilities that when linked together support the complex activities of movement and service operations. The traffic movement through the airport landside consists principally of vehicles containing employees, passengers, visitors and baggage.

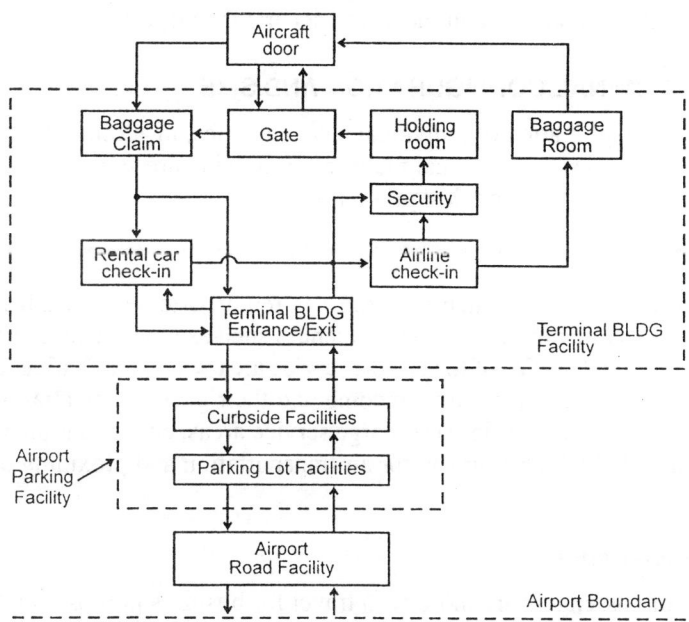

Fig. 15.2. Airport landside functional flow.

Enplaning passenger vehicles entering the airport proceed to a parking lot for long- or short-duration parking, to a rental car check-in area, or to the curbside for unloading. The passengers and visitors then proceed into the terminal. The passenger may wait in the terminal area or proceed to the ticket counter, baggage check-in, car rental check-in counter, or to the airplane gate, where they must pass through a security check before enplaning. Except for enplaning, the order in which these activities can be performed is not necessarily fixed but depends on factors such as the nature and origin of the trip and the terminal geometry.

Deplaning passengers either proceed to another flight or more through the terminal to the airport boundary. Some passengers may need to get their baggage at the baggage claim facilities. Passengers arriving on an international flight must proceed through a customs inspection and an immigration check. Passengers leaving the airport by landside vehicles may require transactions at car rental or intra-airport transit counters or a wait at a curbside or station for a vehicle.

Connecting passengers join in the flow of enplaning passengers in the terminal. The requirements of some transferring passengers are the same as those originating at the airport, and the services are performed at the same facilities. Passengers making interline transfers are generally required to be processed through the security check-point. In general, the baggage is transferred from plane to plane by the airline companies.

Employees are generally assigned parking areas or proceed in public vehicles. The flow of employees is to staffing area or duty stations. At the end of the shift they generally proceed outward beyond the airport boundary.

Visitors are generally well-wishers or greeters. Well-wishers proceed with their respective departing passengers to some point within the airport landside, after which they generally depart the airport. Greeters enter the airport and proceed to the parking areas or to the curbside. The arriving passenger is met at some point within the landside, and the group departs the airport.

15.3 FACTORS IN TERMINAL CONFIGURATION AND SIZE

Each airport has its own combination of individual characteristics, in addition to traffic volumes, to be considered in configuring and sizing terminal facilities. Following are some of the basic factors which need considerations in the planning and design of an airport terminal.

(a) Service area

What percentage of the population which an airport is required to serve, shall require air travel is important for planning. Apart from the obvious influences such as physical size and topography, some of the more significant characteristics of the airport service area which may influence the airport terminal design include population and per capita income and their growth potential, geographic location and distance from other airports with similar or large service areas, concentration of commercial activity that involves a relatively high prospensity for air transportation and proximity of major vacation or recreation areas.

(b) Passenger characteristics

Two basic categories of passengers are those who travel for business purposes and those travelling as tourists or for personal reasons. Business passengers are usually more travel experienced; arrive just prior to flight time; and are more apt to use the full range of public terminal services and concessions. On the other hand, vacation travellers are more likely to arrive much earlier, relative to flight departure time, compared to business travellers; depart from the destination airport later; and, generate a larger number of visitors/greeters. Consequently, significant variations in the characteristics and ratio of these two passenger types can influence space requirements and staffing. A small airport serving a vacation/

resort area with a relatively short season will involve different requirements than an airport handling comparable peak-month volumes of predominantly business travellers.

(c) Airline station characteristics

The route structures of the scheduled airlines serving an airport influence its character and, consequently, its facility requirements. Airports can generally be categorized into three types on the basis of the route structures of the using airlines. These categories and their related characteristics are discussed in succeeding paragraphs. The peak hour movements per gate specified are typical for airports averaging six or more daily departures per gate.

(i) Origination/termination airport

This category of airport usually involves a high percentage (over 70 per cent of total enplanements) of originating passengers and a preponderance of turnaround flights. Ground times range from 45 to 90 minutes, or more. The high flow of passengers between aircraft and ground transportation vehicles generates a relatively high requirement for ticket counter area, curb length, and parking spaces per enplaned passenger. Passengers will usually require maximum baggage-handling services for checking and claiming baggage.

(ii) Through airport

This category has a relatively high percentage of originating passengers combined with a low percentage of originating flights, resulting in the shortest aircraft ground times. Boarding load factors may be lower than originating/termination airports (ranging from 40 to 60 per cent), thereby reducing departure lounge space requirements. Typical domestic peaks will average 1.5 to 2.0 hourly aircraft movements per gate.

(iii) Transfer airport

This category of airport has a significant proportion of passengers, at least 30 per cent of total enplanements, transferring between on-line and off-line flights. Aircraft ground servicing times average 30 to 60 minutes, depending upon connecting patterns and airline operating policies. Compared to the same volume of enplanements at the other two categories of airports, the transfer airport has less ground transportation activity and a lower requirement for curb frontage; less need for airline counter positions serving normal ticketing and baggage check-in (although more positions may be required for flight information and ticket changes); less requirement for baggage claim area; more space for baggage transfer (on-line and/or interline baggage); increased requirements for concessions and public services; and increased need for centralized security control locations.

(d) Aircraft mix

The forecast mix of aircraft expected to use an airport can significantly impact terminal design. For instance, airports serving a large variety of aircraft types and sizes require terminal facilities more flexible and complex than those serving predominantly one class of aircraft. The latter are more conducive to standardizing the area and facilities at each gate position. Terminals at airports serving widebody aircraft require the ability to accommodate the large passenger surges which normally occur when these aircraft load or unload.

(e) Nonscheduled service

In addition to scheduled operations, most airports serve a variety of non-scheduled operations such as charter flights, group tour flights, and air-taxi operations. At some airports, a relatively high volume of airline charter or other nonscheduled operations may warrant consideration of separate, modest, terminal facilities for supplemental carriers. Occasionally, scheduled carriers may desire separate apron hardstands and buildings to serve charter operations which exceed thoroughly, since a separate facility can often create inefficiencies in such aspects as logistics, staffing, and ground equipment utilization.

(f) International service

Airports with international flights may have other characteristics which influence terminal planning and design. One characteristic is a tendency toward higher aircraft activity peaks because of the heavy dependence on schedules for city pairs related to time zone crossing. Another characteristic is the relatively long ground service times (2 to 3 hours for turnarounds, 1 hour for through flights) required for long range aircraft servicing.

15.4 TERMINAL SITING CONSIDERATIONS

For expansion or modernisation of an existing terminal the location is more or less fixed, however, for a new airport or major airport redevelopment, consideration should be given to following important factors :

(a) Runway configuration

The runway configuration at an airport has a significant impact on the location of the apron-terminal complex. The terminal site should be located to minimize aircraft taxiing distances and times and the number of active runway crossings required between parking aprons and runways. At airports with a single runway or very simple runway configuration (for instance, airports with a primary plus cross-wind runway or single set of parallel runways), this may dictate locating the passenger terminal centrally with respect to the primary runway(s). At airports with more complex runway configurations, sitting may require detailed analyses to determine runway use, predominant landing and takeoff directions, location and configuration of exist taxiways, and the most efficient taxiway routings. The runway configuration may also restrict ground access to certain areas of the airport and thus limit alternative terminal sites. Fig. 15.3 depicts the relationship between runway configurations, terminal locations, and ground access facilities.

(b) Access to highway network

The motor vehicle is and will continue to be the major mode of ground transportation to and from the airport. From a cost and efficiency standpoint, the passenger terminal should be located, when possible, to provide the most direct and shortest routing to the access roadway system serving the population center generating the major source of passengers and freight. Adequate area/distance should be provided between the public highway and the primary terminal building to accommodate the ultimate terminal development and necessary future highway interchanges and roadway alignment improvements.

(c) Expansion potential

To assure the long-term success of an airport terminal facility, potential expansion beyond forecast

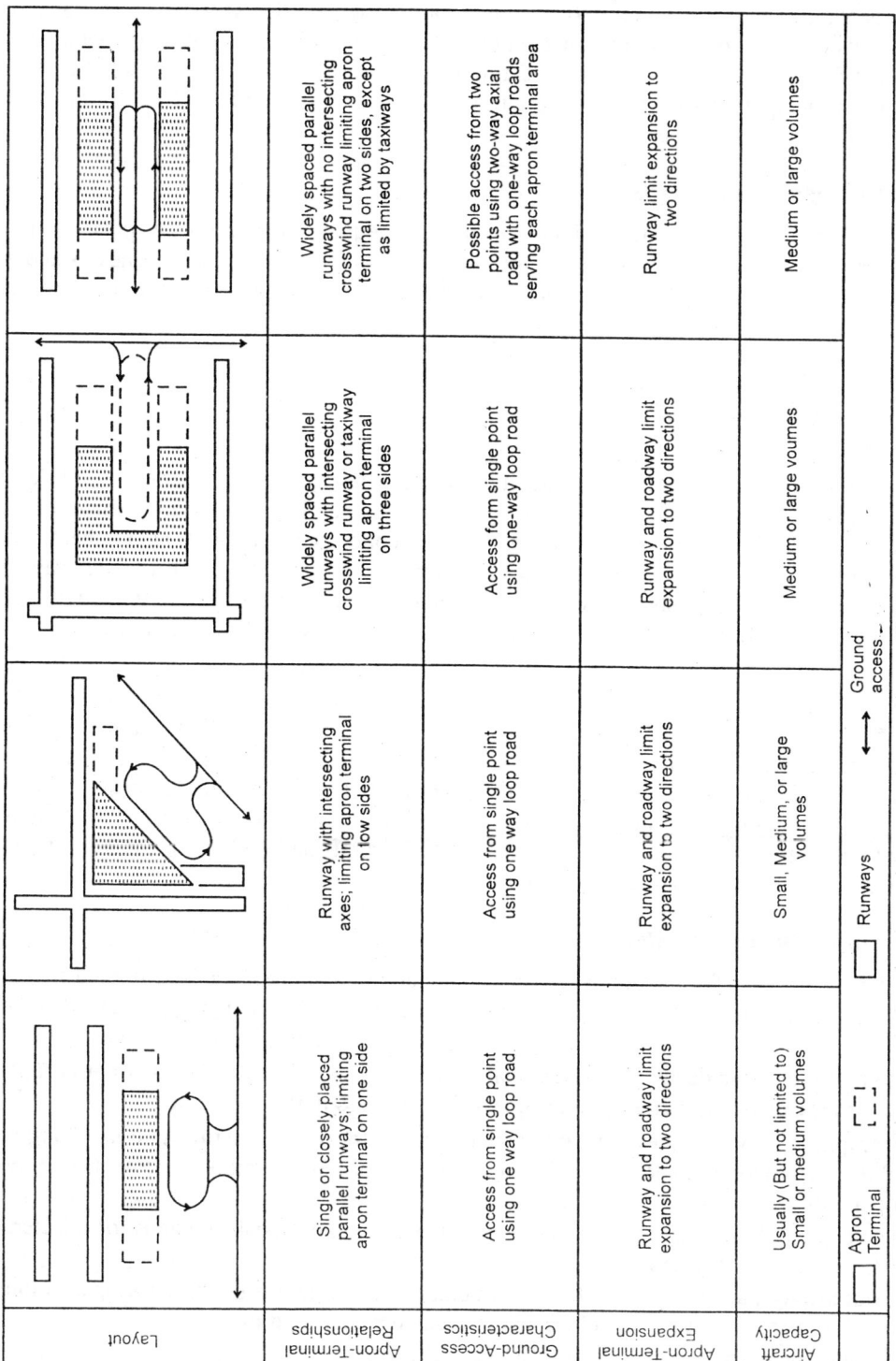

Layout				
Apron-Terminal Relationships	Single or closely placed parallel runways; limiting apron terminal on one side	Runway with intersecting axes; limiting apron terminal on fow sides	Widely spaced parallel runways with intersecting crosswind runway or taxiway limiting apron terminal on three sides	Widely spaced parallel runways with no intersecting crosswind runway limiting apron terminal on two sides, except as limited by taxiways
Ground-Access Characteristics	Access from single point using one way loop road	Access from single point using one way loop road	Access form single point using one-way loop road	Possible access from two points using two-way axial road with one-way loop roads serving each apron terminal area
Apron-Terminal Expansion	Runway and roadway limit expansion to two directions	Runway and roadway limit expansion to two directions	Runway and roadway limit expansion to two directions	Runway limit expansion to two directions
Aircraft Capacity	Usually (But not limited to) Small or medium volumes	Small, Medium, or large volumes	Medium or large voumes	Medium or large volumes

☐ Apron
☐ Terminal   ⌐ ¬
                  ∟ _⌐   ☐ Runways   Ground
                                               access ⌐
                                               ◄——►
                                               Runways

Fig. 15.3. Terminal siting related to runway configuration (FAA).

requirements should always be taken into consideration. In the planning stage, the terminal should be conceived in its ultimate form with reasonable allowance for growth and changes in operation beyond forecasted needs. Use of this principal in selecting a terminal site or expansion scheme will promote the provision of adequate space around the terminal (both on the airside and landside) for orderly construction of succeeding stages.

(d) Geometric design standards

Terminal facilities require a location which will assure adequate distances from present and future aircraft operational areas in order to satisfy airport geometric design standards. These standards include such minimum separation distances as those between a runway centerline and aircraft parking aprons, buildings, and airport property lines; and those between a taxiway centerline and fixed/movable objects and other taxiways.

(e) Existing and planned facilities

Existing and planned structures and utilities should be carefully inventoried and taken into account when planning new or expanded terminal facilities. In some cases, existing facilities or utilities, which are not related to and are restrictive to terminal development, can be demolished, abandoned, or relocated to a more suitable area. In other instances, existing conditions may limit the number of possible alternative terminal sites. In all cases, existing or planned locations of a control tower, navigational aids, weather equipment, etc., should be analyzed to assume that terminal development will not interfere with line-of-sight or other operational restrictions associated with these facilities.

(f) Terrain

Topographical conditions should be considered in the selection of a terminal building site. For instance, potential drainage problems can be reduced if the terrain lends itself to naturally carrying water away from the building. Developing the terminal site on relatively flat land can prove economically advantageous by reducing grading or quantities of fill. An existing terrain feature, such as a grade differential between the landside of the terminal and an aircraft parking apron, can be minimized by use of a multi-level terminal concept.

(g) Environmental consideration

The location of a terminal facility or major expansion of an existing one, should be studied and analysed for its impacts on the environment. An environmental assessment is an important factor to be considered.

Effective planning and design of the terminal area involves active participation of airport planners, managers, airlines, concessionaires and designers. The process normally includes :

- Compiling surveys, questionnaires and forecasts, for short, intermediate and long periods, about passenger enplanements, passenger originations and aircraft movement, size, type and operations.
- Developing design day and peak hour activities to establish peak demands for selected design period of 15 to 20 years.
- Establishing passenger, aircraft and vehicular traffic relationship as passenger volumes depend on aircraft seating capacities and load factors, now and in future.
- Taking inventory and evaluating existing facilities.

- Analysing space requirements for alternative layouts.
- Estimating costs and developing financial plans.

15.5 MAJOR COMPONENTS OF A TERMINAL

To accomplish the interchange between ground and air transportation, following are the major components of a terminal building/complex.

15.5.1 Apron

Located on the airside of the terminal building, the apron comprises the area and facilities used for aircraft gate parking and aircraft support and servicing operations. It includes the following sub-components.

(a) *Aircraft gate parking positions* : Used for parking aircrafts to enplane and deplane passengers. The passenger boarding device is part of the gate position.

(b) *Aircraft service area* : On or adjacent to an aircraft parking position. They are used by airline personnel/equipment for servicing aircraft and the staging of baggage, freight and mail for loading and unloading of aircraft.

(c) *Taxilanes* : Reserved to provide taxiing aircraft with access to and from parking positions.

(d) *Service/file lanes* : Identical sight-of-way on the apron designated for aircraft ground service vehicle and fire equipment.

15.5.2 Connector

The connector consists of the structure(s) and/or facilities normally located between the aircraft gate position and the main terminal building. It normally contains the following elements :

(a) *Concourse* : A passageway for circulation between aircraft gate parking positions and the main terminal building.

(b) *Departure lounge* : An area for assembling and holding passengers prior to a flight departure. In some instances, it may be a mobile lounge also used to transport passenger to a parked aircraft.

(c) *Security inspection station* : A control point for passengers and baggage inspection and controlling public access to parked aircraft.

(d) *Airline operational areas* : Areas set aside for airline personnel, equipment and servicing activities, related to aircraft arrival and departure.

(e) *Passenger amenities* : Areas normally provided in both the connector as well as the terminal components, particularly at the busier airports with relatively long connectors. These amenities include rest rooms, snack bars, beverage lounges and other concessions and passenger services.

(f) *Building maintenance and utilities* : Areas often included in the connector component to provide terminal building maintenance and utilities.

15.5.3 Main terminal building

The following elements comprise this component :

(a) *Lobbies* : Public areas for passenger circulation, services and passenger/visitor waiting.

(b) *Airline ticket counters/office areas* : Areas required for ticket transactions, baggage checkin, flight information and administration backup.

   (c) *Public circulation areas* : Areas for general circulation which includes strairways, escalators, elevators and corridors.

   (d) *Terminal services* : Facilities, both public and non-public, which provide services incidental to aircraft flight operations. These facilities include rest rooms, restaurants and concessions, food preparation and storage areas, and miscellaneous storage.

   (e) *Outbound baggage facility* : A non-public area for sorting and processing for departing flights.

   (f) *Intraline and interline baggage facility* : A non-public area for processing baggage transferred from one flight to another.

   (g) Inbound baggage facility : A non-public area for receiving baggage from an arriving flight and public areas for baggage pickup by arriving passengers.

   (h) *Inspection services* : A control print for processing passengers arriving on international flights.

   (i) *Airport administration and services* : Areas set aside for airport management, operations and maintenance functions.

15.5.4 Airport access system

This component is composed of the functional elements which enable ground ingress and egress to and from the airport terminal facility. They include the following :

   (a) *Curb* : Platform and curb areas (including median strip) which provide passengers and visitors with vehicle loading and unloading areas adjacent to the terminal.

   (b) *Pedestrian walkways* : Designated lanes and walkways for crossing airport roads, including tunnels and bridges which provide access between auto parking areas and the terminal.

   (c) *Auto parking* : Areas providing short-term and long-term parking for passengers, visitors, employees and car rental.

   (d) *Access roads* : Vehicular roadways providing access to the terminal curb, public and employee parking, and to the community roadway/highway system.

   (e) *Service roads* : Public and non-public roadways and fire lanes providing access to various sub-elements of the terminal and other airport facilities, such as air freight, fuel tank stand, postal facility and others.

15.6 PASSENGER BUILDING (TERMINAL)

Activities associated with the transfer of passengers and their baggage from the point of interchange between ground transportation, to the point of connection with the aircraft, and the transfer of connecting and in-transit passengers and their baggage between flights, take place in the passenger building.

15.6.1 General considerations in planning and design of passenger building

Following principles (factors) of design and planning, should be considered for passenger buildings (terminals) :

• Aircraft operations will be less costly and more efficient if the passenger building is as close as possible to the runways, at taxiing distances will be reduced and hence fuel consumption. Expansibility and flexibility, however, should not be compromised.

• Passenger building facilities should be contiguous with general location on the airport, however, in some cases particular functions such as aircraft maintenance may be situated at locations remote from the main passenger building.

- The provision for all of the necessary passenger services should be at an optimum cost while recognising the need for flexibility and expansibility as well as economy of any future passenger building expansion.
- Passenger's physical and psychological characteristics make the passenger area, most sensitive part of the whole air transport system. The facilities, therefore, should provide comfortable, convenient and speedy movement of passengers and baggage between air and ground transportation at the lowest effective cost.
- Passenger building should be designed through close cooperation between planners, architects, engineers and all others concerned, to have a well planned layout.
- Passengers should be forming a homogeneous flow whether constant or intermittent and require indications of what they are expected to do and the flow routes they should follow.
- Facilities for individual needs, disabled and elderly persons, invalids should be suitably accommodate in the plan.
- The plan should be simple with obvious flow routes. Complex buildings are usually costly inflexible and not readily expansible. Fig. 15.4 illustrates each of the important functions of a passenger building and gives passenger and baggage processing interrelationship.
- A walking distance of about 300 m from the centre of the passenger building and the farthest aircraft parking is generally considered reasonable. The large area in the passenger building should be broken down into units or modules, to maintain passenger walking distances within reasonable limits.
- The layout of modular passenger units, within passenger building plan should have the most compact arrangement to minimise transfer distances between the passenger buildings and between associated facilities'within each modular unit.
- The passenger flow should be considered first and baggage movement should then be integrated with the passenger flow, to make it compatible with the best passenger flow. Following points should be considered with respect to passengers :
  - (i) Routes should be short, direct and self-evident. They should not, as far as practicable, conflict with nor cross the flow routes of other passenger, baggage or vehicular traffic.
  - (ii) Change in level of pedestrian routes should be avoided as far as practicable.
  - (iii) The passenger should be able to proceed through a building without anybody's guidance or instructions from staff. The flow system should be for 'trickle flow' rather than controlled movement.
  - (iv) Particular categories of passengers should be diverted from the main flow route to pass through specific controls only at the last point on the main flow route where the character of traffic changes.
  - (v) Departing passengers should have an opportunity to check their baggage at the earliest possible point.
  - (vi) Each flow route, as far as possible, be in one direction only. Where a reverse flow has to be provided, it should be via a self-contained and separate route.
  - (vii) Free flow through all parts of the routes between air and ground transport should be interrupted as little as possible. Every control point in the flow system has a potential to delay and also irritate and confuse passengers, it should, therefore, be planned in best manner to achieve passenger convenience, maximum security, optimum utilisation of staff and minimum cost. Controls should be minimised and concentrate them on few number of points.

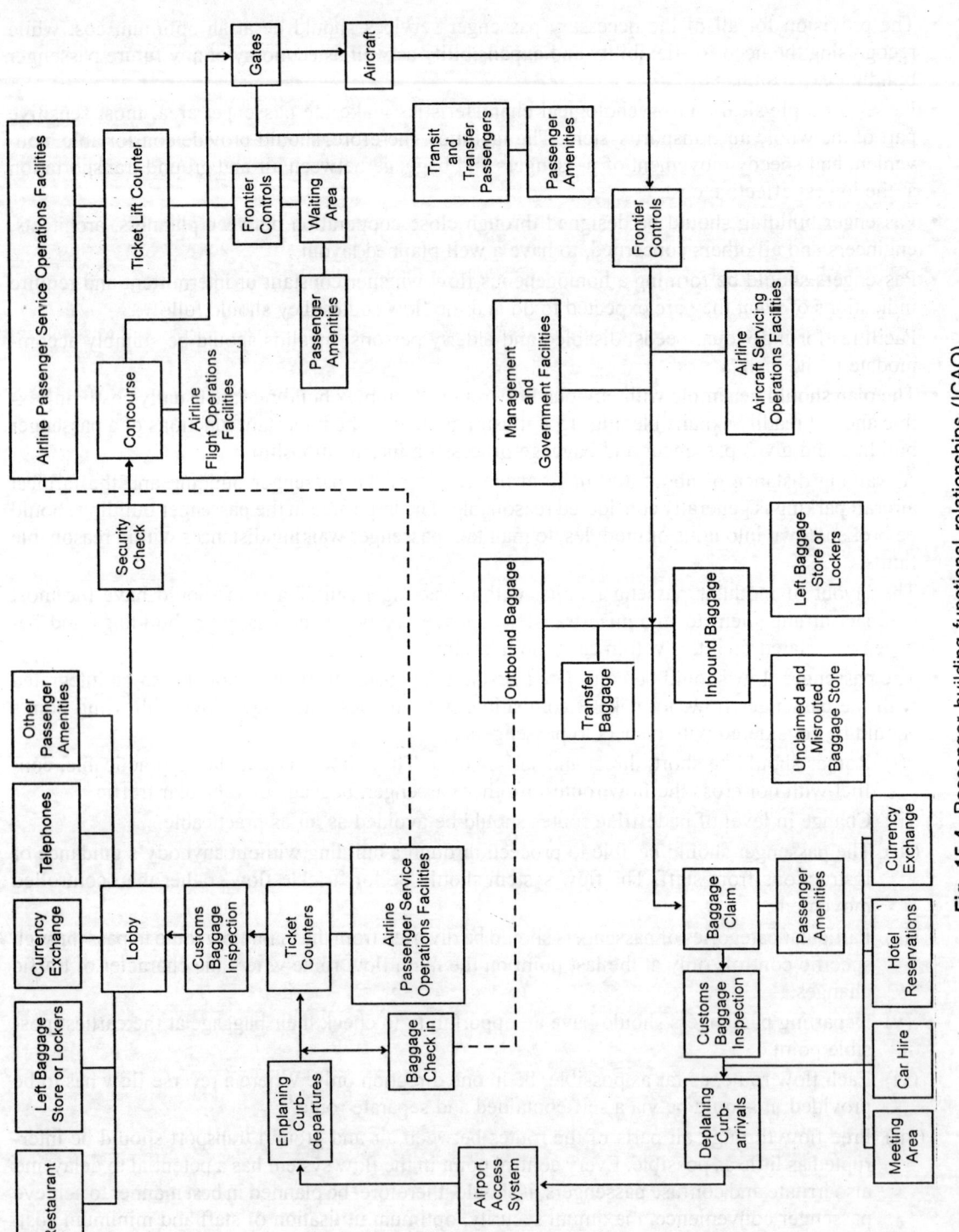

Fig. 15.4. Passenger building functional relationships (ICAO).

(ix) Passenger should not have to pass through the same type of control more than once.

(x) Any control established at an airport for screening of passengers and hand baggage should be sufficiently remote from the boarding gate as to provide maximum restriction of unauthorised access to aircraft. The last control which passenger should pass through is security.

(xi) Flow routes should be planned to give visual continuity to the maximum possible extent, e.g., from the baggage control to customs, or from checkin to immigration. Such continuity assists passenger's understanding of the flow system.

(xii) Features which cause hesitancy, such as ambiguous terminology on signs, flow routes which appear to lead in the wrong directive and multi-directional junctions should be avoided.

(xiii) The speed of flow and capacity of the passenger route should be matched to that of other systems such as baggage flow, aircraft turn-around time etc.

(xiv) Baggage flow routes are those parts of the system which are specifically for baggage handling when it is separated from passengers. Baggage and passengers should be so planned that they come together at expected points.

(xv) In planning, the aim should be to ensure that capacity satisfies demand within practical economic limits and to provide capability for increased capacity as demand increases with traffic growth.

(xvi) The capacity of the public corridor in passenger building depends upon the walking speed, the lateral distance per person and headway distance between persons in the direction of flow. It could be calculated using the following formula :

$$CC = \frac{WS}{Wo \times HD}$$

where :

CC = Corridor capacity (number of persons per minute per metre width)
WS = Walking speed (normally 75 m per minute)
Wo = Width occupancy (0.6–0.8 m per person)
HD = Headway distance between persons (1-2 m)

(xvii) Each component of the passenger facilities should be carefully planned in view of its needs, and required capacity. The appropriate measurement of capacity may not be the same for all individual facility. The capacity required for each facility is determined by the rate of flow to it, the average passenger processing time and the acceptable delay factor.

(xviii) Noise problem is severe in air transportation, and therefore, the terminal building should be almost sound proof towards airside. Control of noise should be achieved through proper design of walls, roofs and other measures.

15.7 PASSENGER BUILDING CONCEPT

The final step in passenger building involves the assessment of the size of facility and their arrangement in optimum relationship to each other and in accordance with the general principles of planning and design described earlier. The requirement and size of facilities vary according to the amount of traffic, type of aircraft operated, nature of operations and number of airlines operating. Information for determining passenger building requirements should be obtained from all present and potential users of the facilities including the airlines, general aviation interests, concessionaires, airport management etc.

The selection of a passenger building concept must be made jointly with the selection of the aircraft parking system discussed in previous chapter on aprons. The following concepts should be considered in the development of passenger building plan :

    (a) Simple concept
    (b) Linear concept
    (c) Pier (finger) concept
    (d) Satellite concept
    (e) Transporter concept.

15.7.1 Simple concept

The simple building concept consists of a single common waiting and ticketing area with several exits into a small parking apron. It is adaptable to airports with low airline activity and is also adaptable to general aviation operations whether it is located as a separate entity on a large airline-served airport or is the operational centre for air airport used exclusively by general aviation. Where the simple building serves airline operations, it will usually have an apron which provides close-in parking for a few commercial transport aircraft; however, due consideration should be given for jet blast effects against the building when a nose-in or nose-out parking configuration is adopted for jet transport aircraft. Where the simple building serves general aviation only, it should be within convenient walking distance of aircraft parking areas and should be adjacent to an aircraft service apron. The simple building concept will normally consist of a single-level structure where access to aircraft is by walking across the apron. The layout of the simple building should take into account the possibility of linear extension for future expansion.

15.7.2 Linear concept

The linear building concept may be regarded as an extension of the simple building concept, that is, the simple building is repeated in a linear extension to provide additional apron frontage, more gates and more space within the building for passenger processing. Passenger and baggage processing can take place in a central area of a terminal (centralization), but when the terminal becomes larger with increased number of aircraft gate positions the problem of long walking distances arises. The problem can be solved by installation of mechanical devices, such as people movers, or by decentralization of some passenger and baggage processing facilities. Complete decentralization would allow passenger and baggage check-in and baggage claim at the individual gate and thus afford very short walking distance between curb-side and aircraft, but construction and operation become costly. The degree of decentralization of processing facilities must be determined after careful study of volume and type of traffic, and of construction and operation costs.

The linear configuration lends itself to the development of adequate close-in public parking. Ample curb frontage for loading and unloading ground transportation vehicles can be provided with each extension of the linear building. Linear buildings can be expanded with almost no interference to passenger processing or aircraft operations. Expansion may be accomplished by linear extension to the existing structure's air-side corridor or by developing two or more linear building units connected by an air-side corridor. The loading of aircraft may be accomplished by nose-in/push-out operations with or without passenger loading bridges.

15.7.3 Pier (finger concept)

The finger or pier concept evolved in the 1950s when gate concourses were added to simple central buildings. Since then, very sophisticated forms of the concept have been developed with the addition of hold rooms at gates, passenger loading bridges, and vertical separation of the ticketing, check-in function from the baggage claim functions. However, the basic concept has not changed in that the main central passenger building is used to process passengers and baggage (a centralized system, although waiting lounges in most cases are dispersed at each gate position along piers) while the pier provides a means of enclosed access from the central building to aircraft gate. Aircrafts are parked at gates along the pier as opposed to the satellite concept where they are parked in a cluster at the end of a concourse.

Walking distances through pier buildings tend to become long. Curb space must be carefully planned since it depends on the length of the central building and is not related to the total number of gates afforded by piers. This is particularly true of deplaning curbs near centralized baggage chain facilities. Although the pier concept has afforded one of the most economical means of adding gate positions to existing buildings, its use for expansion should be limited. Existing piers should not be extended at the expense of taxiway manoeurability nor should new piers be added without providing adequate space for passenger processing in the main building. Most successful additions are effected by extending the main building and then increasing the number of piers.

15.7.4 Satellite concept

The primary feature of the satellite concept is the provision of single centralized terminal with all ticketing, baggage processing, and ancillary services except waiting lounges, which is connected by concourses to one or more satellite structures. The features of the satellite concept are very similar to those of the pier concept except that aircraft gates are located at the end of a long concourse rather than being spaced at even intervals along it. Satellite gates are served either by common or by separate hold rooms. The concourse can be elevated or located underground, thereby providing space for ground service equipment and aircraft taxi operations between the main building and the satellite.

Because the distance from the main building to a satellite is usually well above the average distance to gates found with the pier concept, a people-mover system or some other mechanical devices are often used to reduce walking distances terminal and satellite. There is no direct relationship between the number of gates and curb space so that special care should be taken in the planning of enplaning and deplaning roadways serving the central building to prevent curb overloads.

Building developed under the satellite concept are difficult to expand without reducing apron frontage or disrupting airport operations. Increases in building capacity, are therefore, usually effected by the addition of new units rather than expansion of an existing unit.

15.7.5 Transporter concept

Other passenger terminal concept include the transporter concept (also known as the remote aircraft parking concept) and the unit terminal concept. The former involves the vehicular transport of departing and arriving passengers and may be combined with other concepts to cater for peak hour demands. The latter is one where the individual compact module units are built around a system of interconnecting access and service roads. The buildings are spaced some distance apart under this concept, with each building providing complete passenger processing and aircraft parking facilities. Consideration of the unit terminal concept is usually feasible only for the larger airports.

15.8 LEVELS CONCEPT IN PASSENGER BUILDING

Passenger building concepts can also be considered by the level(s) on which passenger arrival, processing and departure takes place following four typical configurations are common in use as shown in Fig. 15.5.

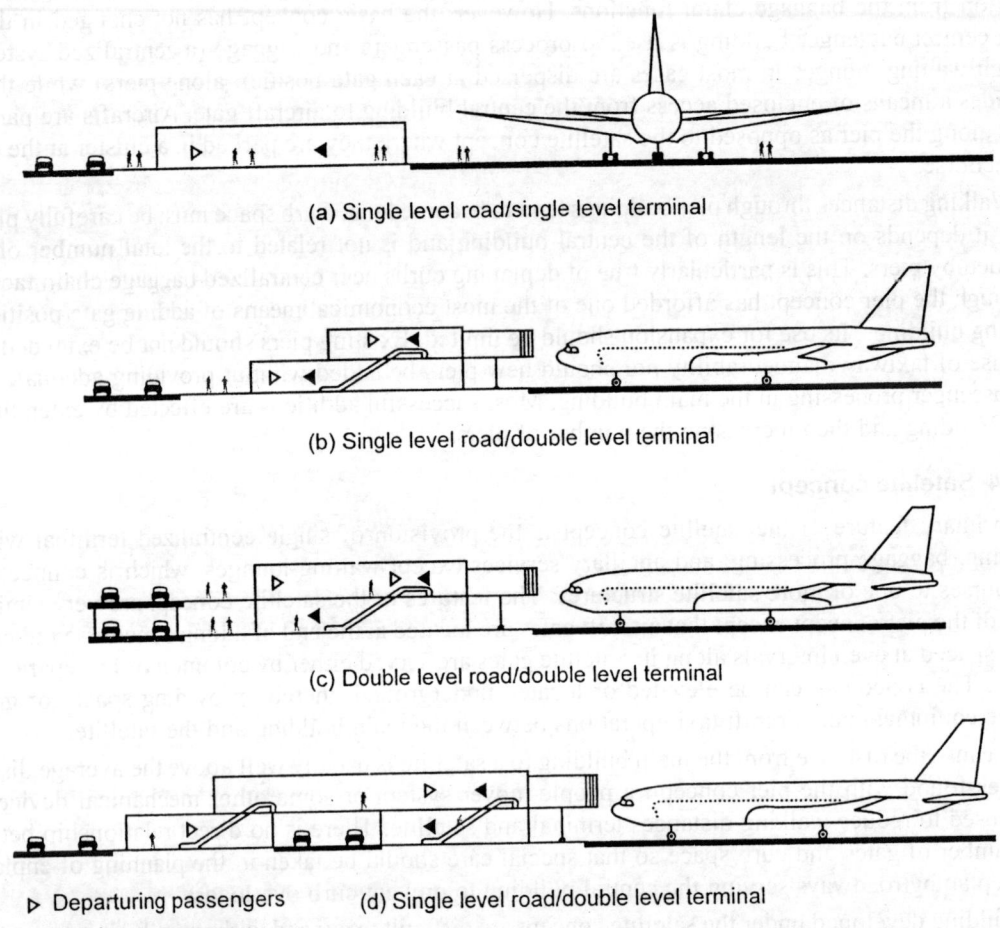

(a) Single level road/single level terminal

(b) Single level road/double level terminal

(c) Double level road/double level terminal

▷ Departuring passengers   (d) Single level road/double level terminal
◁ Arriving passengers

Fig. 15.5. Typical arrangement of levels for processing (ICAO).

15.8.1 Single-level road/single level terminal

In this concept arrival and departure processing in the terminal is done at the same level, but is separated horizontally. Passenger boarding to aircraft is by means of stairs.

15.8.2 Single-level road/double level terminal

In this concept arrival and departure processing in the terminal is normally at grade, with departure lounges on a higher level, permitting the use of passenger loading bridges or of transporters with level change capabilities.

15.8.3 Double level road/double level terminal

Access roads and curb side are on different levels to allow vertical separation of arrival and departure processing in the terminal. Usually, the upper level is for departure and the lower level is for arrival. In the process of developing a terminal concept, the desired degree of centralisation or decentralisation of passenger and baggage processing facilities within the terminal should also be considered.

15.8.4 Single level road/Double level terminal

This is variation of above, with access road and curbside for arrival and departure separated horizontally (laterally), but not vertically.

A centralised design will mean that all passenger and baggage processing facilities are centralised by common use by all gate positions at a terminal, while in a completely decentralised design all of the processing facilities are available at each gate position for the exclusive use. Complete decentralisation affords advantages such as shorter walking distances, efficient passenger and baggage flow, less chances of mishandled baggage etc. but it may turnout to be uneconomical due to under-utilisation of personnel, equipment and space.

The variations and combinations of these two approaches are also used, which often result from the changing conditions experienced at an airport during its life span. In time, the amount of traffic may increase necessitating modification or expansion of the facilities. Growth of aircraft size, a new combination of aircraft types serving the airport or a change in the type of service may affect the suitability of the initial concept, and so also the location and limitation of the site may cause modifications, additions or combinations of basic conceptual form of a terminal building. Fig. 15.6 shows concept combination and variations used in airport terminal designs.

15.9 PASSENGER PROCESSING FACILITIES

15.9.1 Check-in concourse

The area between the passenger building entrance and check-in positions is called the check in concourse. The check-in concourse should be designed so that counters and individual airlines are clearly visible immediately upon entering the passenger building. Provision for seating in this area of the building should be minimum to avoid congestion.

The space between the landside entrances and check-in position should be sufficient to provide free access to check-in and other facilities. Check-in concourse sizing is a function of total length of airline ticket counter, frontage, queuing at counters and allowance for lateral circulation without congestion. It also depends upon visitor/passenger ratio. Higher ratio may require bigger size.

Although check-in is the primary activity is this area, other allied functional activities such as ticket sale counter, standby registration and information counter, currency exchange counter, airport tax and passenger service charge counters etc. are also located in this area clear of the primary flow streams.

It is necessary to ensure that passengers arriving just before the designated final check-in time can be processed without delay.

The type and number of counter positions required depend upon, design hour enplanement, contact ratio, passenger arrival distribution patterns, average process time for each counter activity and service goals of an individual airline for specific types of counter position.

The minimum facilities will be required if the aircraft operators share traffic, frequency of operation and check-in positions homogeneously and any passenger can check-in at any position for any flight. Homogeneous use of all check-in positions provides the greatest passenger convenience and ensures highest utilisation of check-in positions, and therefore, requires minimum provision of these facilities and building space.

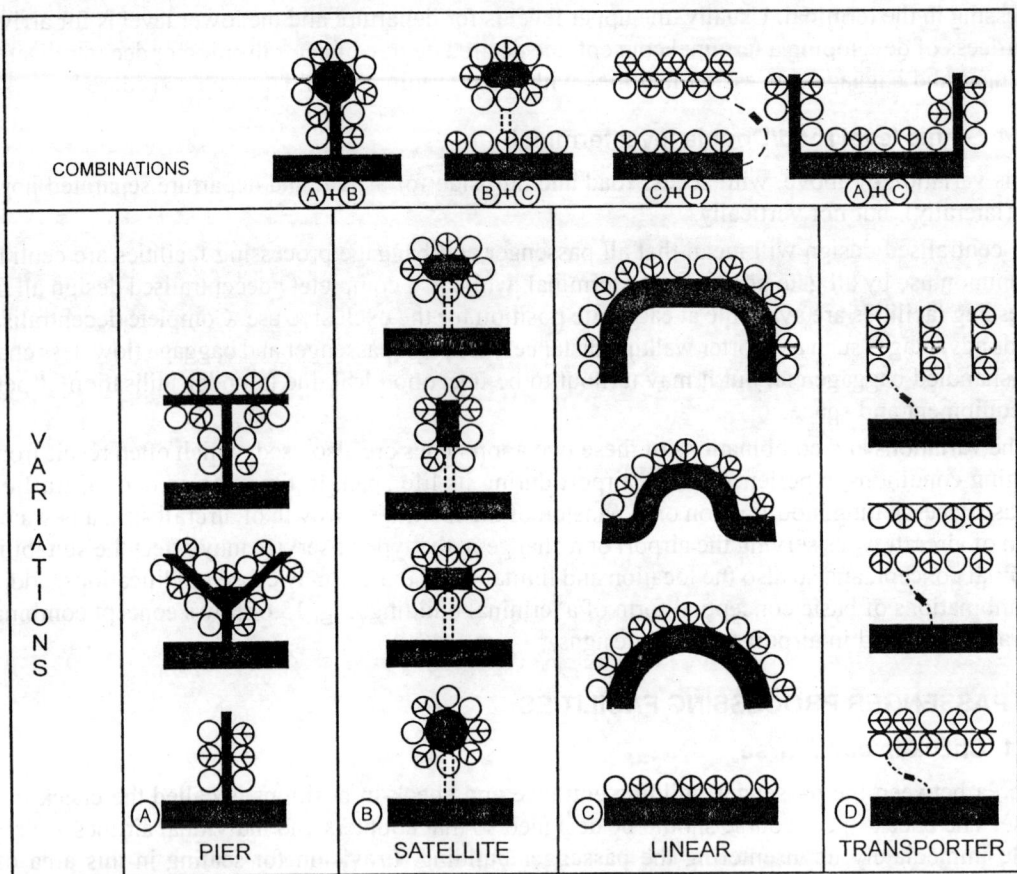

Fig. 15.6. Concept combination and variations.

The conventional check-in system of manual ticket control and baggage weighing labelling is used on small airport these days. Many airport now-a-days use computer check in system and already computerized departure control systems. The airport planner should be aware of common use terminal equipment (CUTE) which is a generic airline industry term for a facility which allows individual airlines to access their host computers and to share passenger terminal handling facilities. A current development is the elimination of baggage weighing and thus the need for scales. This already applies to some routes where the passenger baggage entitlement is fixed as a specified number of pieces of defined size. This new system can affect passenger building by imposing different space requirements for the check-in positions.

15.9.2 The check-in concepts

Following three concepts are adopted in check-in system :

(a) Centralized check-in concept

Passengers and baggage are processed at check-in counters located in a common central area, usually the departure concourse of the terminal. The counters may be of different configurations and may be divided into sections specially designated for individual airlines (airline base) or flights (flight base) or alternatively passengers may be free to check-in at any counter positions (common base).

(b) Split check-in concept

The check-in function is split between two or more locations within the terminal complex, e.g., baggage may be accepted at check-in counters on the lower level and seat assignment takes place at the waiting lounge on the upper level of the terminal.

(c) Gate check-in concept

Gate check-in is normally directly related to the decentralized passenger terminal concept. By this system passengers and baggage are processed at check-in counters located very close to an aircraft gate position (or a few positions in case of semi-decentralized passenger terminal concept) and its waiting lounge. This concept can afford advantages such as short distances, simple check-in handling for both passengers and baggage, etc. However, economic aspects should be well taken into consideration, since the facilities and personnel tend to be under utilized during off-peak hours.

The check-in counters, may be divided into following three types of configurations :

(a) Linear counter

This is most frequently used ticket counter configuration at low volume airports, where an agent can perform ticket transactions, check in baggage and other operational services. Multipurpose positions reduce number of servicing stops and afford flexibility in staffing, in non-peak periods.

During the peak-periods and at high volume airports, single function positions become more common, to expedite processing of passengers.

(b) Flow-through counters

This concept is successful when specialised for baggage check-in, where passengers queue along the baggage input, complete their transactions and walk through to a lobby or circulation area beyond. The advantages are reducing the cross-circulation and increasing baggage take-away capability. With a flow through counter, the out-bound baggage system becomes complicated, because of the greater number of individual inputs and the difficulty of merging multiple inputs into a single transport conveyor or sorting device, thereby increasing investment and maintenance cost for baggage systems.

(c) The island counter

This concept combines some features of both the flow-through and linear system arrangements. The agent positions form a "U" around a single conveyor belt (or pair of belts) providing interchangeability between multipurpose or specialised functions.

Fig. 15.7. shows typical passenger check-in positions. Check-in facilities should be located so as to enable passengers to check in at the earliest possible moment and permitting the latest possible arrival at the airport before flight departure. This will enable passengers to be relieved of their baggage at the earliest opportunity.

The layout of the check-in facilities is influenced by the two considerations. Preservation of the straightness of the parallel flow and minimum distance to the airside.

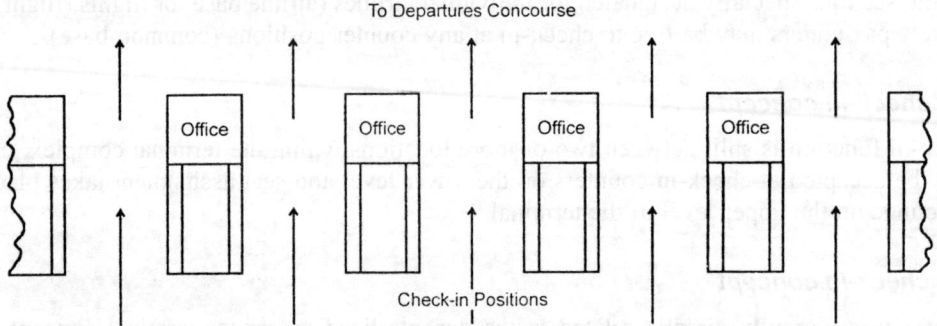

Fig. 15.7. Passenger check-In positions and check-in concourse (ICAO).

In the check-in concourse and waiting areas passengers have to be informed when their aircraft is ready for boarding and when delay occurs. This is generally done by loud speaker announcements and visual presentation of flight information. They should be located so that flight information is visible from all principal parts of these areas.

15.10 BAGGAGE PROCESSING FACILITIES

Baggage flow refers to these parts of the system which are specifically for baggage handling – the baggage which is separated from passengers. Various factors to be taken into account are :
- baggage and passenger flow should be matched in speed and capacity;
- flow routes should not conflict with passenger or vehicular flows;
- flow routes should be accessible so that baggage can be recovered at various stages;
- the flow system should involve a minimum number of individual handling operations, e.g. transfers between different types of vehicles, etc., and the flow should be steady and uninterrupted;
- passengers should have an opportunity to check their baggage at the earliest possible point;
- baggage claim systems should provide continuous presentation to passengers and permit them to recover their baggage personally;
- flow routes may be influenced by the type of handling system adopted, e.g., manually or mechanically propelled trucks, conveyor belts, etc.;
- palletized systems should be compatible with aircraft baggage holds and loading systems;
- security checks of baggage should be performed prior to flight check-in;
- The departing baggage, after being checked in must be sorted with flight groups as the destination or particular aircraft etc. At some airports baggage may have to be submitted for custom inspection. Such provisions should be made.

Fig. 15.8 shows a typical passenger/baggage flow.

The choice and layout of baggage handling system depends upon the size and nature of the traffic space available and local considerations such as the cost, availability of labour and equipment for operation. The baggage flow system should be designed so that persons not concerned with processing of baggage, will be denied access to the baggage. Under certain circumstances, however, it may be necessary for passengers personally to identify and search their own baggage prior to loading in order to ensure that nothing has been placed surreptitiously in the baggage. Facility for this has to be provided.

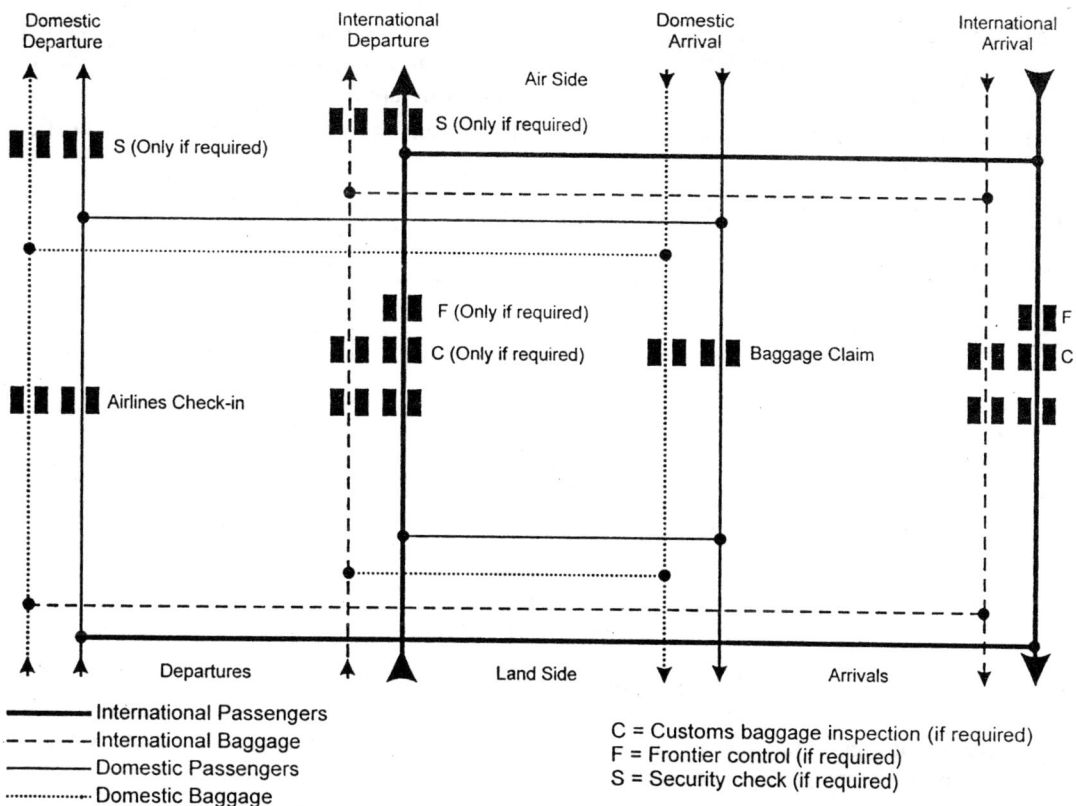

Fig. 15.8. Passengers/baggage flow (ICAO).

Baggage is transported between passenger building and aircraft by road vehicles or special trolleys formed into trains and towed by prime movers. The large aircrafts are equipped for the carriage of baggage in containers which are loaded and emptied in the passenger building. The type and size of container vary between aircrafts.

Baggage claim on arrival

A variety of manual, mechanical and semi-automatic baggage claim systems are available for use. Manual systems which rely on passengers moving to their bags are satisfactory, where number of baggages or passengers in claim area is small.

As passenger flow rate and aircraft size increases, a baggage claim system should be arranged, by having passenger remain in the principal flow streams and presenting their baggage to them, on equipment arranged in a 'comb' across the line of the flow. For high flow rates, baggage is presented on a moving display, such as a revolving turntable or belt which passes infront of the passengers. Facilities should be provided for storage of baggage of delayed passengers or misrouted and unclaimed baggage.

At international airports, passengers proceed from baggage claim to custom baggage inspection. Flow streams through the custom control should be arranged so that passengers with goods to declare do not hold up passengers without dutiable goods to declare. At many airports two separate channels are provided :

(a) one green channel for passengers having with them no goods or only goods which can be admitted free of import duty and tax and which are not subject to import prohibitions or restrictions. Random checks to these passengers may be applied.

(b) The other red channel of other passengers. This stream should flow through custom officer in normal way.

Fig. 15.9 shows a typical custom inspection system.

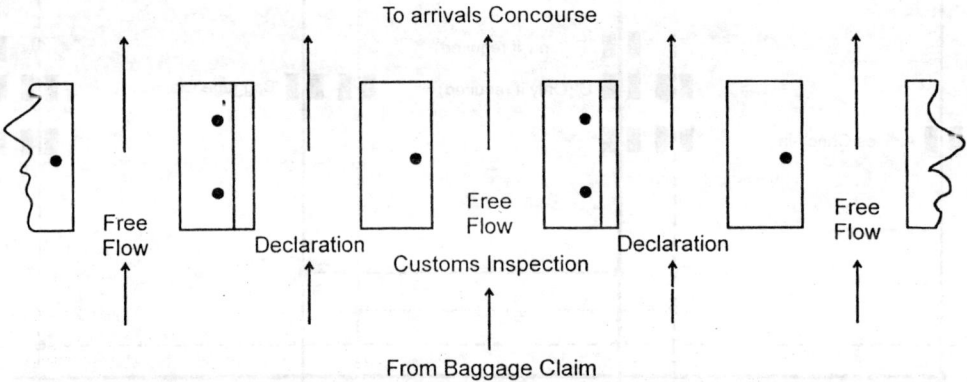

Fig. 15.9. Customs inspection (ICAO).

Delivery of baggage to the claim area, at the rate comparable with the passenger flow is one of the most important elements of airport operation.

Transferring passengers between international flights do not claim their baggage until they reach their final destination. The baggage of such passengers should be identified in the baggage vehicle loading area and transferred directly to the departures baggage sorting area for integration with all other departure baggage.

15.11 PASSENGER WAITING AREAS

In some passenger processing system, waiting areas at or close to the aircraft gates may be required. The capacity of such areas is determined by number of passengers to be accommodated at any one time, levels of comfort considered appropriate, average time spent in the area, climate and local customs. The capacity should be sufficient to absorb the difference in flow rates between check-in and aircraft boarding.

Fig. 15.10 shows a typical layout for waiting areas. To maintain the straight possible flow routes, the waiting area should be of the same general length as the departure concourse. Entrances should be provided for each main flow stream. Space for passenger circulation within waiting area should be adequately provided.

15.12 GOVERNMENT CONTROLS

The most commonly applied government exit controls are immigration, health and police, however, at some airports, custom inspection of passengers and their hand baggage is also imposed. For such inspection and controls, following factors should be considered :

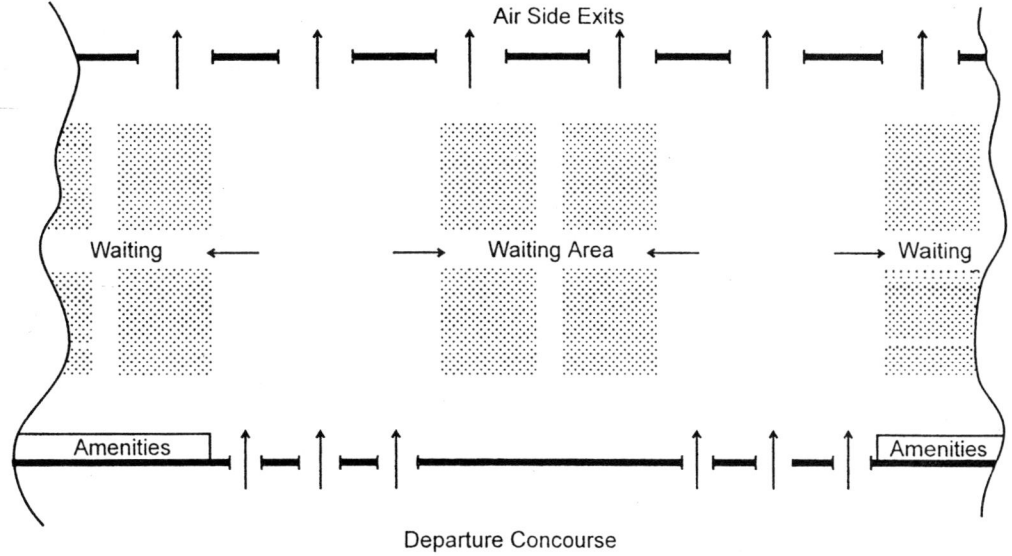

Fig. 15.10. Waiting areas (ICAO).

- Passenger flow from aircraft to the control facilities should be as short and direct as possible and unimpeded by obstructions.
- Passenger circulation should be designed so that there is no cross-circulation between international passengers and domestic passengers.
- It should be so located that there is no possibility for passengers to bypass the control facility and thereby avoid inspection.
- Physical contact between international passengers and visitors to the airport should not be permitted, once they have cleared controls.
- The capacity of control counters should be such to maintain minimum service time maintaining good flow rate in view of number of passenger to be inspected.
- Health controls are often integrated with immigration controls. The location of passenger medical inspection facilities should be immediately adjacent to, but at the side of other controls.
- Passport clearance often includes or is associated with, police inspection. This control requires considerable time in opening passports, searching visas and entry stamps etc. It is preferable for inspections requiring the longest time to be placed first, thus the delays at the second control will not cause obstructions at the first control, and minimum distance may be provided between the two controls.
- At some governments, varying degree of inspection documents, depending on the category of traffic and nationality of passenger are required. A faster overall flow and some economy in number of counters can be achieved if some counters are allocated for the use of only of these categories of passengers who are subject to minimal inspections.
- The control authorities may require offices for search and interviews at control/inspection points.

Search and interview rooms require absolute privacy of both sound and vision. In providing these, it is most important that visual continuity through the passenger flow route is not obstructed.

Figs. 15.11 and 15.12 show typical control facility for outgoing and incoming passengers.

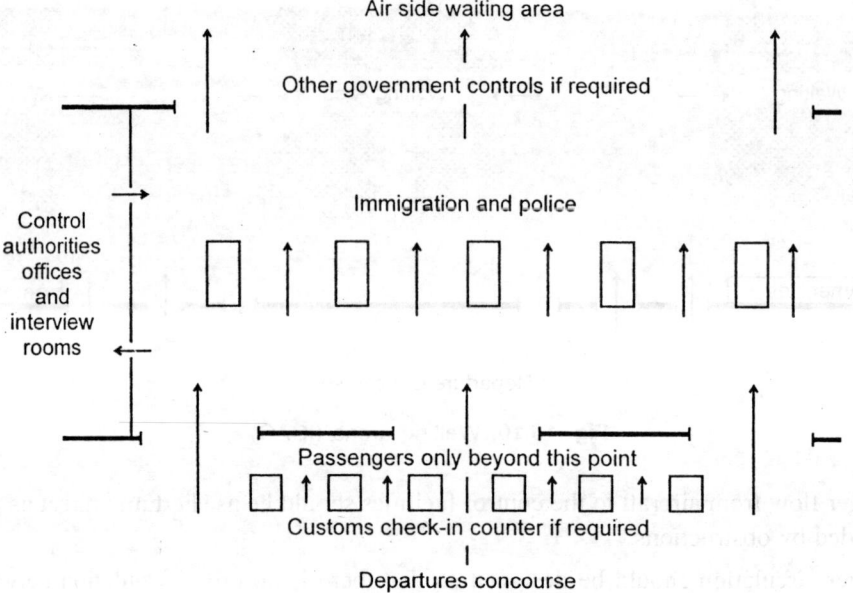

Fig. 15.11. Departure controls (ICAO).

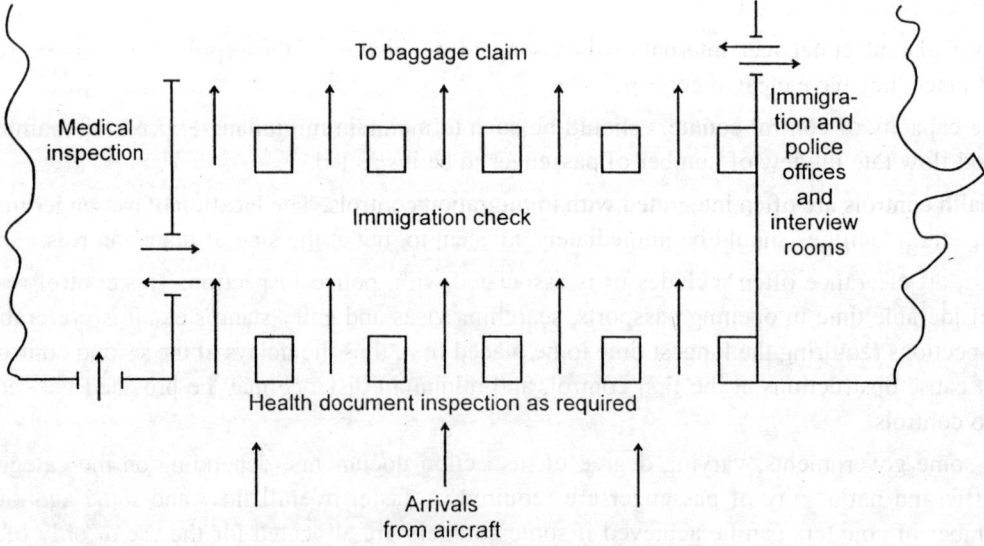

Fig. 15.12. Arrival controls (ICAO).

15.13 PASSENGER CONNECTION WITH AIRCRAFT

15.13.1 Airside exits

The connection between the passenger building and aircraft should be arranged to form the passenger flow into a linear pattern of a narrow front compatible with the size of the aircraft or apron and passenger vehicle door. Some control is necessary to ensure that only authorised persons and bona fide passengers are allowed to pass to the air side and board the aircraft.

15.13.2 Aircraft gates

The gates may be no more than doorways giving access to the aircraft stand, but they can also accommodate a number of departure facilities including gate check-in. The type depends upon nature of traffic, flow rate and processing system adopted for the passenger building.

In a trickle flow system the passengers flow freely, at their own speed, through all parts of the route, ideally, in the case of departures, a flow directly into aircraft. A waiting area for passengers is often necessary, for servicing the aircraft cabins on arrival.

For quick aircraft turnaround it is necessary for passengers to be at the gate ready to board the aircraft immediately when it is available.

Provision of waiting areas at the get reduce the requirement of the waiting areas in passenger building. The size and layout of waiting areas depend upon functions to be performed, location of waiting area, climate etc.

15.13.3 Forward waiting areas

Forward waiting areas generally serve three purposes namely, the provision of passenger lounge, passenger processing area and passenger deplaning area.

(i) Passenger lounge area

The lounge comprises seating, processing and circulation areas. The area is required to serve passengers to be in lounge 15 to 30 minutes prior to boarding the aircraft. All passengers will not have to sit in the lounge and some may want standing area. Based on experience, as aircraft arrivals and departures and distributed overtime so that full passenger population is never experienced at one time. It is, therefore, common to combine lounge area for a number of aircraft gate positions, the total area can be reduced to 20 to 30 per cent based on four to six gates.

(ii) Passenger processing area

The passenger processing area, depends upon the number of airlines and standard of service that specifies minimum waiting and processing time for the passengers. The average depth of lounge area, generally considered reasonable is 7.5 to 9 m. Length will depend upon airline agents, public facilities and means for delivering baggage.

(iii) Deplane area

The deplaning area is a corridor for deplaning passengers leading from the building entrance door (from the apron, loading bridge or transporter) to the public corridor without interfering with the passengers waiting in the departure lounge.

To allow for two passengers and baggage side by side or one passenger and baggage with passing space, an acceptable width is 1.5 to 1.8 m. The length is a function of the depth of the departure lounge itself and may include transition area for visitors.

15.13.4 Passenger security check

The location of the passenger security check is dependent upon traffic characteristics and the terminal concept. The security check can be centralised, partly decentralised or completely decentralised. A centralised security check can be located at the point in the terminal where passengers are separated from the general public i.e., after immigration control. In case of a fully decentralised system, the security check is made at the entrance to the waiting lounge.

15.13.5 Connection between passenger building and aircraft

The system for moving passengers between the passenger building and the aircraft is an integral element in the choice of aircraft parking system and apron plan. As described earlier, the systems used for this purpose include having passengers walk up to boarding stairs or along a passenger loading bridge or conveying them in a transporter. The route may be over the open apron, through enclosed routes at or below apron level or at passenger building and aircraft floor levels. Any specifically defined route over which passengers walk, other than over an apron is a "pier". Thus a pier can be at, above or below apron level.

For boarding stairs, the passenger flow rate for aircraft in the 40–210 seat capacity range is approximately 20 to 22 persons per minute and 25 persons per minute for the 220–420 seat capacity aircrafts. In the later case the rate of boarding or disembarking can be increased by the use of more than one door.

Passenger loading bridges can provide quicker and more even passenger flow between aircraft and passenger building and protect passenger from weather, noise and fumes. The installation of passenger loading bridges, however, should be justified by traffic volumes and other considerations. The size and form of passenger loading bridges should be choosen to provide sufficient flexibility to serve different type of aircraft other than those forcasted. The width of passenger loading bridges should be sufficient for at least two people to walk side by side. The floor slopes should generally not exceed one in ten.

The type of passenger loading bridges are fixed pedestal, apron drive or suspended and its length are functions of variables including apron dimensions, wing span, door locations, fixed aircraft services and economics. Normally only one passenger loading bridge is required to serve any aircraft up and including the B–747. If two passenger loading bridges are to be used there should be a separate tunnel to the terminal for each bridge or alternatively, a double door corridor from the junction of two bridges to the terminal building. The minimum width of double corridor should be 3.2 m.

Transporter vehicles are used when aircrafts are parked remote from the terminal. They range from a specially designed bus in combination with stairs to a specifically designed vehicle with an elevating capability For these vehicles with elevating capability, special attention should be paid to their relatively low speed, lack of easy manoeuverability and potential hazard, they may be to aircraft operation. These vehicles have high capital, operating and maintenance costs and require highly skilled drivers.

15.3.6 Airside entrance (transit and transfer passengers)

From piers or transporter loading positions passengers go into the passenger building. In no circumstances should arrival flow routes pass through departure areas in the passenger building. The airside entrance should therefore, give access directly to the arrival areas of the building. These may be on a lower floor in multi-story building or by the side of the departure areas in single story building. In multilevel buildings the descent should be direct, obvious and easy.

The airside entrance should arrange to separate passengers into appropriate flow streams i.e., transit and transfer passengers as well as passengers ending their air journey, as shown in typically in Fig. 15.13.

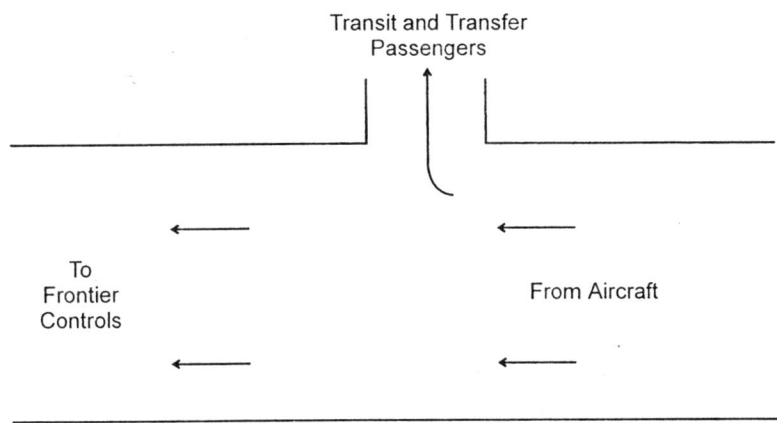

Fig. 15.13. Air side entrances (ICAO).

Transit passengers stay at airports only for the duration of the aircraft turnaround. They should follow the main arrivals route, until being diverted into the departures airside waiting area or into a 'in transit' waiting hall, where all aminities which they may require are provided on departure of their flight, they follow the normal routes and procedures of departure passenger, including security check if required.

The flow route for transfer passengers depends upon whether the transfer is between flights of the same or different categories, i.e., domestic to domestic, international to international or between international to domestic. When it is between international and domestic, transfer passengers are subject to the normal arrival controls and should follow the main arrival route to the landside, where they then pass through the main departure flow route for domestic departures.

When traffic is entirely domestic or international, transfer passengers should not pass through arrival controls. They should be segregated from the main arrivals flow and pass directly to the departures airside waiting area, usually following the same route as transit passengers.

Unlike transit passengers who leave the airport on the same flight on which they arrive, the transfer passengers change flights and it may be necessary for them to check in for connecting flights. This can be undertaken either at the gate or preferably on the route to the departures waiting area.

15.14 PASSENGER AMENITIES AND OTHER SERVICES

In the passenger terminal building considerations must be given to planning, design and location of passenger amenities, concessions and other services as described below.

15.14.1 Passenger amenities

Amenities should be located to ensure that they do not interfere with the primary flow stream of pas-

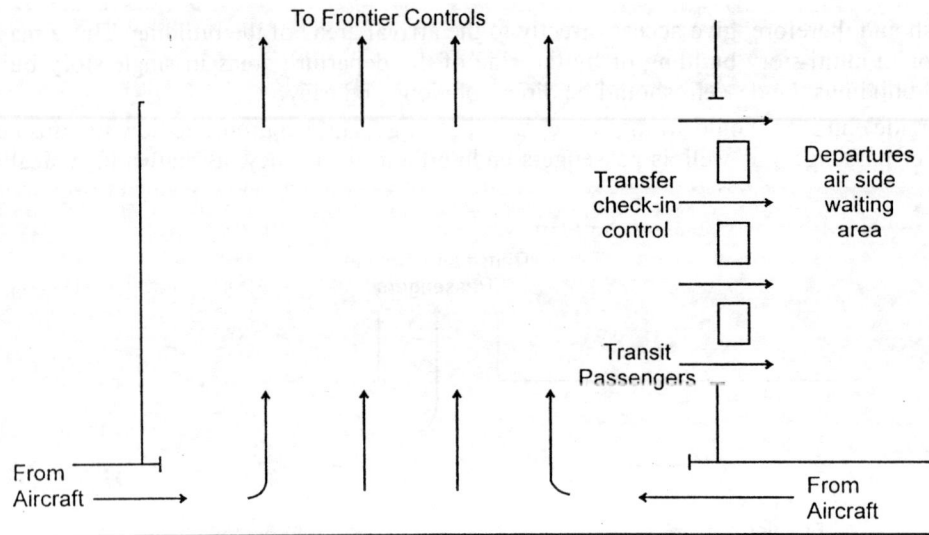

Fig. 15.14. Transfer passengers (ICAO).

sengers through the building and they are appropriately sited relative to each other. For example duty free and liquor shop can be located adjacent to the main flow routes for easiest access and provision of fast services. Amenities are provided in departure concourse and waiting areas.

Food and beverage services include snack bars, coffee shops, restaurants and bar-lounges. They may be provided at one or several locations depending upon the size of airport, building concept surrounding community, and amount of passenger traffic. Vending machines for beverages and other items should be considered to supplement staffed facilities at small airports, where traffic volumes might not justify operating during all hours.

Other amenities to be provided include newspaper/book stores, gift and apparel shops, barber and shoe shine, car hire counters, insurance office, left luggage locker, public telephones, displays etc.

15.4.2 Other passenger building services

Other passenger facilities which are common to most public buildings include public toilets, airport management offices, airport police and security office, medical aid facilities, travellers aid, building maintenance and storage facilities, space for heating and refrigerating plants, public address system room, government offices, post offices etc. Most of these facilities are required to be provided regardless of traffic volume.

15.4.3 Considerations for disabled and elderly people

Due to the speed and comforts, the air travel is expected to be patronised, increasingly by physically handicapped and elderly people, including the chair bound. Their special problems and different needs should be recognised, to be accommodated in planning and design for safety and convenience.

Ramp should be provided for wheel chairs with easy gradients such as 1 in 6. The transition between air and surface transport needs special considerations. It is desirable to provide identified reserved parking areas for physically disabled people, with proper signs and direction indicators. The route from reserved parking to terminal should be free of obstructions.

Wheel chairs should be available to people to move to taxi, bus or private car loading areas. The service should be clearly advertised.

Hand rails should be provided wherever required. There should be at least one main entrance without steps usable by people in wheel chairs. Automatic opening of doors may be highly desirable.

All floors should be maintained in a non-skid conditions. The only really effective way of moving chairbound people from floor to floor is by elevator which should be large enough to accommodate a wheel chair and one or two standing persons. An audio description of the floor reached may be desirable. Directional signs to the lift should be placed at various points in the building.

It is good to provide sign posting of "special" facilities for disabled people. A pictorial symbol effectively advertises the availability of facilities for the disabled.

Toilet facilities should be accessible to wheel chair users. Audible and visual means should be used liberally to provide guidance to various areas and facilities.

Various design standards developed for such special facilities are beyond the scope of this book.

15.15 PASSENGER BUILDING CONNECTION WITH ACCESS SYSTEM

The components of this system are the following :

15.15.1 Vehicular traffic lanes, through, bypass, curb and manoeuvring lanes

The required curb lengths and the vehicular traffic lanes will influence the passenger building configuration. The vehicular curb areas at the passenger building are required for efficient unloading of departing passengers and their baggage and for the efficient on loading of arriving passengers and their baggage.

15.15.2 Signs for direction and identification

Direction and identification is most necessary at this point of connexion to facilitate an orderly flow of passengers to their desired locations.

A set of uniform signs is developed by ICAO for use at international airports throughout the world. These are meant to facilitate air travellers in locating various facilities and services such as telephones, check-in-counters, baggage claim areas, post offices, toilets, banks etc.

15.15.3 Curb-side layout

To make provision of sufficient space of a suitable shape for vehicle unloading is an important element of passenger building planning. For shortest flow route the unloading points should be as close as possible to the first processing positions in the passenger building. For straight and direct flow, it should be possible to enter the building directly from the unloading points at any point along its frontage.

Proper planning and analysis should be done to establish relationship between the possible number and location of terminal openings, the terminal functions which they connect, and the total required curb length, to achieve economy, efficiency and passenger convenience. The curb spaces for buses, courtesy car are usually designated areas and thus controlled ones. Similarly queue lines for taxis are designated and controlled. The loading and unloading of passengers by private vehicles and unloading by taxis can not be rigidly controlled, however, each vehicle should occupy a curb space only for the time it takes to load or unload passengers and baggage.

The required curb length can be calculated as follows :
(a) Determine design hour passengers enplaning and deplaning. Identify the design period for deplaning passengers within the peak hour (10 or 20 minutes).
(b) Determine the percentage of transfer passengers of the total, and deduct from the total design hour requirement to find the number of passengers entering the airport using the road system.
(c) Determine the modal preference by vehicular type.
(d) Determine the percentage that go directly to parking facility and do not use the curb system.
(e) Determine the visitor ratio of passengers to visitors, and apply to the percentage of passengers using private vehicles.
(f) Determine occupants per vehicle and the average curb dwell time for that type of vehicle.

The total calculated curb lengths need to be related to actual terminal layouts for both enplaning and deplaning.

15.16 CARGO FACILITIES (CARGO TERMINAL)

The same considerations which influence the sitting of the passenger facilities (terminal) also apply to the cargo area. In most airports at present, the amount of cargo traffic is considerably less than passenger traffic, and therefore, movement of cargo aircrafts is not high. However, for planning and design of cargo facilities, forecasts for future demands for cargo traffic must be considered.

The rapid growth rate of increase in cargo traffic, the advent of very high capacity aircraft capable of accommodating increased quantities of cargo and new developments in cargo handling methods including the use of containers and automated equipment, need for flexibility and expansibility is an important aspect of cargo facilities planning.

Uniform standards for air cargo facility design are not possible to develop, due to requirements of storing, customs, types of cargo, mix of national and international carriers, warehouse needs and the degree of containerised traffic. A single design concept cannot meet the varying needs of all carriers or all locations. However, some general guiding principles to be followed in planning, sitting and designing of a cargo terminal listed below :

• Collect all possible information related to past, present and future cargo traffic from airline sources;
• Determine the impact of cargo, mail and company stores upon the facility;
• Determine the desired material handling system based upon the nature and volume of the forecast traffic and the operating method best suited to the particular locality;
• Design the terminal to accommodate the ultimate material handling system and with the ability to be progressively expanded within the confines of the building or site;
• Ensure that the site area will accommodate required aircraft stands, truck loading areas and customer/employee parking areas and desired access/egress roads, and will allow for future expansion;
• Site the terminal with the consideration for the type of operations (all-cargo or combination) and to provide the shortest possible time for the movement of on-line, interline and in-bound cargo;
• Provide sufficient space to accommodate the maintenance of fixed or mobile equipment, and the maintenance, parking and refuelling of powered ground equipment;
• Maximize interval overhead dimensions, as well as clear floor space, to allow optimum utilization of available cubic capacity and to accommodate multi-level unit-load handling and/or storage;

- Limit the amount of administrative area occupying warehouse floor space to the absolute minimum and consider second level administrative accommodations wherever feasible;
- Consider means to prevent the unauthorized removal of air cargo and equipment;
- Provide adjustable or flexiblke connexions at air side and land side to accommodate fixed loading bridges, mobile ramp equipment and variable delivery/collecting vehicle heights;
- Provide terminal building bypass means to transfer unitized loads or large single pieces between air side and land side; and
- Make adequate provision for holding or staging areas for unitized loads, including cargo containers and lower deck containers, both of which have specialized handling requirements. In the case of lower deck container, particular attention should be paid to the necessity of ensuring that such facilities will enable the containers to be handled expeditiously at all times, including periods of unavoidable multi-operations, as it is essential to keep aircraft ground times to a minimum.
- The site selected for cargo terminal and facilities should be flexible and expansible to take care of future needs over a period of 20 years;
- The cargo terminal and facilities should be easily accessible from ground transportation. The taxiing distances for aircraft between the terminal and the runway should be as short and direct as possible.
- The cargo terminal should take into account the prevailing winds and location of aprons and cargo building should not obstruct or interfere with imaginary surfaces and navigational aids.
- Provisions should be made for handling of large containers and pallets between trucks and cargo terminals and between cargo terminal and aircraft.
- In planning air cargo, both types of flows should be considered namely :
  (i) flow of documents, and
  (ii) flow of cargo itself.
  Fig. 15.15 shows the principles of achieving a continuous and direct flow of air cargo within the cargo terminal.
- Cargo processing facilities must be capable of supporting export (outbound) and import (inbound) functions and activities.
- The aprons planned should suit the cargo handling methods to be used. Nose-in or tail-in parking with fixed mechanical loading systems minimise the size of the apron.
- To ensure efficient cargo handling, the cargo apron must be treated as a continuation of cargo terminal, with sufficient reserve space for future needs.

15.16.1 Cargo facility requirements

The basic facility requirements in a cargo terminal are the following :
- Separate areas for inward and outward consignments.
- Adequate space for presentation, opening and examination of air cargo for customs.
- Adequate space, near to the final delivery area, for repacking of air cargo after customs examination.
- Enough warehousing areas.
- Weighing facilities.
- Cold storage for perishables and foodstuffs.
- Strong rooms for valuables and bullion.
- Storage space for human remains.
- Holding areas and accommodation for animals and livestock.

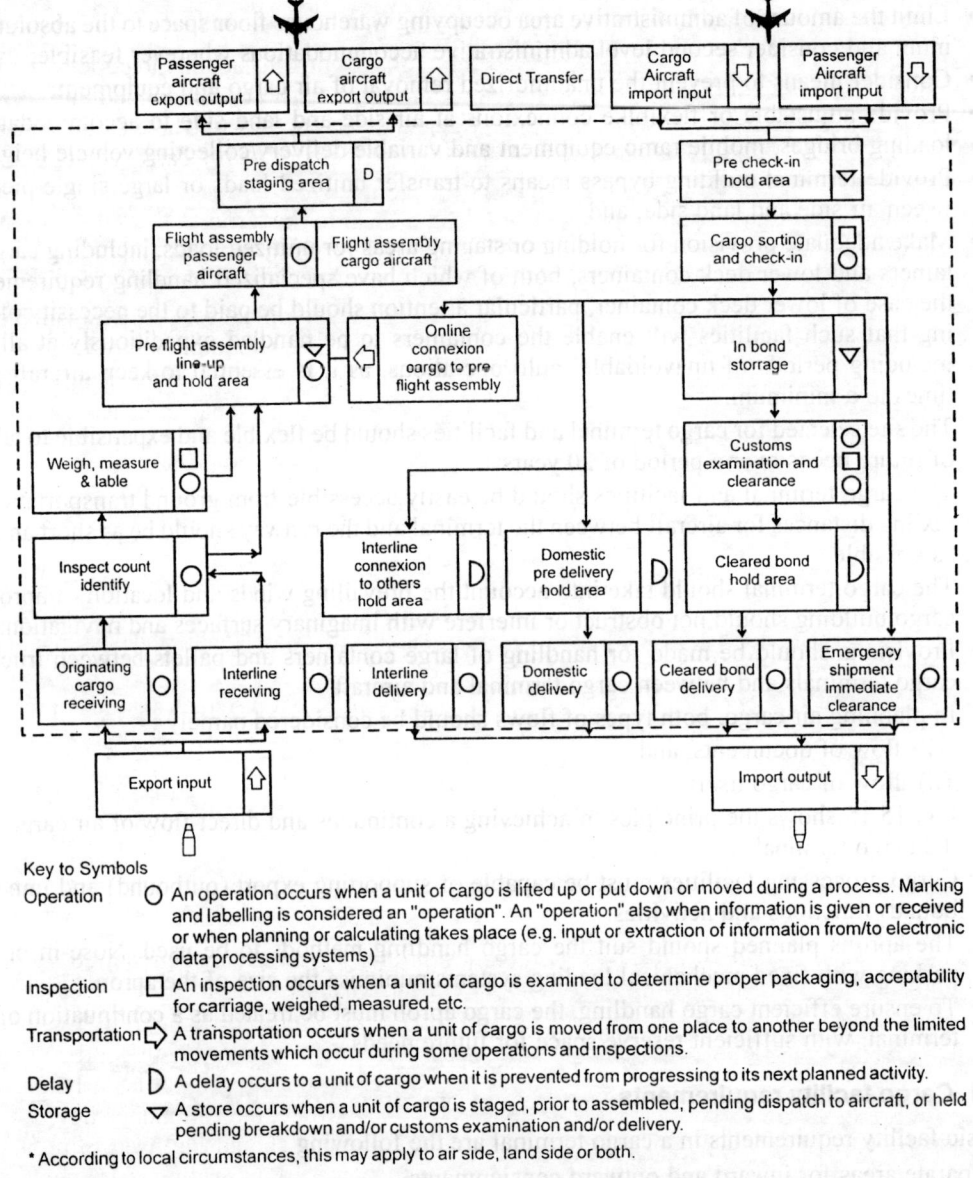

Fig. 15.15. Flow in cargo terminal (ICAO).

Key to Symbols

Operation ○ An operation occurs when a unit of cargo is lifted up or put down or moved during a process. Marking and labelling is considered an "operation". An "operation" also when information is given or received or when planning or calculating takes place (e.g. input or extraction of information from/to electronic data processing systems)

Inspection □ An inspection occurs when a unit of cargo is examined to determine proper packaging, acceptability for carriage, weighed, measured, etc.

Transportation ⇨ A transportation occurs when a unit of cargo is moved from one place to another beyond the limited movements which occur during some operations and inspections.

Delay D A delay occurs to a unit of cargo when it is prevented from progressing to its next planned activity.

Storage ▽ A store occurs when a unit of cargo is staged, prior to assembled, pending dispatch to aircraft, or held pending breakdown and/or customs examination and/or delivery.

* According to local circumstances, this may apply to air side, land side or both.

- Parking and storage space for loading vehicles and other equipment.
- Public reception counters.
- Space and offices for control authorities, management, accounting functions, security etc.
- Aircrew functional space and rest rooms.
- Storage space for aircrafts and servicing facilities.
- Storage space for empty containers, pallets and ULD (unit load device).
- Workshop facilities for cargo handling equipment and battery charging facilities.

15.16.2 Cargo terminal area access for road system

Road system linking to the cargo terminal complex should have enough capacity to cope with the peak hour volume of pickup and delivery vehicles along with other traffic considering future growth. A truck access road, separate from passenger vehicle roads may be necessary, where cargo volumes are heavy. The pavements should have enough strength to carry loads of heavy cargo vehicles and container transporters.

There should be an easy access from the major road system to the airport. A connection from the road system direct to the cargo terminal should also be provided for the use of authorised airlines or commercial vehicles.

Convenient and direct access road should be provided to cargo terminal for vehicles which operate on the airside of the airport, ensuring the safety of personnel, vehicle and equipment from aircraft blast.

Adequate vehicle parking space should be provided on the landside of cargo terminal, considering present and future needs, for pickup and delivery vehicles as well as personal vehicles.

15.17 AIRPORT OPERATION AND SUPPORT FACILITIES

Following facilities for a variety of operational purposes are required at an airport :

   (i) *Administration and maintenance building* : For facilities such as offices for airport management, aircraft operators, government control authorities, police station, telephone exchange, maintenance depot etc.

  (ii) *Medical centre* : For treatment of medical emergencies (first aid) to staff and passengers should be at walking distance from passenger area.

 (iii) *Ground vehicle fuel stations* : For landside ground vehicles, it provides a good source of revenue for airport authority. A separate station for airport vehicles may also be justified.

 (iv) *Generating stations* : They may be required for heating, electricity etc. Standby generators may be necessary at some airports, independent of main airport power system, as a secondary power supply.

  (v) *Water supply and sanitation* : Adequate water supply properly processed and chlorinated and a sewage disposal system for handling and treating waste must be provided. Dump for refuse should not create a bird hazard problem.

 (vi) *Flight catering kitchens* : Fairly large facilities, properly located may be required for preparation and storage of food, drinks and other aircraft cabin stores.

(vii) *Meteorological services* : Meteorological information is useful to flight crew, weather radar for forecasting may be provided some times, at convenient location, with good communication with meteorological station. An unobstructed view of the airport, particularly the runway complex and good communication with meteorological office, the communication centre and local air traffic services units are essential. Considerations should be given to provision of necessary ducts to allow the satisfactory siting of sensors, distant reading instruments such as thermometer, anemometers etc.

(viii) *Aircrew briefing and reporting* : Before a flight can depart, aircrew may be required to undertake certain briefing, for which facilities should be provided in administration building.

 (ix) *Aircraft maintenance area* : Size of this area will depend whether the airport is a base for major maintenance or only for line maintenance, or for some combination of both.

  (x) *Rescue and fire fighting services* : Airport fire station should be located so as to ensure that response time for airport accident and incident is 2 to 3 minutes, to the end of each runway in optimum conditions of visibility and surface conditions. More than one fire station may be necessary some times.

(xi) *General aviation facilities* : General aviation includes many different type of facilities for aircraft like personal flying, private aeroplanes, air taxi and instructional flying. General aviation may involve fleet ranging from single-engine aircraft to multi-engine turbo jet. Facilities for general aviation activities both national and international should be provided, without causing unacceptable delays to normal departure and arrival of commercial flights. When possible, provide a separate runway and taxiway system and other facilities to serve general aviation type aircrafts exclusively, if general aviation operations are substantial in volume.

15.18 AIRCRAFT FUEL FACILITIES AND SAFETY

As handling of fuel at airports is an important, it deserves attention in planning and design of the airports specially because of safety, minimum gate occupancy times.

The storage capacity of fuel is estimated based on types of operating aircraft, frequency of operation, fuel uplift per aircraft and different types of fuel required, for a period of time, determined as a matter of policy based on distances to source of delivery and risks of disruption. In supply storage areas should be located as close to the aircraft fuelling area as practical.

15.18.1 Fuelling of aircrafts

Aircrafts are fuelled at their parking positions either in stands close to terminal building or at remote ones, by fuel servicing tank vehicles, fuelling pits or hydrant systems. Care should be exercised in locating hydrant outlets at the stands to ensure that they provide optimum flexibility and capacity. Sometimes combination of hydrants and tankers is used with advantage as large jet aircrafts require a considerable amount of fuel (nearly 70,000 litres for the Boeing 707–120 and DC–8 to almost 115,000 litres for Boeing–747).

Another method fuelling is to install pipelines running from a central fuel storage area located adjacent to the landing area to pits located at the aircraft stands on the apron. Fuel is transferred to the pit by pumps located at the storage tanks. The pits must be located near the fuel intake in the wings of the aircraft. The advantages of fuel pits are that a continuous supply of fuel is available at all times, it is carried safely underground and trucks are eliminated from the apron. At large airports trend is towards the hydrant system which requires a simpler installation than the pits while providing similar advantages. Essentially, the hydrant system consists of the same elements as the fuel pits, except that the pit is replaced by a special valve mounted in a box in the pavement and flush with the surface. The hose reel, meter, filter and air eliminator are contained in a mobile self-propelled or towed hydrant dispenser. In the hydrant system duplication of hose reel, meter, filter etc. which are required for each pit is eliminated. Number of hydrants required, per gate position depends upon type of aircraft and number of grades of fuel required, as each grade of fuel requires a separate hydrant.

Considerations should be given to the need for accessibility of emergency fire equipment when establishing aircraft fuel servicing locations and laying out of airport fixed fuelling systems. The fuel supply system at airports should be in accordance with standards prescribed by competent authorities. Fuel-resistant pavements should be used on aprons where refuelling operations and engine shut-downs are likely to take place. The surface must slope away from the face of the terminal building for proper drainage and safety in case of fuel spillage. Fuel piping should not run under buildings or passenger loading fingers. The apron surface materials should be graded to form a gradual slope away from the rim or edge of fuelling hydrants or fuelling pits to prevent flooding.

15.19 BLAST FENCES

Blast fences are used at airports to reduce or eliminate the detrimental effects of blast by deflecting the high air velocities, heat, fumes and noise associated with blast. The application of either fences or screens becomes necessary when it is impractical to provide safe, reasonable separation between the aircraft engines and people, buildings and other objects on the aircraft. Fig. 15.16 shows the typical airport locations requiring blast fences.

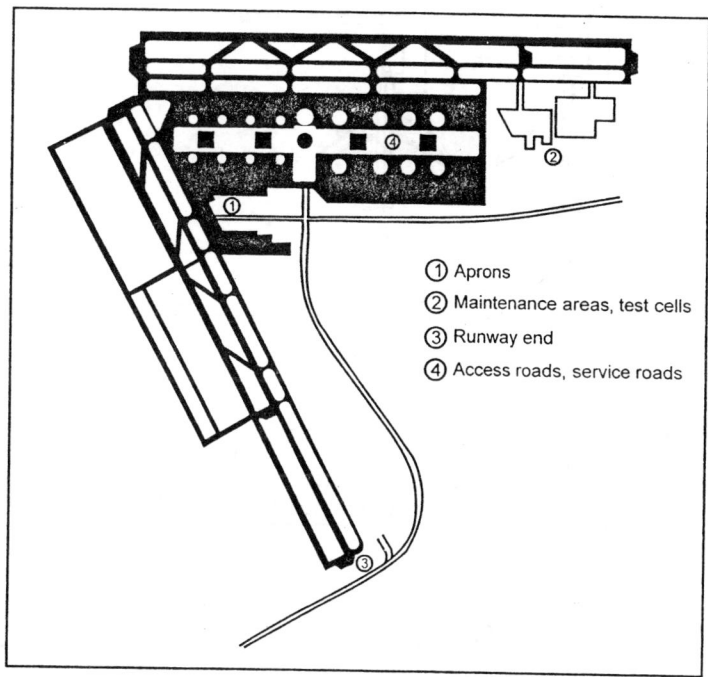

① Aprons
② Maintenance areas, test cells
③ Runway end
④ Access roads, service roads

Fig. 15.16. Locations for blast fences (ICAO).

15.19.1 Planning and location of blast fences

A complete study of the aircraft types, their movement pattern should be undertaken at a new or existing airport, for planning a system of blast fences. The whole airport movement area including the aprons, taxiways, holding bays and runways must be analysed for possible magnitude and orientation of blast.

The type of apron movement pattern used by aircrafts for entering or leaving a stand is critical factor in deciding the need and location of blast fences. Fig. 15.17 shows an example of the fence requirements in apron areas. Because an aircraft on self-manoeuvring stand must make a full 180° turn within the apron area under breakaway blast conditions, all areas along the public access road, service road and between aircraft parking positions can be subjected to excessive blast, requiring fences.

In off-apron areas also the blast fences should be used, where blast could cause a danger to personnel, building, equipment or other aircraft. They are often used along taxiways, to protect hangers or terminal facilities where the aircraft turns through 90 or 180 degrees.

Another critical location is the end of the runways, centred above the runway centreline, which should be examined closely because this area is subjected to the aircrafts maximum continuous thrust on take-off. The use of fences should not cause a hazard to movement of aircrafts or ground vehicles.

The blast protection can be provided by specially manufactured blast fences which are effective or by using other methods and materials. Tall hedges and other natural or manmade obstructions may be used with advantage in some cases such as around engine run-up areas.

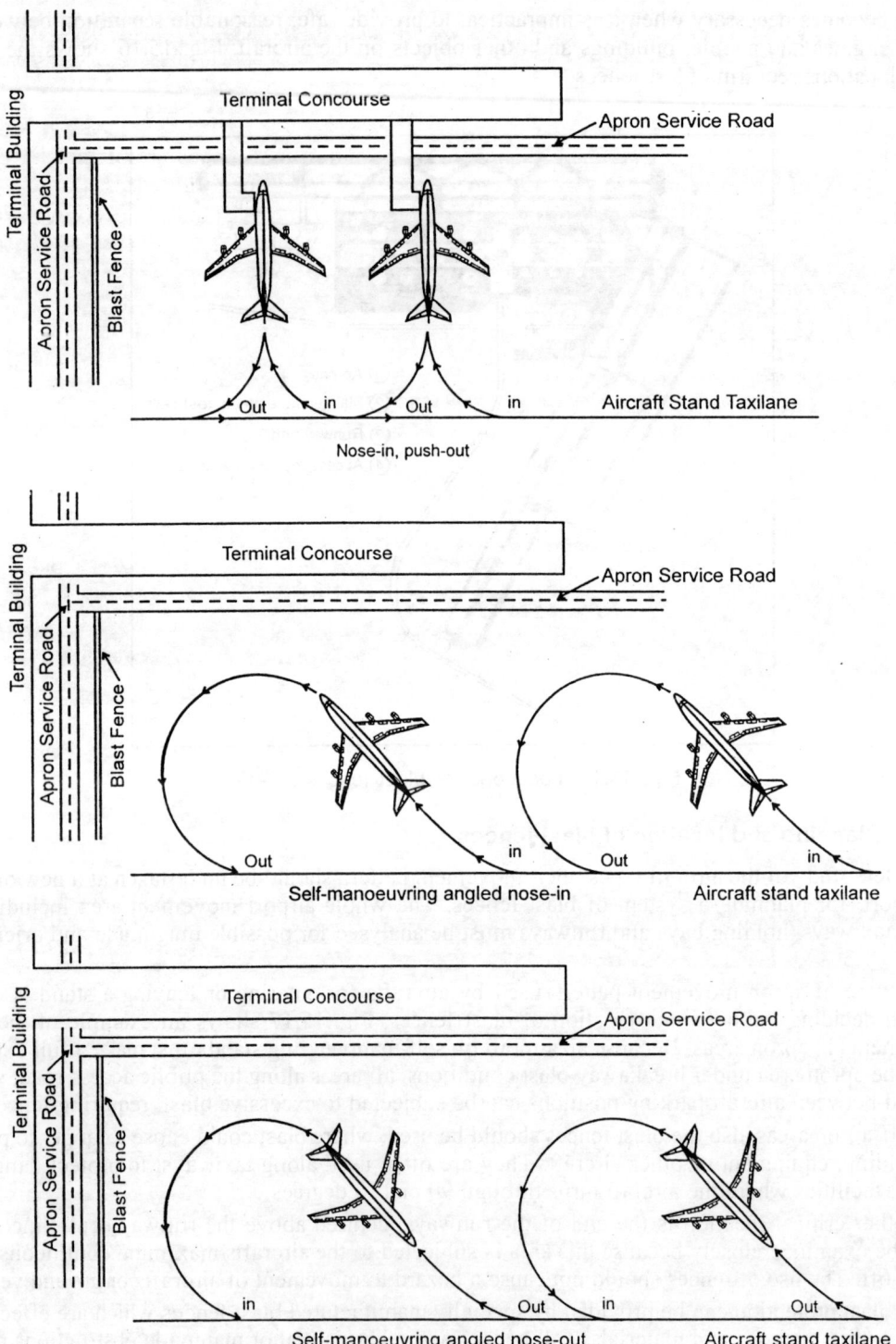

Fig. 15.17. Blast fences in apron areas (ICAO).

Blast fences are usually located only after the basic airport layout is planned. The design of blast fences depends more on architectural considerations to give a good appearance.

15.19.2 Types and design of fencing

The fencing material is either concrete or metal. Pre-manufactured fences are of metals. Concrete deflectors require lesser maintenance. They may be louvered fences or solid fences. Baffles, perforations, louvers and corrugations can be used singly or in combination. Fig. 15.18 shows several types of blast fences.

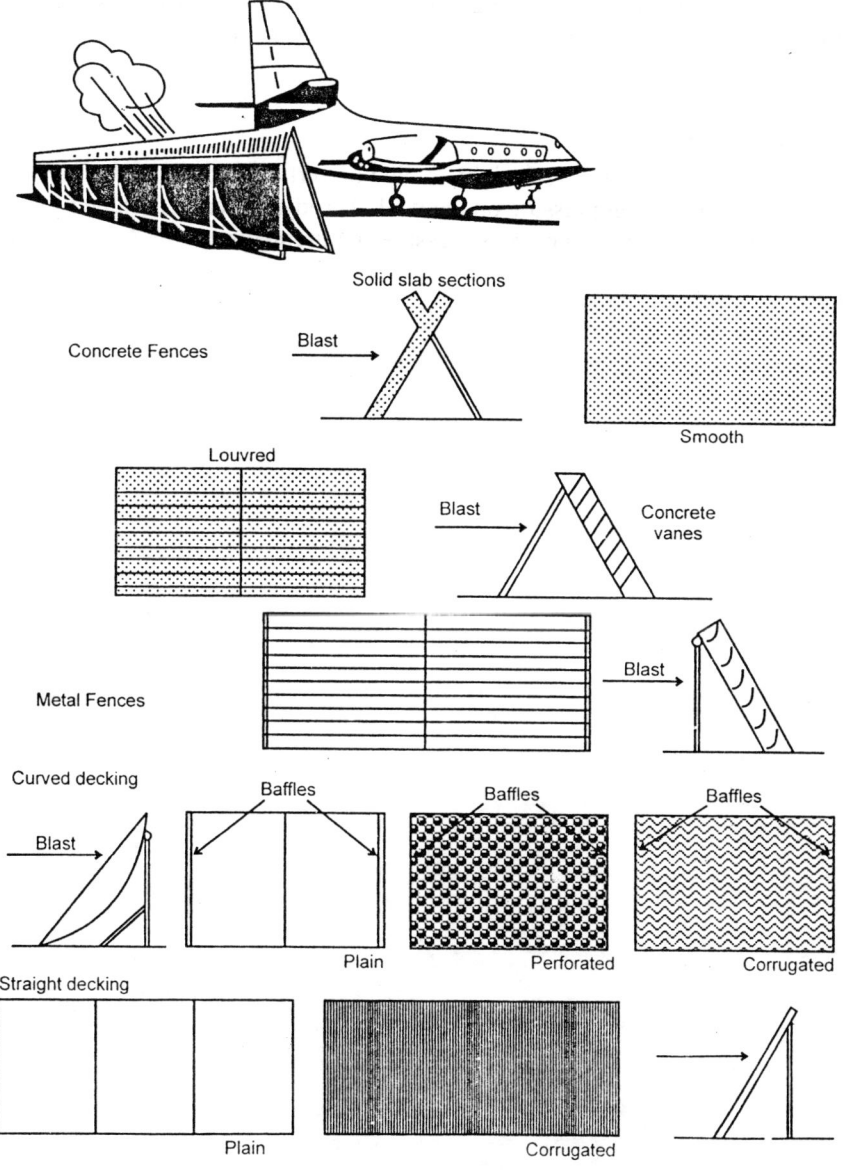

Fig. 15.18. Types of blast fences (ICAO).

The blast fences of all types should be structurally strong to carry the wind forces. The following factors need attention for structural design.

(i) Gross wind pressure

For a given fence location, the worst possible blast velocity should be determined. Wind velocity can be converted into pressure using standard curves.

(ii) Height of fence

The blast fence should be high enough to deflect the centre portion of blast. This height depends upon aircraft type and should be used in conjunction with the calculation of pressure to establish the critical fence section.

(iii) Shape and type of fence

The shape of the fence whether curved, straight, angled or vertical and the type of fence such as solid or louvered, will determine the net pressure on the wall. Aerodynamically designed shapes and the use of openings in the fence will reduce the gross pressure requirements.

CHAPTER 16

Maintenance, Evaluation and Rehabilitation of Airport Pavements

16.0 INTRODUCTION

Careful construction control, quality management and some degree of maintenance will be required to produce a pavement which will achieve the intended design life. Poor construction and lack of preventive maintenance will usually shorten service life of even best designed pavements. A systematic procedure should be developed for determining priorities, schedules, and techniques for maintenance and rehabilitation of airport pavements, making best use of the available funds.

Formation of cracks, patches, waves, corrugations, ruts and unevenness etc. are the causes of deterioration in flexible pavements. Cracks due to temperature stresses, uneven joints, mud pumping etc. are common causes of deterioration of rigid pavements. If immediate attention is not paid to these defects in pavements, they may result in failure of the pavements in serving the intended purpose.

16.1 MAINTENANCE MANAGEMENT SYSTEM

A "pavement management system" (PMS) is a systematic approach for providing consistent and objective evaluation of the pavement condition and establish the priorities and schedules for preventive pavement maintenance and rehabilitation. It involves maintaining continuous records of pavement condition and develop procedures for maintaining the pavement network to acceptable condition within budgeting restraints, thus making a cost effective decision regarding maintenance and rehabilitation.

As shown in Fig. 16.1, the rate of deterioration of a pavement accelerates with time. By adopting a maintenance and rehabilitation strategy, to upgrade the pavement condition at proper time, overall costs can be minimised. FAA observed that the total annual cost to maintain or rehabilitate a pavement in relatively poor condition can be 4 to 5 times that of maintaining or rehabilitating a pavement in relatively good condition.

An effective pavement management system (PMS) should include the following :

- A systematic technique of regularly collecting, storing and retrieving condition of the pavement and extent to which it is used, i.e., a record of traffic data using the pavement.
- Minimum acceptable standard for the maintenance of the airport pavement e.g., runways, taxiways etc.

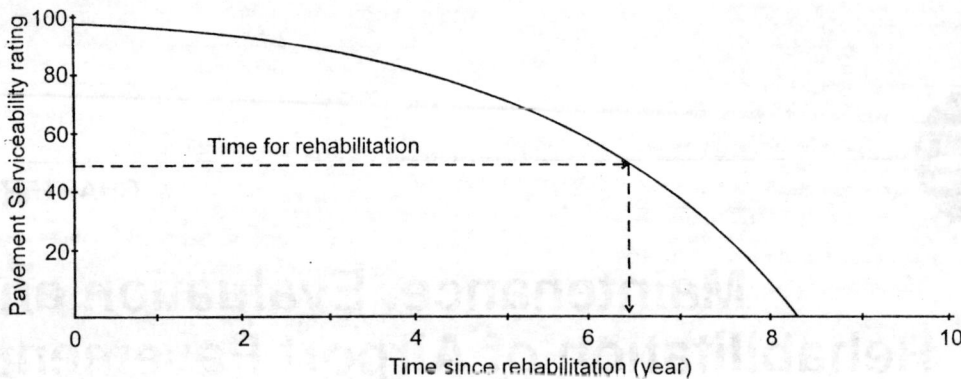

Fig. 16.1. Pavement serviceability rating versus time.

- Developing alternative maintenance and rehabilitation strategies.
- Mechanisms for predicting and evaluating impacts of different strategies on pavement condition, serviceability and life of pavement.
- Estimates of costs and different proposed strategies and alternatives.
- Comparing the costs of various strategies and adopting the one which meets the stated requirements.

A pavement rating system should be developed upon the quantity, severity and type of distress affecting the pavement surface condition. The periodic collection of condition-rating data is essential to tracking pavement performance.

FAA has developed a computerised PMS programme called Micro-Paver covering various aspects of pavement structural integrity and surface conditions.

16.2 DEFECTS IN AIRPORT PAVEMENTS

Defects in airport pavements are developed mainly due to poor quality control in materials and construction techniques or due to poor drainage. Through proper workmanship and technique of construction, and good quality material, pavement defects can be minimised. Several factors combine in development of defects, which deteriorate the pavement performance, and lead to the pavement failure. Defects in the pavements should be immediately rectified through proper maintenance/rehabilitation procedures. If adequate steps are not taken against development of defects in initial stages itself, it may result in failure of the pavement to perform the intended purpose.

Flexible pavement require greatest care in consolidation and selection of material for subbase, base and surface courses, as failure of any course, ultimately results in formation of defects and failure of the whole pavements.

Rigid pavements develop cracks and defects due to poor quality of cement, aggregates, inadequate compacting, curing and pavement thickness.

Excessive water due to poor drainage retained around and below the pavement, increases the detrimental effects on the pavements. Table 16.1 shows various defects causing factors in pavements.

Defects as mentioned above are removed or repaired through regular maintenance work for keeping the airport pavements in good condition for performing their function. If pavement is properly

Table 16.1. Common defects in flexible pavements

Defect/distress	Description	Possible causes
1. Pot holes	Formation of holes and patches	• Inadequate bond between layers
2. Alligator cracks	Cracks developed on the surface, like a map or alligator, in pattern	• Movement of materials in pavement layers • Fatigue failure of base • Moisture variations
3. Ruts	Depression/deformation of pavement longitudinally	• Cumulative deformation due to excessive wheel loads, applied repeatedly • Poor consolidation of one or more layers
4. Longitudinal cracks	Cracks developed in longitudinal direction throughout the thickness	• Differential volume changes in subgrade • Frost action • Settlement of fills, slopes etc.
5. Shear cracks	Upheaval of pavement resulting in fracture or cracking	• Excessive loading • Weak material in pavement mix
6. Reflection cracks	Cracks appearing on bituminous surface	• Poor bond with underlying layer, often seen in overlays on concrete pavements
7. Waves and corrugations	Waves and corrugation are developed on the surface, resulting in uneven surface	• Failure of one or more pavement layers • Localised depression due to excessive loads
8. Frost heaving	Localised heaving up of pavement portion	• Due to heavy frosts • Water and soil conditions
9. Broken pavements	Localised areas of the pavement are broken	• Deficient pavement thickness • Unstable subgrade • Poor drainage

Table 16.2. Common defects in rigid pavements

Defect/distress	Description	Possible causes
1. Structural cracking	Cracking on the surface of concrete slab, spalling of joints, poor riding quality, shrinkage cracks, disintegration etc.	• Structural inadequacy • Poor quality of cement, aggregate, filter and sealer • Poor workmanship • Inadequate curing
2. Shrinkage cracks	Longitudinal and/or transverse cracks	• Improper/inadequate curing
3. Mud pumping	Ejection of soil slurry through joints/cracks, or edges	• Slab deflection • Subgrade type/clayey • Presence of water, below slab
4. Warping cracks	Cracks at the edges of the slab	• Joints not well designed/reinforced
5. Spalling of joints	Excessive cracking and subsidence at joints	• Filler material in joints not placed properly
6. Scaling of cement concrete	Overall deterioration of concrete. Surface of the pavement becomes rough and ugly	• Poor mix design • Impurities in concrete mix • Excessive vibrations of mix
7. Temperature cracks	Cracks in interior region	• Due to temperature variations

designed and well constructed the maintenance work is considerably lesser, during the early years of pavement life.

When pavement totally fails to perform its functions in spite of regular maintenance, overlaying of layers, is considered economical by strengthening it through well designed and constructed overlays. If damages to the pavement are due to ponding of water, suitable surface and subsurface drainage system should be designed and implemented to remedy the defect permanently.

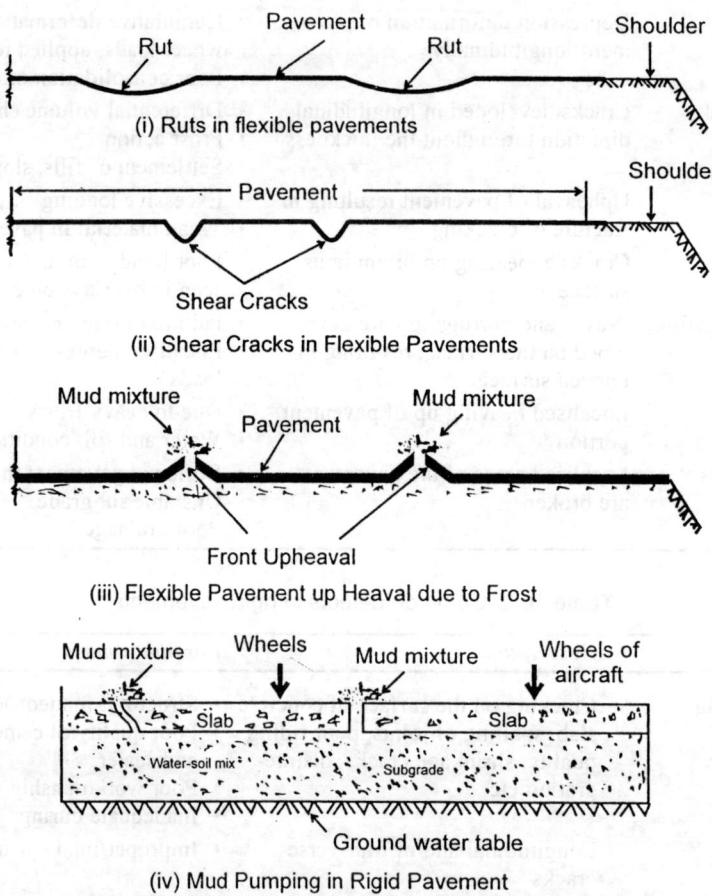

(i) Ruts in flexible pavements

(ii) Shear Cracks in Flexible Pavements

(iii) Flexible Pavement up Heaval due to Frost

(iv) Mud Pumping in Rigid Pavement

Fig. 16.2. Defects in airport pavements.

16.3 EVALUATION OF AIRPORT PAVEMENTS

Airport pavement evaluations are necessary to assess the ability of an existing pavement to support different types of weights and volumes of aircraft traffic. Evaluation procedures are the reversal of design procedures. In design, pavements are designed to support the expected load, while in evaluation pavements are tested for their ability to support the prevailing and expected aircraft traffic. It would, therefore, be appropriate to evaluate existing pavements by the same method, by which they were designed. Design begins with the aircraft loading to be sustained and the subgrade strength provides the necessary thickness and quality of material for the needed pavement structure. While evaluation begins with existing subgrade strength, finds thickness and quality of each component of pavement and uses design procedure pattern to determine the thickness of layers, their strengths and subgrade strength for the purpose of evaluation.

16.3.1 Purpose of evaluation

Pavement are evaluated for several reasons :

- To establish load carrying capacity of existing pavements.
- To assess ability of the pavement to support expected volume of air traffic.
- To plan and design improvements for upgrading the pavements to meet greater/expected volumes of traffic.
- Residual life of the existing pavement can be assessed through evaluation.

The load carrying capacity of existing bridges, culverts, drains and other structures should also be considered in evaluation, to assess overall capability of the airport.

16.3.2 Evaluation process

The evaluation of airport pavements is a methodical step by step process, involving the following :

(a) *Study of records* : A review of construction, design and maintenance records, testing methods, drawing and design details, traffic history is done, through available records and files.

(b) *Site inspection* : Site of the pavement is inspected and the condition of the pavement is noted by visual inspection. Evidences of adverse climatic, soil or other factors are also noted.

(c) *Sampling and testing* : Depending upon the study of records, site inspection and type of evaluation, physical tests are planned to be conducted. Two types of testing procedures are commonly used :

  (i) *Direct sampling procedure* : Laboratory or field tests like CBR, soil classification tests etc. are done. The thickness of various layers of flexible pavements must be known to evaluate the pavement. The flexural strength of concrete and the modulus of subgrade reaction are determined to know the thickness of rigid pavements for evaluation. Boring and test-pits may also be required to verify thickness of pavements. Plate bearing tests are performed on the subgrade to know the modulus of subgrade reaction (K). Consolidation test and moisture tests are performed in laboratory or in situ.

  (ii) *Nondestructive testing* : Several methods of non-destructive testing (NDT) of pavements are available. NDT eliminates subjective judgement needed in other testing procedures. The major advantages of non-destructive testing are :
  - The pavement is tested in place under actual conditions of moisture, density etc.
  - The disruption to traffic is minimum.
  - The need for other testing is minimised.

(d) *Other evaluation tools* : There are number of other equipments available for evaluation. These equipments are useful in determining pavement condition index (PCI). Subsurface conditions through ground penetrating radar – non-destructively and physical properties of pavements through infrared thermography, are determined.

(e) *Evaluation report* : The findings and analysis of the test results is incorporated in an evaluation report. Load carrying capacity of pavement, including an adjustment inview of the soil, moisture, weather conditions etc. must be included in the evaluation report.

16.3.3 Flexible pavement evaluation procedure

After all the parameters of evaluation process are established, the evaluation process is essentially the reverse of the design procedure for flexible pavements. The design curves are used to determine the load carrying capacity of the existing pavement, using the inputs of subgrade and subbase CBR values, thickness of surfacing, base and subbase, and annual departure levels. Several checks must be made to determine the load carrying capacity of the flexible pavements. The calculation which gives the lowest allowable load will control the evaluation.

The deficiencies in base course, surface course should be considered and plans for overlaying the pavement to correct the deficiencies are worked out.

16.3.4 Rigid pavement evaluation process

The evaluation of rigid pavements for aircrafts requires concrete flexural strength, 'K' value of the foundation slab thickness and annual departure levels as inputs. The rigid pavement design curves are used to establish load carrying capacity, as is done during the design phase.

If in the evaluation the existing pavement is found to be deficient, considerations should be given to the corrective action, through reconstruction or overlaying.

ICAO has developed a standardised method of reporting airport pavement strength known as ACN/PCN airport classification number/pavement classification number described earlier.

16.4 ELEMENTS OF PAVEMENT EVALUATION

Elements of the pavement for evaluation are the subgrade, its structure including all layers up to surfacing, and the weight and frequency of using aircraft.

16.4.1 The subgrade

Strength of the subgrade is a significant element and this must be characterised for evaluation. Subgrade strengths is dependent on soil moisture and must be evaluated in situ beneath the pavement structure. Subgrade strength established for a particular pavement may fall anywhere within the range indicated by four subgrade categories used in ACN-PCN method.

16.4.2 Pavement structure

A rigid pavement responds 'stiffly' to the surface loads and distributes the loads by bending or beam action to wide area of subgrade. The strength of the pavement depends on the thickness and strength of the portland cement concrete and any underlying layers above the subgrade. The pavement must be adequate to distribute surface loads so that the pressure on the subgrade does not exceed its evaluated strength.

A flexible pavement consists of a series of layers increasing in strength from the subgrade to the surface layer. Pavements meant for aircrafts usually have a bituminous bound wearing course. At each level from the surface to subgrade, the layer must have strength sufficient to tolerate the pressures at their level. A flexible pavement must also have thickness of structure above each layer to reduce the pressure to a level acceptable by the layer. In addition, the wearing course must be sufficient in strength to accept without distress tire pressures of using aircraft.

16.4.3 Aircraft loading

The aircraft mass is transmitted to the pavement through the undercarriage of the aircraft. The number of wheels, their spacing, tire pressure and size determine the distribution of aircraft load to the pavement. The pavement must be strong enough to support the loads applied by the individual wheels, not only at the surface and the subgrade but also at intermediate levels. With a subgrade strength category the relative effects of two different aircraft types on pavements can be uniquely stated with good accuracy.

16.4.4 Repetition of load and traffic composition

There is a fatigue or repetition of load factor which should be considered. Thus magnitude and repetition must be treated together and the pavement which is designed to support one magnitude of load at a defined number of repetitions can support larger load at fewer repetitions and a smaller load at greater number of repetitions. It is thus possible to establish the effect of one aircraft mass in terms of equivalent repetitions of another aircraft mass (and type). This concept permits the determination of a single (selected) magnitude of load and repetitions level to represent the effect of the mixture of aircraft using a pavement.

16.4.5 Pavement condition survey

An important part of the evaluation is a critical condition survey of the pavement. Pavement should be closely examined for evidences of deterioration, movement, defects of any kind. Observable effects of traffic along with an assessment of the magnitude and composition of that traffic can provide an excellent basis for defining the bearing capacity of a pavement.

16.5 EVALUATION BY PCN (PAVEMENT CLASSIFICATION NUMBER)

Pavement classification number (PCN) is an index rating (1/500th) of the mass which an evaluation shows can be borne by the pavement when applied by a standard (1.25 MPa tire pressure) single wheel load. The PCN rating established for a pavement indicates that pavement is capable of supporting aircraft having an ACN (Aircraft classification number) of equal or lower magnitude. The ACN for comparison to the PCN must be of the aircraft ACN established for the particular "pavement type" and "subgrade category" of the rated pavement as well as for the particular aircraft mass and characteristics.

Pavement type could be rigid, flexible or composite (PCC overlay on a flexible pavement or bituminous overlay on concrete pavement).

Subgrade category : ACN-PCN method uses four subgrade strength categories termed as high, medium, low and ultra low with prescribed ranges for the categories. Commonly one subgrade category may be appropriate for an airport. Subgrade strength evaluated, must be 'in situ' beneath the pavement.

Tire pressure categories : Directly at the surface the tire contact pressure is the most critical element of loading with little relation to other aspects of pavement strength. This is the reason for reporting permissible tire pressure in term of tire pressure categories. Rigid pavements do not require tire pressure restrictions. Tire pressure category is determined as low, very low, medium or high.

The ACN of the most critical aircraft should be reported as the PCN of the pavement. Thus any aircraft having ACN no higher than this PCN can use the pavement facility.

Pavements for light aircrafts : In evaluating pavements for light aircrafts – mass 5700 kg or less, it is necessary to consider the geometry of the undercarriage of aircraft or how the aircraft load is distributed among the wheels. Only maximum mass and allowable tire pressure is reported. Subgrade strength and pavement type are not necessary.

16.6 TECHNIQUES FOR EVALUATION

Process for defining or qualifying the bearing capacity of a pavement and its behaviour under load is called technical evaluation. Some of the evaluation methods are :

16.6.1 The early methods

Early methods for design and evaluation of "flexible pavements" were based on experience and extended by theory. Index type methods like CBR test, plate bearing tests and many others were used to assess subgrade strength and strengths of pavement layers, especially in highway design. They are still the methods in primary use for aerodrome pavements.

Early methods of "rigid pavement design" and "evaluation" made use of westergaard's model, including various extension to treat fatigue, ratio of design stress to ultimate stress, strengthening effects of sub-base or base layer etc. Westergaard's methods for loading at the centre of the pavement slab (width unlimited) and loading at the edge of the slab (otherwise unlimited) were used. Plate bearing tests are used to characterise subgrade or sub-base. Early method, further developed still remains the primary basis for aerodrome pavement design, and evaluation.

16.6.2 The newer (more fundamental) methods

Methods using stress-strain response of materials and rational theoretical models are now developed and continuing to develop. The advanced computer technology have made computer-oriented developments possible.

Most common newer design method is 'elastic layered system'. Layers are of finite thickness and infinite extent laterally except that the lowest layer (subgrade) is also of infinite extent downwards. Responses of each layer is characterised by its modulus of elasticity and poisson's ratio. Values of these parameters are determined by laboratory tests of several types, by field tests of several types with correlations.

These model permit the stresses, strains and deflections from imposed loads to be computed.

Finite element models, through interactive computional techniques are also being used, but these largely remains for research applications.

16.6.3 Direct load response methods

Theories applied to pavement behaviour indicated a proportionality between load and deflection, thus implying that deflection should be an indicator of capacity of a pavement to support load. Pavement deflection determined for a particular load could be adjusted proportionately to predict the deflection which could result from other load. This is the basis for pavement evaluation.

This established strong relationship between load and deflection has been used in pavement evaluation of flexible pavements. Methods based on plate tests were commonly used using 76.2 cm diameter plates. The Bankleman Beam method is used for evaluation of light aircraft pavements.

A variety of dynamic load equipment has been developed, using sensors which could merely be positioned on the pavement or load plate and would measure deflection. When dynamic plate load testing is carried out on existing pavements it is possible to measure the velocity of propagation of stress waves within the pavements. Modulus of elasticity of each layer can be deduced.

16.6.4 Evaluation of inversion of design

To design a pavement, one selects a design method, to determine the thickness and acceptable characteristics of materials for each layer and the wearing course, taking into account the subgrade on which pavement will rest and intensity of loading (traffic) pavement must support. For evaluation, the process must be inverted since the pavement is already existing. Character of subgrade and thickness and character of each structural layer and surface must be established, from which the maximum allowable magnitude and frequency of allowable aircraft loading can be determined by using a chosen design method in reverse. The elements of the design method in the existing pavement must be evaluated in accordance with the selected design method.

16.6.5 Dynamic methods

These methods involve a dynamic loading device which is mounted on vehicle or trailer and which is lowered, in position onto the pavement. These methods make use of inertial instruments (sensors) which when placed on the pavement surface or on to loading plate can measure deflection (vertical displacement). The dynamic loading is determined, usually by a load cell through which the load is passed on to the axle plate. Comparison of the load applied and displacements measured provide load deflection relations for the pavement tested.

Some dynamic methods can evaluate the strength or the stress-strain relationship of the subgrade and overlaying pavement layers for use in various design methods.

16.6.6 Non-destructive evaluation method

Non-destructive techniques (NDT) are use to evaluate the load carrying capacity of airport pavement, for making rapid evaluation with a minimum of interference to normal airport operations. At present NDT evaluation procedures are applicable to conventional rigid and flexible pavements. Work in USA, is in progress to extend NDT procedures to other types of pavement systems which incorporate thick bituminous surfacing and stabilised layers. A conventional rigid pavement consists of a non-reinforced concrete surfacing layer on non-stabilised base and/or subgrade materials. A conventional flexible pavement consists of a thin 15 cm or less thick bituminous surfacing layer on non-stabilised layer of base, sub-base and subgrade materials.

16.7 REPORTING OF PAVEMENT STRENGTH

For reporting information on pavement bearing strength of pavement, ICAO specifies following four elements and the PCN (pavement classification number) :

1. *Pavement type* :
   Rigid pavement (code-R)
   Flexible pavement (code-F)

2. *Subgrade strength* :
   High strength (A) – CBR > 15, K > 150 MN/m^3
   Medium strength (B) – CBR × 10–15, K > 80 MN/m^3
   Low strength (C) – CBR × 6–10, K > 40 MN/m^3
   Ultra low strength (D) – CBR > 3, K < 20 MN/m^3

3. *Tire pressure* :
   High tire pressure (W) – Unlimited
   Medium tire pressure (X) – 1.5 MPa
   Low type pressure (Y) – 1.0 MPa
   Very low tire pressure (Z) – 0.5 MPa

4. *Evaluation method* :
Technical evaluation (T)
Experience based (U)

PCN to be reported can be determined from the aircraft loads (masses) which the evaluation has established as maximum allowable for the pavement.

The classification is thus indicated by 50/F/A/X/T.

16.8 AIRPORT PAVEMENT OVERLAYS

Overlay means putting a new layer over the top of existing pavement. This may be required due to variety of reasons such as :

- A pavement may have been damaged by overloading and overlay is required to maintain service-able level.
- A pavement may require strengthening by overlaying to serve aircraft heavier than those for which pavement was originally designed, or increased repetitions.
- A pavement which is worn-out after serving its design life. may require an overlay.
- An overlay may be required sometimes to merely improve the riding quality of the pavement.
- Overlay may also be used to improve skid resistance and provide better safety.

In this section, we shall mainly deal with overlays required for structural purposes. Overlay pavements generally consists of either portland cement concrete or bituminous concrete, i.e. rigid or flexible type placed on an existing pavement.

16.8.1 Type of overlay pavements

(i) *Bituminous overlay* : Bituminous concrete pavement placed on an existing pavement.

(ii) *Concrete overlay* : Portland cement concrete pavement placed on an existing pavement.

(iii) *Bonded concrete overlays* : An overlay which is bonded to existing rigid pavement.

16.8.2 Design of bituminous overlays

Bituminous overlays can be used on either flexible or rigid pavements.

(a) *Bituminous overlay on existing flexible pavement* :

(i) use the appropriate basic flexible pavement design curved to determine the thickness requirement for a flexible pavement for the desired load and equivalent design departures, using the CBR value of subgrade and subbase, assuming that existing pavement does not exist. Thickness of all pavement layers must be determined.

(ii) The thickness of pavement required over the subgrade and sub-base must be compared with existing pavement to determine the overlay requirements. The thickness of the bituminous overlay is equal to the difference between the computed thickness and the thickness of the existing pavement.

(iii) Adjustment to the various layers of the existing pavement is then done, rather than using the entire overlay of bituminous type. Layer conversion i.e. converting base to sub-base etc. is largely a matter of engineering judgement.

CBR values of subgrade and sub-base are determined by conducting field in place CBR tests, at the equilibrium moisture content.

(b) *Bituminous overlay on existing rigid pavement* :
   (i) To establish the required thickness of bituminous overlay for an existing rigid pavement, determine the single thickness of rigid pavement required to satisfy the design conditions.
  (ii) This thickness is then modified by a factor 'F' which controls the degree of cracking which will occur in the existing rigid pavement. It is a function of the amount of traffic and the subgrade strength.

   The 'F' factor in effect indicates that entire concrete slab thickness as determined by design is not needed, because a bituminous overlay pavement is allowed to crack and deflect more than a conventional rigid pavement. Appropriate 'F' value is selected from the Fig. 16.3.
 (iii) Another factor 'C_b' is applied to represent the condition of the existing rigid pavement. 'C_b' factor is an assessment of the structural integrity of the existing pavement.

   A C_b value = 1 is used when existing slabs contains nominal cracking.

   C_b value = 0.75 when the slab contains multiple cracking.
  (iv) After the 'F' factor, condition factor 'C_b' and single thickness of rigid pavement have been worked out the thickness of the bituminous overlay is computed from the following formula :
$$t = 2.5\,(Fh - C_b h_e)$$
where :
   t = thickness of bituminous overlay in cms
   F = factor which controls the degree of cracking in the base pavement
   h = single thickness of rigid pavement required for design conditions in cms
   C_b = condition factor for base pavement ranging from 1.0 to 0.75
   h_e = thickness of existing rigid pavement in cms

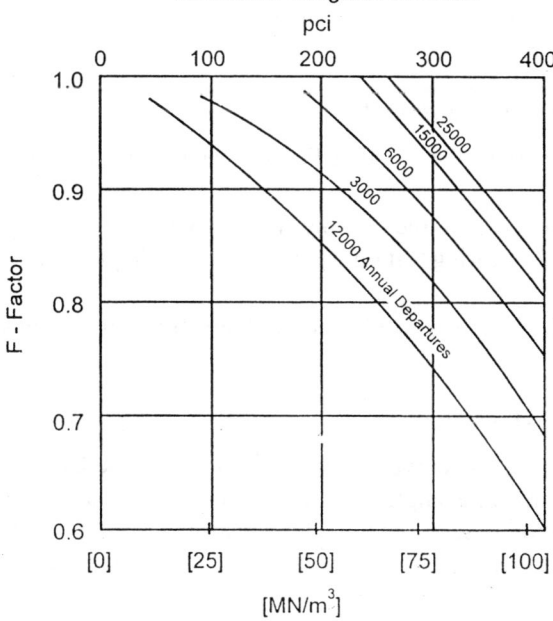

Fig. 16.3. Graph of F factors vs. modulus of subgrade reaction for different traffic levels (FAA).

For an existing bituminous overlay on a rigid pavement, the procedure is to assume as if existing overlay was not present, as explained in Example 16.6.

Example 16.1

An existing pavement consists of 25 cms rigid pavement with 7.5 cms bituminous overlay. The existing pavement is to be strengthened to be equivalent to a single rigid pavement of 36 cms. Assume an F factor of 0.9 and C_b of 0.9 for existing conditions.

Solution

(i) Calculate the required thickness of bituminous overlay as if existing 7.5 cm overlay was not present.

$$T = 2.5 (Fh - C_b h_e)$$
$$= 2.5 (0.9 \times 36 - 0.9 \times 25)$$
$$= 23 \text{ cms}$$

(ii) This overlay of bituminous would have been required if there was no existing overlay, however, since an overlay of 7.5 cms is existing, an allowance is to be made.

The effective thickness of existing overlay can be assumed as only 6 cms rather than 7.5. It is a matter of engineering judgement.

The required overlay thickness would be 23 – 6 = 17 cm.

16.8.3 Design of concrete overlays

Concrete overlays can be constructed on existing rigid or flexible pavement. As per ICAO, the minimum allowable thickness for concrete overlay is 13 cm, when placed on a flexible pavement, directly on a rigid pavement, or on a levelling course. The minimum thickness of a concrete overlay which is bonded to an existing rigid pavement is 7.5 cm.

(a) Concrete overlay on flexible pavement

For design the concrete overlay, the existing flexible pavement is considered a foundation for the overlay slab.

For design of the rigid pavement, the existing flexible pavement is assigned a 'K' value using Fig. 16.3 or by conducting a plate bearing test on the existing flexible pavement. In either case the K value assigned should not exceed 500.

In frost areas, additional thickness will be required. For frost protection stabilised material must be provided.

(b) Concrete overlay on rigid pavement

(i) The rigid pavement design curves are used to find the thickness of concrete required to satisfy the design conditions for a single thickness of concrete pavement. The K-value to be assigned to existing foundation is determined by field bearing tests conducted in the test pit cut through the existing rigid pavement or may be estimated from construction records of existing pavement.

(ii) Structural integrity of the existing rigid pavement is assessed, on the basis of engineering judgement or through non-destructive testing (NDT) method. Following values are adopted to provide uniform assessment of condition factor Cr.

Cr = 1.0 for existing pavement in good condition except minor cracking

Cr = 0.75 pavement having initial corner cracks due to loading but not progressive cracking or joint faulting.

Cr = 0.35 pavement in poor existing condition – badly cracked.

Within these ranges intermediate value of Cr may be adopted.

(iii) There may be two situations : (a) concrete overlay without levelling course and (b) concrete overlay with levelling course. Different formulas are used for the above two cases, based on research on test track pavements and observations of in-service pavements.

(a) *Concrete overlay without levelling course* : The thickness of the concrete overlay slab applied directly over the existing rigid pavement is computed by the formula :

$$h_c = 1.4 \sqrt{h^{1.4} - C_r\, h_e^{1.4}}$$

where :

h_c = required thickness of concrete overlay

h = required single slab thickness determined from design curves

h_e = thickness of existing rigid pavement

C_r = condition factor

Due to the inconvenient exponents in the above formula graphic solutions are provided in Fig. 16.4 and Fig. 16.5 for Cr values of 1 and 0.75. Lower value of condition factor is not recommended because of the likelyhood of reflection cracking.

(b) *Concrete overlay with levelling course* : In some cases it may be necessary to apply a levelling course of bituminous concrete to an existing rigid pavement, prior to application of concrete overlay. When the existing pavement and overlay pavement are separated, the slab acts more independently than when slabs are in contact with each other. Following formula is used for the thickness of overlay slab, when levelling course is used:

$$h_c = \sqrt{h^2 - C_r\, h_e^2}$$

where :

h_c = required thickness of concrete overlay

h = required single slab thickness determined from design curves

h_e = thickness of existing rigid pavement

C_r = condition factor

Levelling course must be constructed of highly stable bituminous concrete.

Graphic solutions of the above equation are given in Fig. 16.6 and Fig. 16.7, for condition factor of 0.75 and 0.35. Other C_r values can be interpolated.

(c) *Bonded concrete overlays* : In some circumstances overlays are bonded to existing rigid pavements. In such cases the new section behaves as a monolithic slab.

The thickness of bonded overlay is computed by substracting the thickness of the existing pavement from the thickness of the required slab thickness determined from design curves.

$$h_c = h - h_e$$

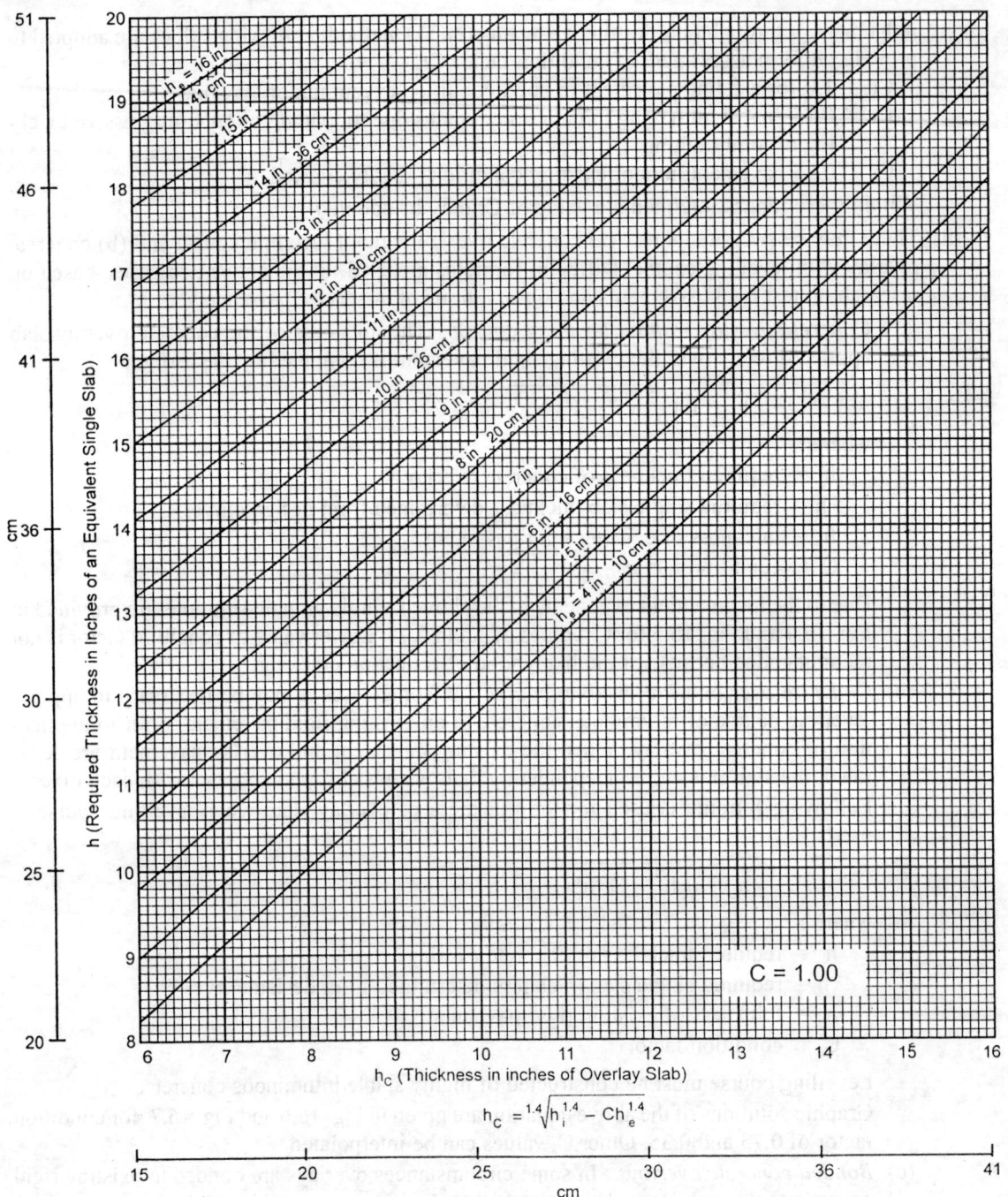

$$h_C = {}^{1.4}\sqrt{h_i^{1.4} - Ch_e^{1.4}}$$

Fig. 16.4. Concrete overlay on rigid pavement. C = 1 (FAA).

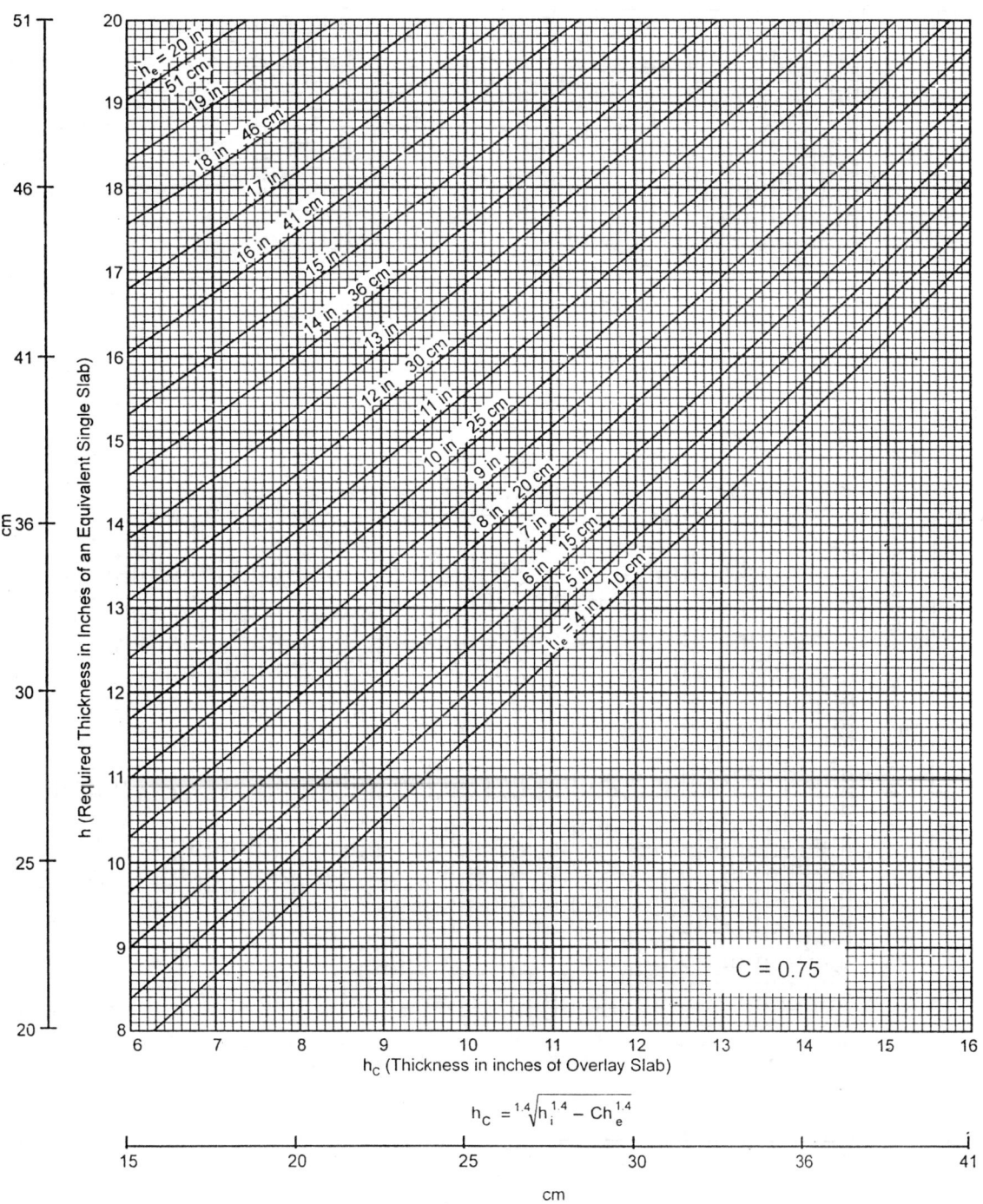

$$h_C = \sqrt[1.4]{h_i^{1.4} - Ch_e^{1.4}}$$

Fig. 16.5. Concrete overlay on rigid pavement. C = 0.75 (FAA).

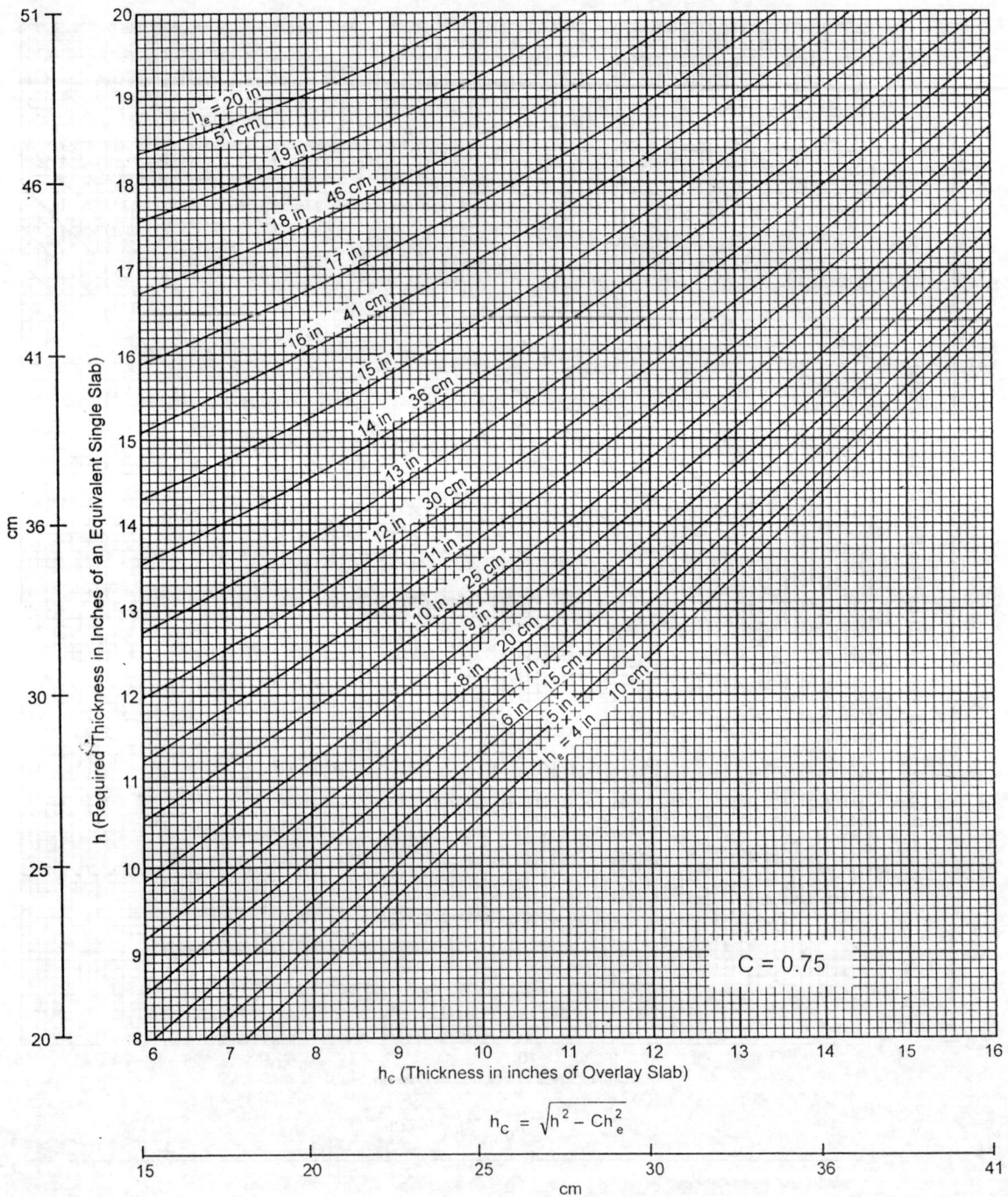

$$h_C = \sqrt{h^2 - Ch_e^2}$$

Fig. 16.6. Concrete overlay on rigid pavement with levelling course. (C = 0.75) (FAA).

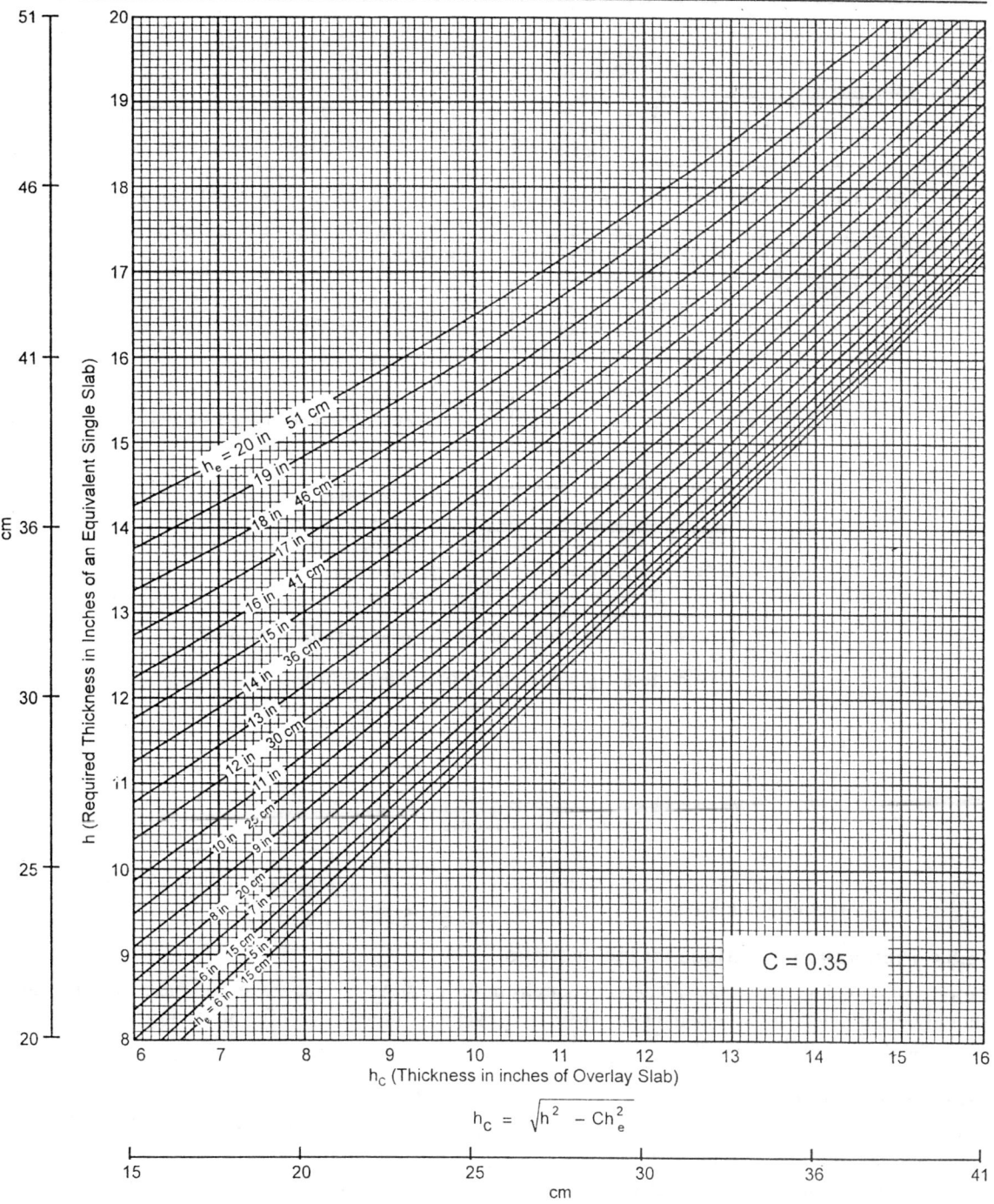

$$h_c = \sqrt{h^2 - Ch_e^2}$$

Fig. 16.7. Concrete overlay on rigid pavement with levelling course. (C = 0.35) (FAA).

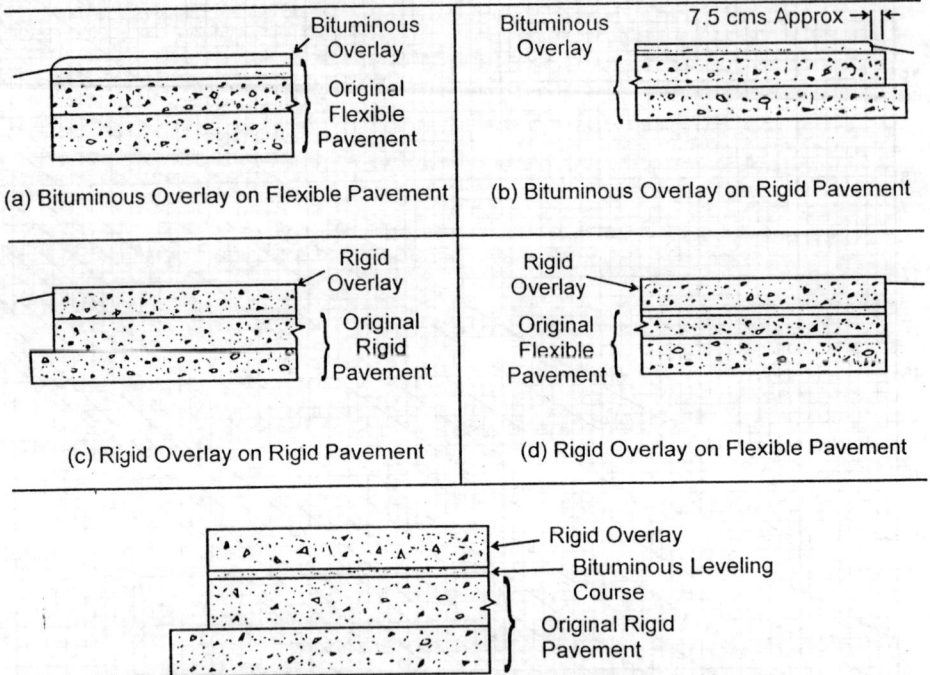

Fig. 16.8. Typical overlay pavement (FAA).

where :

h_c = required thickness of concrete overlay

h = required single slab thickness determined from design curves

h_e = thickness of existing rigid pavement

Bonded overlays should be used when existing rigid pavement is in good condition. Adequate bond should be achieved through elaborate surface preparation and precise construction techniques.

16.8.4 Preparation of surface for the overlays

Before constructing an overlay, all defective areas in the existing surface, base, sub-base and subgrade should be corrected, to have satisfactory overlay pavement, otherwise deficiencies in the base pavement may reflect in the overlay.

Flexible pavements

Pavement breakups, patches, surface irregularities and depression etc. in the existing flexible pavements should be removed, before putting the overlays.

- Broken pavement areas are repaired by improving the subgrade, or drainage facilities. After correction of the subgrade condition sub-base, base and surface courses of required thickness should be placed and thoroughly compacted layer by layer.
- Surface irregularities and depression like shoving, rutting, depressions should be levelled by rolling and/or by filling with suitable hot mix bituminous mixtures.

- Bleeding surface, i.e., hot mix bituminous material accumulated at one place on the surface of the pavement is bladed off/or blotted up.
- Old joints or cracks are removed by putting the crack filler – a lean mixture of sand and liquid bituminous material. The mixture is well tamped in place, levelled with the pavement surface and any excess removed. Material place should dry and become hard, before placing the overlay.
- Pot holes should be thoroughly cleaned and filled with suitable bituminous mixture and tamped in place.

For the hot mix overlay, a light tack coat is put immediately, on the pavement after cleaning. The overlay should not extend to the edge of the pavement, but should be cut off approximately 75 mm from the edge.

In recent years, new pavement overlay materials such as fibrous concrete, roller compacted concrete and rubberised asphalt are used, with varying degree of success.

Rigid pavements

In rigid pavements, narrow transverse, longitudinal and corner cracks will generally need no special attention, unless there is displacement and faulting between the separate slabs. If the subgrade is stable and no pumping has occurred no corrective measures are required as overlay will take care of such defects. If subgrade support is not stable and pumping has occurred, pumping cement grout or bituminous cement is used to fill the voids, to provide even support for the overlay.

- Slab may be removed and replaced if the pavement slabs are broken and subject to racking because of uneven bearing on the subgrade.
- Bad broken slabs are also removed and replaced before putting the overlay.
- If the existing pavement is rough due to slab distortion, faulting or settlement, a provision should be made for a levelling course of hot bituminous mix concrete, before overlay is placed.
- Cracks and joints 1 cm or more in width, should be filled with a lean mixture of sand and liquid bituminous material, and tamped firmly in place and levelled with pavement surface.
- After repairing and prior to placing of the overlay, the surface should be swept clean of all dirt, dust, and foreign material. Any extruding joint-sealing material should be trimmed from the rigid pavement.
- Bonded concrete overlays require special attention to insure bond with the existing pavement. Shot peening or mechanical texturing by cold milling are two techniques which have been used to provide a surface which will allow bonding. Bond can also be achieved by placing a neat cement grout on the prepared surface, before placing overlay.

16.9 STAGE CONSTRUCTION OF AIRPORT PAVEMENTS

In many instances it may be necessary to construct the airport pavements in stages, that is to build up the pavement profile, layer by layer, as the traffic using the facility increases in weight and number. Planning for future widening of the pavement may be necessary to accommodate larger aircraft. If stage construction is to be undertaken, sound planning should be done. Complete pavement should be designed prior to start of any stage, and each stage must provide an operational surface. Planning of a stage constructed pavement requires considerations of economics, material, labour and equipment cost.

CHAPTER 17

Heliport, Stolport and Vertiport

17.1 INTRODUCTION

Air transportation vehicle which can "hover" i.e., stand still while flying is called a "Rotorcraft". Helicopter is a rotorcraft vehicle which can take-off from and land in a nearly vertical direction. It is a rotary-wing aircraft that can lift vertically and takes forward flight by power-driven rotor blades turning on a vertical axis. Helicopter is thus vertical takeoff-landing aircraft, known as "vertical takeoff and landing" (VTOL). Facilities provided for helicopters and other rotary-wing aircrafts on the ground are called "Heliports".

In the recent years other vehicles capable of "short take-off and landing (STOL) are developed for local intercity transportation, and used as commuter aircrafts feeding major outlaying airports. These aircrafts are unable to lift vertically, but take-off and land on very short runway. They climb and descend at steeper angles than conventional fixed wing aircrafts. STOL aircrafts require "Powered Lift" for their operation rather than relying entirely on "mechanical lift". Powered lift involves propulsion from the engines which are used to pass air over parts of the aircraft body, improving the life, particularly at low speeds. Facilities on the ground for STOL aircrafts are called "Stolports". STOL aircraft is less noisy and less costly to operate than VTOL aircraft, requiring less length of runway (about 1000 m) than conventional aircrafts. Another type of vehicles "Tilt-rotor aircraft", "tilt-wing aircraft" and "fan-in-wing" aircraft are also developed.

A tilt-rotor aircraft has fixed wings but can take off and land vertically and can fly in cruise as a conventional fixed wing aircraft. The axis of propeller blades on the engines of these aircrafts are capable of being pivoted to vertical takeoff and landing, and to horizontal to drive lift from the wings while cruising.

In a tilt wing aircraft, the wing chord and the axes of the power driven propeller blades are capable of pivoting from vertical, for vertical take-off and landing to horizontal, to drive lift from the wing in cruise.

Fan-in wing aircrafts are fixed wing aircrafts with rotor fans to permit vertical operations. *Vertiports* are facilities provided for take-off and landing of these aircrafts.

HELIPORTS

Heliport is the area on land, water or structure used or intended to be used for the takeoff and landing of helicopters, together with appurtenant buildings and facilities.

Helidock is a heliport located on a floating or an off-shore structure.

Helistop is an area developed minimally and used for helicopter takeoffs and landings to drop of or pick up passengers and cargo.

Helipad is a paved or other surface used for parking helicopters at a heliport. It is a load bearing area on which helicopters can land and takeoff.

17.2 ADVANTAGES OF HELICOPTER TRANSPORTATION

The modern helicopter is one of the most versatile transportation vehicle. Its ability to operate from minimal area has given it the capability of providing important services. In addition to serve the transportation of people helicopters are useful to the community in following ways :

- *Disaster relief* : When natural ground transportation system breaks down due to natural disasters like floods, earthquakes etc. helicopters are used to bring rescue teams, supplies and to evacuate injured and other people during the critical period before situation is normalised.
- *Air ambulance services* : Injured or critically ill people are transported at high speed, from point to point in a helicopter.
- *Police departments* : Search, rescue, chase and surveillance operations are conducted by the police in helicopter. They provide vital force.
- *Moving high value assets* : For newspaper, perishable goods, taking photos etc. helicopters provide fast and flexible mode.
- *Other uses (benefits)* :
  - Politicians frequently travel on helicopters to visit disaster sites, to local airports and other sites.
  - Areas are developed where helicopter can land and take-off.
  - Helicopter are used for reconnaissance of areas which are in accessible.
  - For military operations and defence purposes helicopters provide best services.
  - In many advanced cities, helicopters provide taxi services for point to point transportation within a city.
  - Helicopters are used for crop dusting and sprays.
  - For transportation to offshore oil well locations, helicopters are commonly used.
  - Helicopters also provide for inter-city transportation, between cities.

17.3 HELICOPTER TERMINOLOGY

Following terms used commonly for developing a heliport are defined below :

Approach/Take off path : The flight track which helicopters follow when landing at or taking off from a heliport.

Design helicopter : A generic rotorcraft which reflects the maximum weight, overall length, rotor diameter etc. of all helicopters expected to operate at the heliport.

Final approach and takeoff area (FATO) : A defined area over which the final phase of the approach to a hover or a landing, is completed and from which the takeoff is undertaken.

Final approach reference area (FARA) : A 45 m × 45 m long obstacle-free area with its centre aligned on the final approach course. It is located at the end of a precision instrument FATO.

Heliport imaginary surfaces : The imaginary planes, centered above the FATO and the approach/takeoff path that identify the objects, which should be removed, lowered, and/or marked and lighted otherwise the approach/takeoff path should be realigned.

Protection zone : An area off the end of the FATO and under the approach/take off path to enhance the protection of the people and property on the ground.

Taxi route : An obstruction free corridor in which helicopters hover and taxi above the surface at airspeeds less than approx. 20 knots.

Taxiway : A defined path established for the ground taxi of helicopters from one part of a heliport to another.

Touchdown and lift area (TLOF) : A load bearing area, generally paved, normally centred in the FATO, on which the helicopter lands or takeoff. It is also called helipad or helideck.

17.4 CLASSIFICATION OF HELIPORTS

Heliports are classified according to its use :

17.4.1 Private use heliport

These heliports are used by the private owners. Their use is restricted. They may be owned by police, fire departments, oil companies, manufacturing firms. With their diverse uses they vary considerably both in size and complexity.

17.4.2 Public use general aviation heliports

A public use heliport is available for use by the general public. It accommodates helicopters used by individuals, corporations and air-taxi services. Scheduled passenger services may be available if sufficient demand exists. The extent of facilities depend upon the demand, present and in future.

17.4.3 Transport heliports

Transport heliports are available to general public and are intended to accommodate, scheduled service with large helicopters. Facilities are designed for commercial passenger or cargo services.

17.4.4 Hospital heliports

Hospital heliports are for transportation of injured persons from scene of accident to a hospital and for transferring patients in critical conditions, from one hospital to another. They are as simple as a cleared area on the lawn with a wind indicator and a clear approach/take-off path-aligned with dominating winds.

17.4.5 Military heliports

These heliports are owned by armed forces and prohibited for non military use.

17.4.6 Government heliports

These heliports are used by central and state government officials, governors and politicians. They are owned by government departments.

17.4.7 Personal use heliports

Facilities of these heliports are used exclusively by the owner.

17.4.8 Helicopter facilities on airports

Helicopters operate on an airport without interfering with airplane traffic. Provisions are therefore, made at an airport for landing and takeoff of helicopters, such that maximum separation is achieved from fixed wing aircrafts in taking off and landing as well as in taxiing.

Passenger walking distances should not be large, and ample separate parking for helicopter should be provided.

Heliports can also be located at the roof of the terminal building.

17.5 CHARACTERISTICS OF HELICOPTERS

17.5.1 Operating characteristics

Helicopter gains its lift from the rotary motion of airfoil surfaces and is able to hover through application of power to the rotating air foils. Because of this characteristics helicopters can develop higher flight speeds and flight altitude and can takeoff and land from comparatively small areas than conventional aircrafts. Helicopters are able to taxi on ground under their own power.

The initial vertical lifting up of the helicopters is aided by the ground cushion built up by the pressure of air directed against the ground by the revolving rotors. After few meters of vertical ascent, horizontal acceleration starts until the climb-out speed is reached. Just before touchdown, the helicopter hovers momentarily 2 to 4 m above the landing pad.

For a single engine helicopter, from the safety reasons, it is necessary that emergency landing facilities are provided along the entire flight route, so that in engine failure case, safe landing using autorotation is possible. This problem is avoided in twin-engine helicopters.

Helicopters are either single-rotor or tandem rotor and are powered by one or two engines. The landing gears could be pontoons for landing on water, skids or wheels with rubber tires. When wheels are used landing gear consists of two main wheels and a single nose wheel or tail wheel, or four wheels as shown in Fig. 17.1.

17.5.2 Physical characteristics

Physical characteristics and dimensions of the helicopters needed for heliport design are given in Table 17.1 and illustrated in Fig. 17.2. Dimensions of helicopters in Table 17.1 are given in ft-lb units as provided by the manufacturers. To convert them in metric units, multiplying factors can be used.

Mechanical and gas turbine (jet) systems are used in driving the rotors. Gas turbine engines are lighter in weight and provide greater energy. Thus helicopters are piston driven (P) or turbine driven (T).

Helicopters may be single rotor type or twin roters fitted side by side. Number of blades vary from 2 to 4 or 5. Helicopters equipped with conventional landing gear wheels are normally supported by two main wheels and one tail wheel, or nose wheel. For large helicopters, each main landing gear consists of two wheels. Each main gear typically support 40 to 45% of the weight of the helicopter, and the tail wheel or nose wheel supports the 10 to 20 per cent weight. If the helicopter is supported by tubular skids, 50 per cent of the weight is supported by each skid.

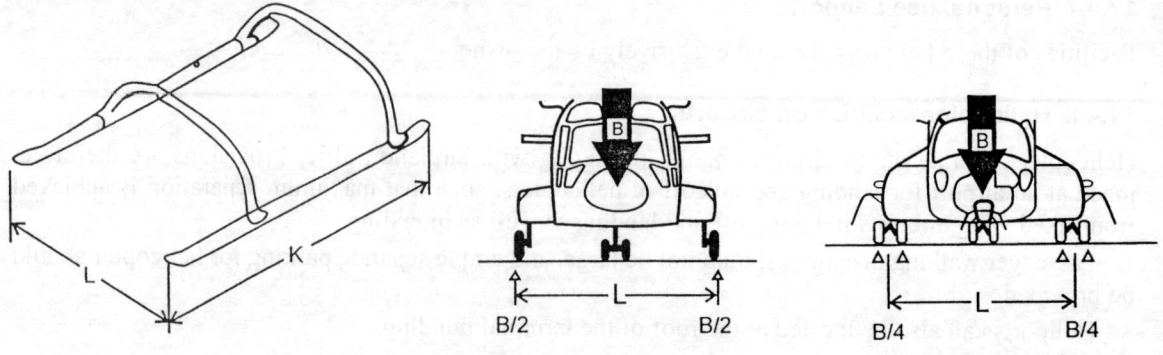

Typical Skid Configuration   Typical Single Wheel Configuration   Typical Dual Wheel Configuration

B = Gross weight; B/2 = Gross weight/gear/skid; B/4 = Gross weight/tire; K = Wheel base/skid length; L = Tread

Fig. 17.1. Helicopter landing gear configuration (FAA).

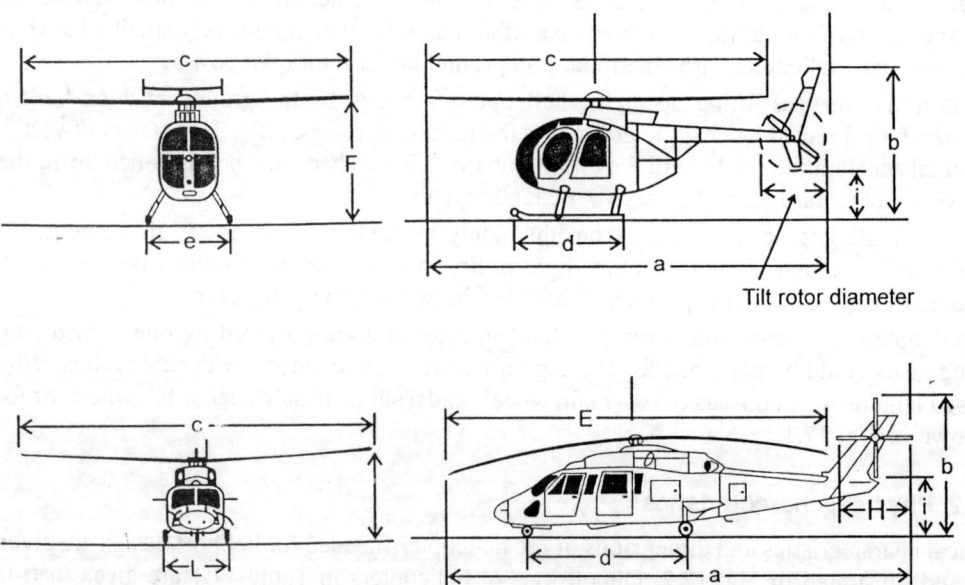

Tilt rotor diameter

Fig. 17.2. Helicopter dimensions (FAA). (a = overall length in feet; b = overall height in feet; c = rotor diameter in feet/no. of blades; d = undercarriage length in feet; e = undercarriage width in feet).

17.6 SITE SELECTION FOR HELIPORTS

Heliports are usually provided in large urban areas between central business districts and airports, for commercial transportation. Sometimes more than one heliports may be required in a big city. Therefore, a heliport must be accessible through good road and public transport facilities so that passengers can easily reach. In addition, following important points for the selection of the site for a heliport in an urban area need considerations :

Table 17.1. Dimensions of typical helicopters (Source : FAA)

Manufacturer/Model	Overall (ft)		Rotor dia./ no. of blades ft	Under-carriage length ft	Under-carriage width ft	Number and engine type	Crew/ passengers	Max. take of weight lbs
	Length ft	Height ft						
1. Eurocopter 330 Puma	60	17	50/4	13.3	9.8	2-T	2 and 20	16315
2. Eurocopter 332 Super Puma	62	17	52/4	17.3	9.8	2-T	2 and 24	18960
3. Eurocopter 341 Gazelle	40	11	35/3	6.4	6.6	1-T	1 and 4	3970
4. Eurocopter 365 Douphin	45	14	40/4	11.9	6.2	2-T	1 and 13	9369
5. Eurocopter BK-177	43	13	37/4	6.2	8.2	2-T	1 and 10	7385
6. Augusta A-109	43	11	37/4	11.6	7.5	2-T	1 and 7	5997
7. Bell 205	58	15	48/2	12.1	8.6	1-T	1 and 14	9500
8. Bell 212	58	13	49/2	12.1	8.7	2-T	1 and 14	11200
9. Bell 214	63	16	52/2	12.1	8.6	2-T	2 and 18	17500
10. Bell 412	57	15	46/4	7.9	8.3	2-T	1 and 14	11900
11. Boeing 107	84	17	50/3	24.9	12.9	2-T	3 and 25	20000
12. Boeing 234	99	19	60/3	25.8	10.5	2-T	3 and 44	48500
13. Boeing 360	84	20	50/4	32.7	13	2-T	3 and 30	36160
14. MDX explorer	39	12	34/5	7.3	7.3	2-T	1 and 7	5800
15. Schweizer 330	31	10	27/3	–	6.5	1-T	1 and 3	2200
16. Sikorsky S-61	73	19	62/5	23.5	14	2-T	3 and 28	20500
17. Sikorsky S-64 Skycrane	89	26	72/6	24.4	19.8	2-T	3 and 0	42000
18. Sikorsky CH-53	100	29	79/7	27.3	13	3-T	3 and 55	69750
19. Sikorsky S-76	53	15	44/4	16.4	8	2-T	2 and 12	11700
20. Westland 30-100-60	53	16	44/4	17.9	10.1	2-T	2819	12800
21. Brantly-305	33	8	29/3	6.2	6.8	1-P	1 and 4	2900
22. EH-101	75	22	61/5	22.9	14.9	3-T	3 and 30	31500
23. Enstrom 480	37	10	5/2	–	8.1	1-T	1 and 3	2680
24. Robinson R-44 PS	38	11	4.8/2	4.2	7.2	1-P	1 and 3	2400
25. Hiller RH-1100	42	10	6/2	7.9	7.2	1-T	1 and 6	3500

T : Turbo; P : Piston

- It should be located where traffic demand is maximum.
- There should be well connected surface transportation and space for parking should be available.
- It should be an area where there are no or minimum obstructions for approach and departure of helicopters.
- The disturbance from the noise generated by helicopters to the adjacent land use should be minimum.
- Cost of acquiring the land and its development for an heliport should be as low as possible.
- The prevailing winds should be favourable so that two approach paths separated by at least 90° can be developed with respect to prevailing winds.
- Visibility of the site should be good. Not restricted by tall buildings, fog etc.

- It should be located such that no interference is created with other existing air traffic in the area.
- For emergency landing of the single engine helicopter, along the whole route, it should be possible to develop facilities.
- For small heliports, facilities for landing and takeoff of a helicopter are provided on the roof of buildings, hospitals etc. The building designed for such purpose should take into account vibration and impact effects of a helicopter landing and takeoff.
- On large airports, an area is provided separately for landing and take-off of helicopters, developed as an heliport. It should involve minimum taxiway and free from interference with other air traffic.
- Heliports are sometimes located on terminal buildings of an airport, providing easy facilities for processing of passengers and cargo.

All the above factors, may not be satisfied at a particular site, and therefore, a compromise is required to be done, to select the most appropriate site considering the economics and safety.

Fig. 17.3 shows typical layout of a heliport.

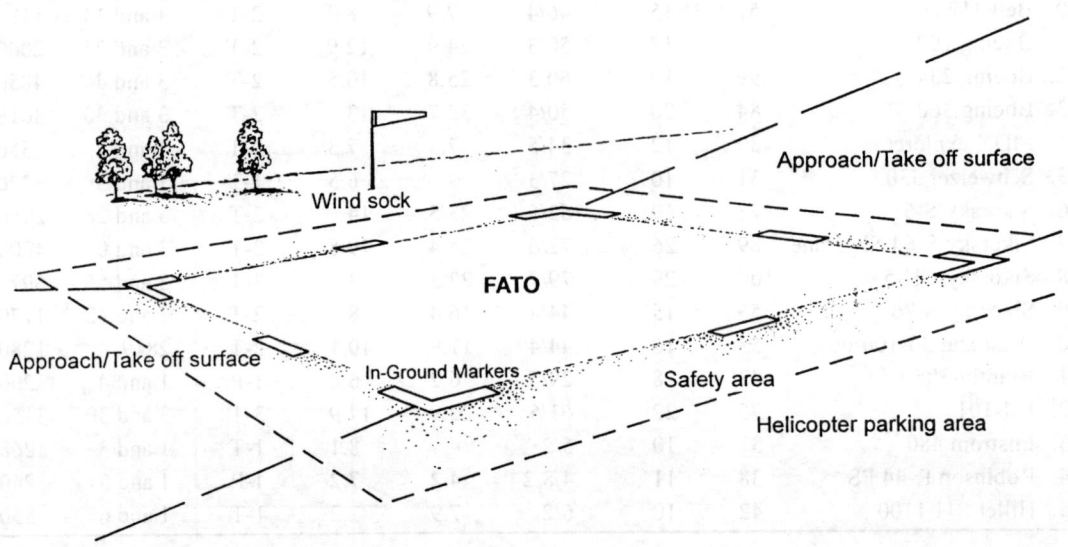

Fig. 17.3. General features of a heliport (FAA).

17.7 COMPONENTS OF A HELIPORT

The main components of a heliport are :

1. Final approach and take-off area (FATO)
2. Touch down and lift-off area (TLOF)
3. Safety area
4. Approach/Takeoff surface
5. Taxi route and taxiways
6. Helicopter parking areas
7. Heliport markings

8. Heliport lighting
9. Wind direction indicator
10. Visual glide path indicator
11. Terminal facilities (terminal building)

17.7.1 Final approach and take-off area (FATO)

A general avaiation heliport must have atleast one FATO. The FATO may contain one or more touchdown–liftoff areas within its borders at which arriving helicopters terminate their approach in a hover or a landing, and from which departing helicopters takeoff.

The FATO may be at ground or water level or a roof-top level. The least dimensions (length, width or diameter) of the FATO should not be less than 1.5 times the overall length of the design helicopter. FAA recommends that for heliports at elevations of 300 m or more above MSL (mean sea level), elongation in dimensions of FATO should be done as shown in Fig. 17.4. The elongation should be in the direction of take-off. Gradient for drainage purposes of FATO may range from 0.5 per cent (minimum) to 5.0 per cent maximum depending upon site.

FATO grades should not exceed 2 per cent in landing area. To make it dust free FATO should be paved or provided with turfing as helicopter rotors create dust on unprepared surface.

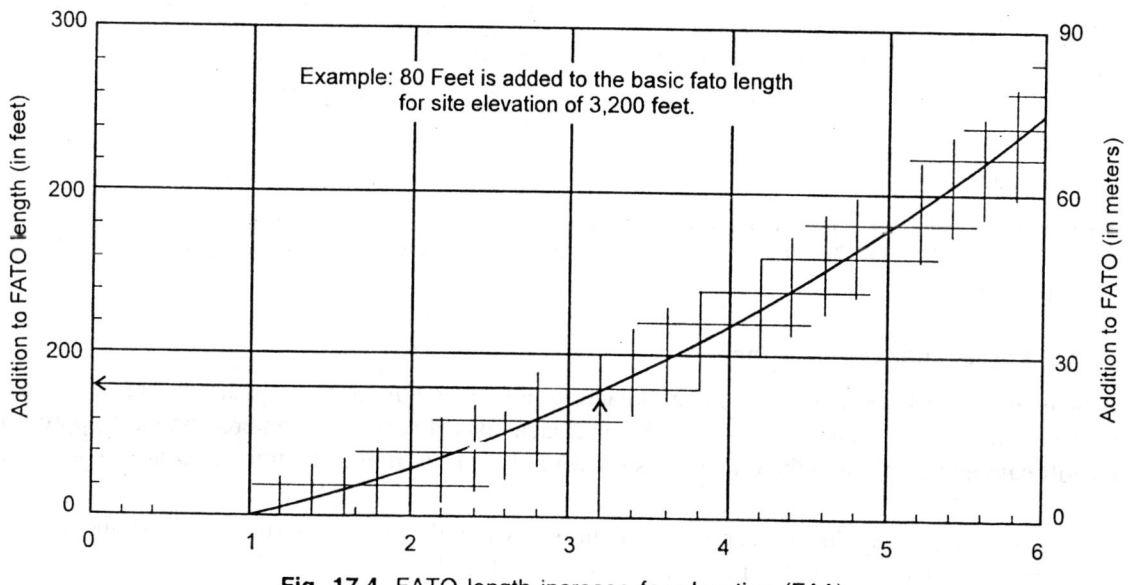

Fig. 17.4. FATO length increase for elevation (FAA).

17.7.2 Touch down and lift-off (TLOF)

When entire FATO is load bearing, an identifiable TLOF may not be necessary. A TLOF may have any shape. It is centred in FATO and has a hard or paved surface. An elongated FATO, may have an elongated TLOF of either cement or bituminous type. Rough or broomed surface of concrete is considered better.

The FAA recommendation is that minimum dimension of TLOF should not be less than the rotor diameter of the design helicopter. Pavement should be designed to support 1.5 times the maximum takeoff weight of the design helicopter. Gradient of TLOF should be minimum of 0.5 per cent and maximum of 2 per cent. To ensure drainage gradient of shoulders for rapid run-off may range from 2 to 5 per cent. General FATO/TLOF relationship for a heliport is shown in Fig. 17.5.

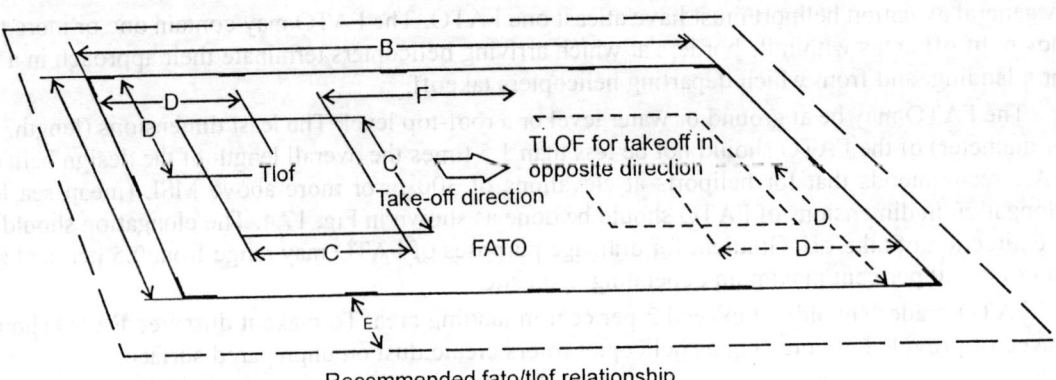

Recommended fato/tlof relationship
for a general aviation heliport

A--Fato width, 1.5 x overall length, **B**--Fato length, 1.5 x overall length plus, addition from figure 3-2.
C--Tlof length and/or width, 1.0 x rotor diameter, **D**--Distance from fato edge to center of tlof,
0.75 x overall length, **E**--Safety area, 0.33 rotor diameter, minimum of 20 feet (6 m).
F--Recommended minimum distance betweer, tlof edges developed for, 1.0 x rotor diameter.

Fig. 17.5. Recommended FATO/TLOF relationships (FAA).

17.7.3 Safety area

A safety area, equal to 1/3 the rotor diameter of the design helicopter, but not less than 6 m in width should surround FATO. The FATO and the safety area must be free and clear of objects such as parked helicopters, fences etc.

17.7.4 Approach/take-off surface

A heliport should have more than one approach/takeoff path. One of these paths should be oriented with the direction of predominant wind. An approach/takeoff surface is centered on each approach/takeoff path and conform to the dimensions shown in Fig. 17.6. It must be free of hazards to air navigation.

Property underlying the approach/takeoff surface where the surface is 10.5 m, above the heliport elevation is considered as protection zone, as shown in Fig. 17.7.

At private-use heliports, it is recommended that these flare out in horizontal plane from the final approach and take-off area at the rate of 1 to 20 and slope upward at the rate of 8 to 1.

For the heliports, the principal surfaces are :

- the approach and departure surfaces
- the transitional surfaces, and
- the heliport protection zone.

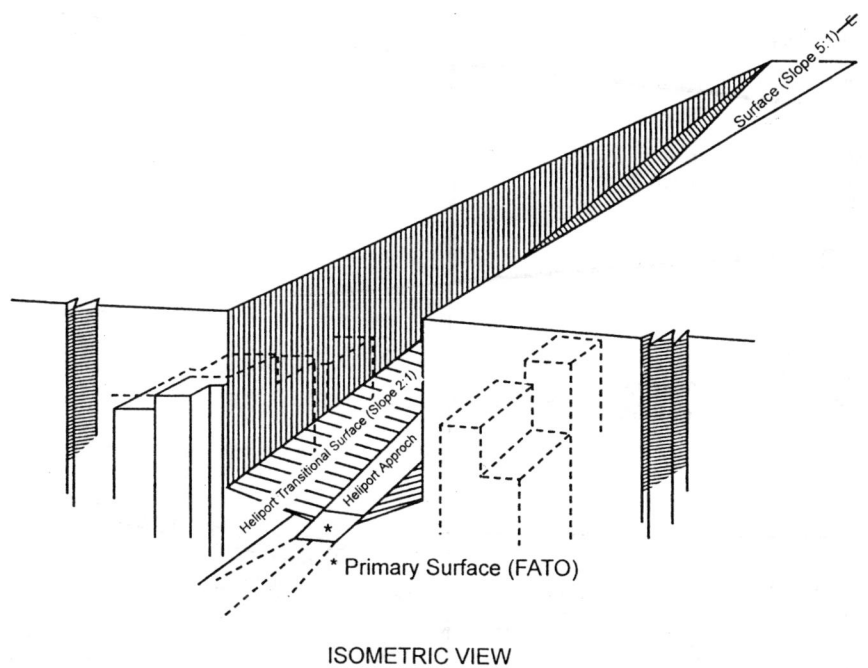

ISOMETRIC VIEW

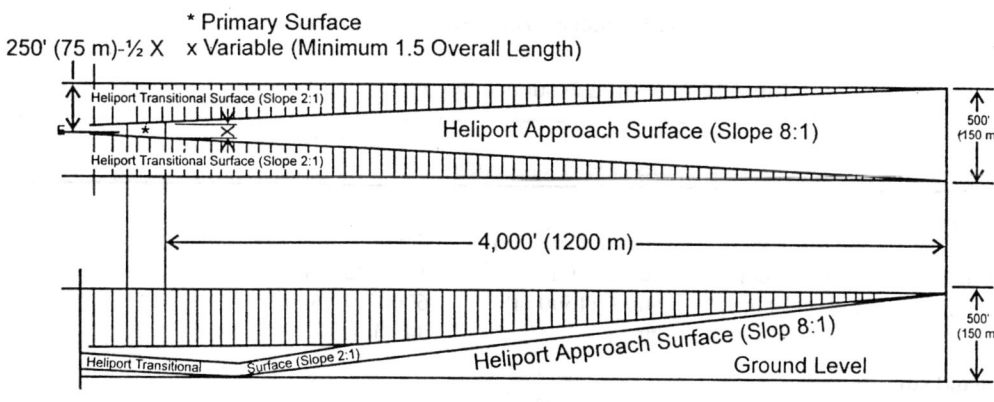

PLAN VIEW

Fig. 17.6. Imaginary surfaces for heliports (FAA).

The horizontal surface, as required for airports is not necessary for heliports. The approach surface requirement for visual, non-precision-instrument and precision-instrument operations specified by FAA is given in Table 17.2.

17.7.5 Taxi route and taxiway*

A taxi route is an object free right of way connecting the FATO to a parking area or apron. It is also a manoeuvering aisle on the parking area or apron. Taxiways are paved areas (surfaces) used by wheel equipped helicopters in ground manoeuvering.

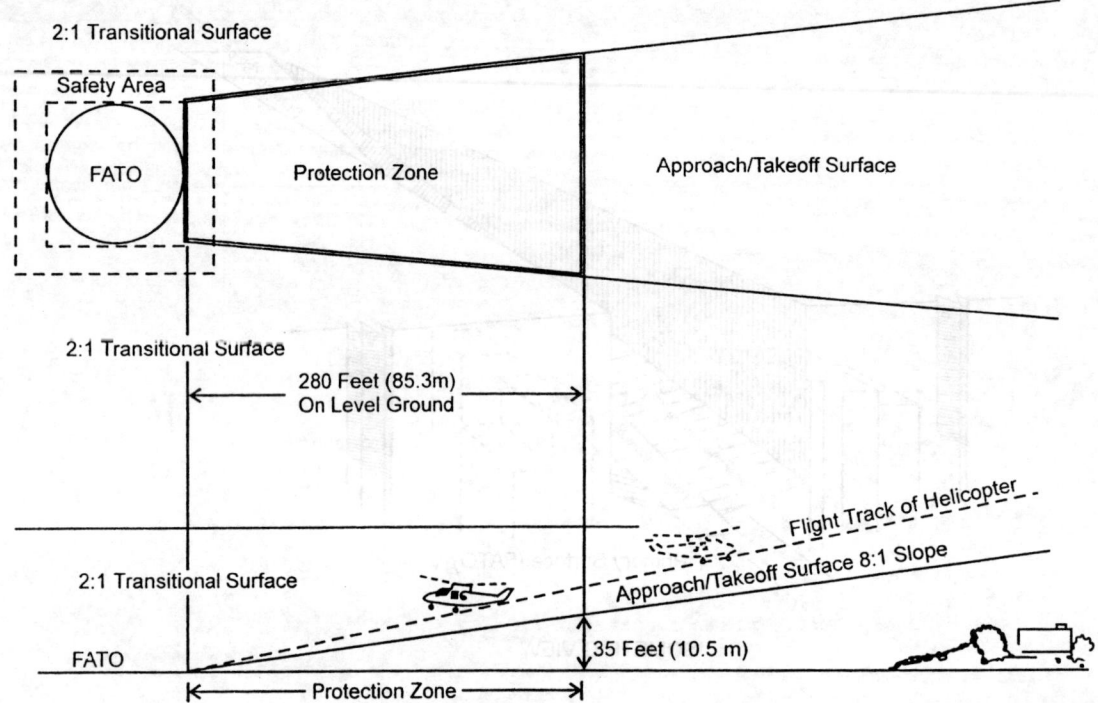

Fig. 17.7. Protection zone for a heliport (FAA).

Table 17.2. Approach surface dimensions for heliports

Surface	Visual approach	Non-precision instrument approach	Precision instrument approach
Primary surface			
Length	4000′ (1200 m)	10000′ (3000 m)	25000′ (7500 m) begins 367 m from FATO end
Inner width	FATO width	500′ (150 m)	1000′ (300 m)
Outer width	500′ (150 m)	5000′ (1500 m)	6000′ (1800 m)
Slope	8 : 1	20 : 1	34 : 1
Transitional surface			
Inner width	FATO width	FATO width	600′ (180 m)
Outer width	250′ (70 m)	600′ (180 m)	1500′ (450 m)
Slope	2 : 1	4 : 1	7 : 1

Taxi routes are required to have a width equal to maximum rotor diameter of the helicopter that will hover or ground + a clearance (6 m).

Paved taxiways widths should be designed to be equal to twice the undercarriage width of the design helicopter.

If taxiways are not paved, they should be turfed or treated so that dirt, debris are not raised by taxiing helicopter's rotor. Taxiway pavements should be able to take maximum gross weight of the design helicopter under all weather conditions.

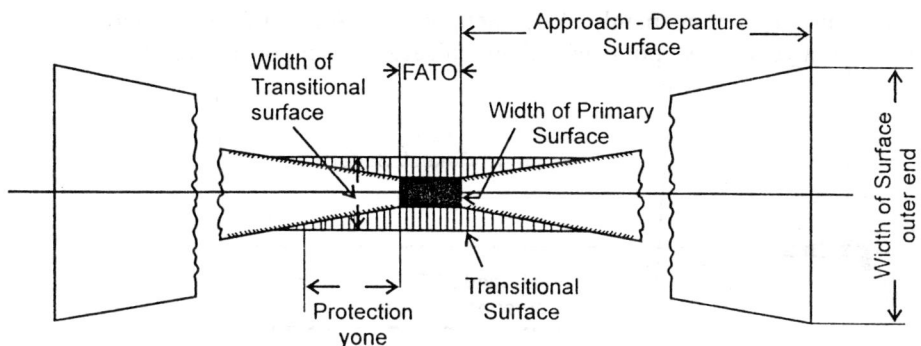

Fig. 17.8. Operational clearance requirements for heliports.

Taxiway longitudinal gradient should not exceed 2.0 per cent. Transverse gradient should not be less than 0.5 per cent nor greater than 2 per cent.

17.7.6 Helicopter parking areas

Heliports should have area for parking helicopters. The size will depend upon number of helicopters to be parked, and the intended path for manoeuvering. The least dimension of a parking pad should be a minimum of 1.5 times the undercarriage length or width of the design helicopter.

17.7.7 Heliport markings

Perimeter of the FATO and/or TLOF should be defined by markers and/or lines. The purpose of markings is to identify areas clearly for the use of helicopters.

- Surface markings may be point or preformed material.
- FATO's and TLOF's are marked with white lines on the surface or in-ground markers (30 cm by 1.5 m) located at corners and along the FATO. 30 cm wide dashed lines define paved FATO perimeter.

The corners must be defined and the edge segments should be 1.5 m in length.

- Taxi routes are defined with raised edge markers – yellow-blue-yellow that are not more than 20 cm in height and not less than 10 cm in diameter.
- Taxiways and aprons are defined with yellow lines/markings. The centre line is continuous 15 cm wide yellow line.
- An "H" is painted in the centre of the touchdown and lifeoff area. In the hospital heliports, a white cross is also inscribed within the square along with letter 'H'.
- A continuous 30 cm wide solid white line defines the perimeter of a paved or hard surfaced TLOF. A continuous 30 cm in ground marking shows perimeter of a load bearing TLOF.
- The edges are marked with two continuous 15 cm wide yellow lines spaced 15 cm apart.
- The yellow taxiway centreline continues into the individual parking position to mark the centreline of the parking positions. A parking position is further identified by a 30 cm wide yellow line defining a circle. The diameter of circle is equal to rotor diameter of the largest helicopter to be parked.

- All markings of a permanently closed heliport i.e. FATO and TLOF markings should be removed, or a yellow "X" put over "H" to indicate that heliport is closed.

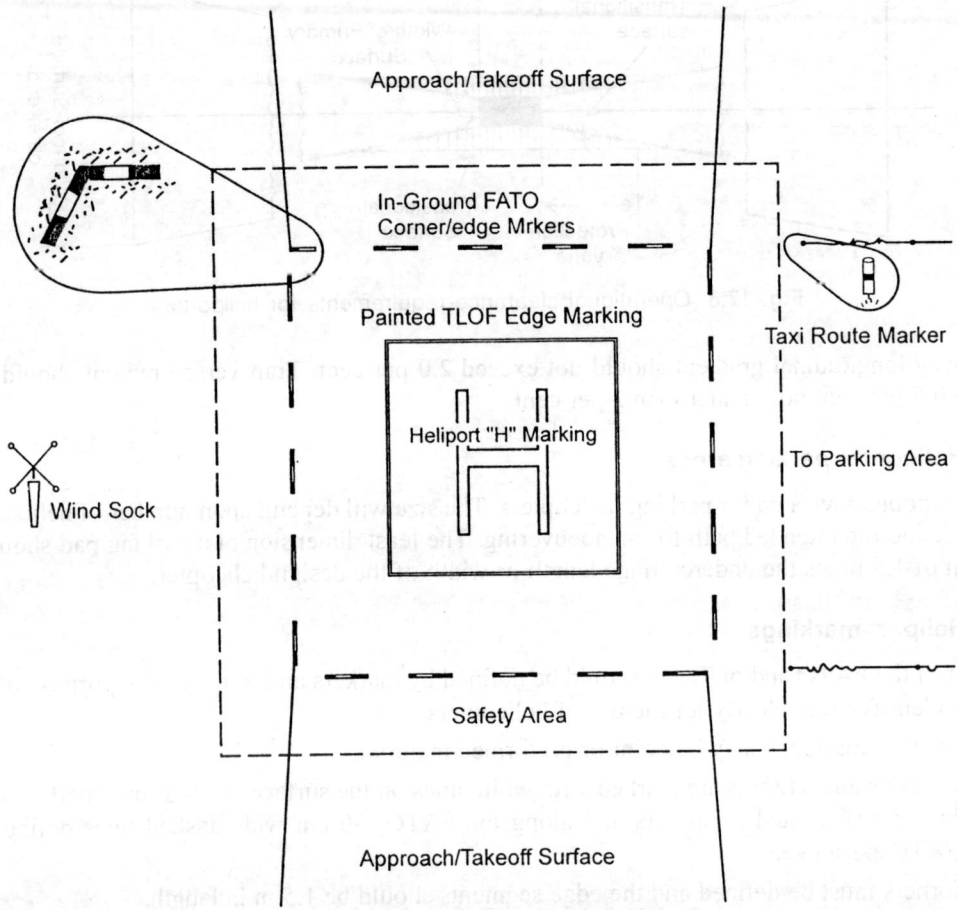

Fig. 17.9. Typical heliport with markers and markings (FAA).

17.7.8 Heliport lighting

For operation during night time or hours of darkness various types of lights are used at the heliports. The lighting system depends upon the volume of traffic and the type of the heliport. Lights help in identification of helicopters, direction of wind, landing area, taxiways and obstructions. Either FATO or TLOF, not the both, and the taxiways (or taxi routes) need to be lighted. FAA recommendations for lighting at heliport are summarised below :

- Yellow lights show the limits of FATO or TLOF, spaced 7.5 m centre.
- Flush green lights indicate taxiway centre-line, spaced at 15 m interval on straight section and 7.5 m on curves.
- Blue omni-directional lights or reflectors define taxi route edges spaced 15 m on straight section and 7.5 m on curved section.
- Flush lights are located on the TLOF edge to provide illumination.

- Landing direction lights are a configuration of five lights with omni-direction yellow lenses, spaced at 4.5 m intervals.
- A heliport beacon is recommended to aid pilots in locating the heliport. The beacon flashing white/green/yellow flashes 30 to 45 times per minute.
- Flood lights are used to illuminate the apron. They should be aimed down providing a minimum of 3 foot candle (32 lux) of illumination on the apron surface.

Fig. 17.10 shows typical lighting at heliport, for night operations.

17.7.9 Wind direction indicator

A wind direction indicator, as used on airports is used to show the direction and magnitude of the wind. Wind cone assemblies are placed at a suitable place clear of safety area, the approach/take off surfaces and the heliport transitional surfaces. The wind cone must be lighted for night operations. It should provide best possible colour contrast to its background. Multiple wind cone may be necessary on large heliports as wind direction and speed may differ significantly from one part of the heliport to another.

17.7.10 Visual glide path indicator

A visual glide path indicator, such as "Heliport Approach Path Indicator" (HAPI), "Visual Approach Slope Indicator" (VASI) or "Precision Approach Path Indicator" (PAPI), provides pilots with visual course and descent cues. The lowest on course visual signal must provide a minimum of 1 degree of clearance over any object in the approach path that lies within 10 degrees of the approach course centreline.

The optimum location of visual glide path indicator is on the extended centreline of the approach path at a distance that brings the helicopter to a hover 0.9 to 2.5 m above the TLOF centre.

17.7.11 Terminal building (facilities)

For low traffic volumes, loading and unloading of the passengers can be accomplished on the final approach and takeoff area itself. As traffic increases or on large heliports serving heavy volumes, the heliport terminal will require curb side access for passengers using private vehicles, taxies or public transport.

Waiting areas for passengers, number of gates for loading/unloading, space for processing of passengers etc. may be necessary, similar to an airport terminal and its facilities, discussed earlier. The planing concept is same.

A helipad is connected to final approach and takeoff area – FATO by taxiways and taxilanes.

17.8 ELEVATED (ON THE ROOF) HELIPORTS

Heliports can be located on piers, roof tops or other structures on water, if ground level sites are not available or when they are to be located close to central business area. The dimensions of touchdown and lift of area remain the same as for ground heliports, while FATO could be smaller. Peripheral areas are not required. For roof top heliports air currents developed by adjacent building and environmental factors, noise etc. should be carefully considered. Heights of the heliport above ground level may also result in visibility problems due to clouds.

The strength of the floor of the elevated heliport should be higher than the stresses produced by the design helicopter. FAA recommends that for design purposes the touchdown and liftoff area be ca-

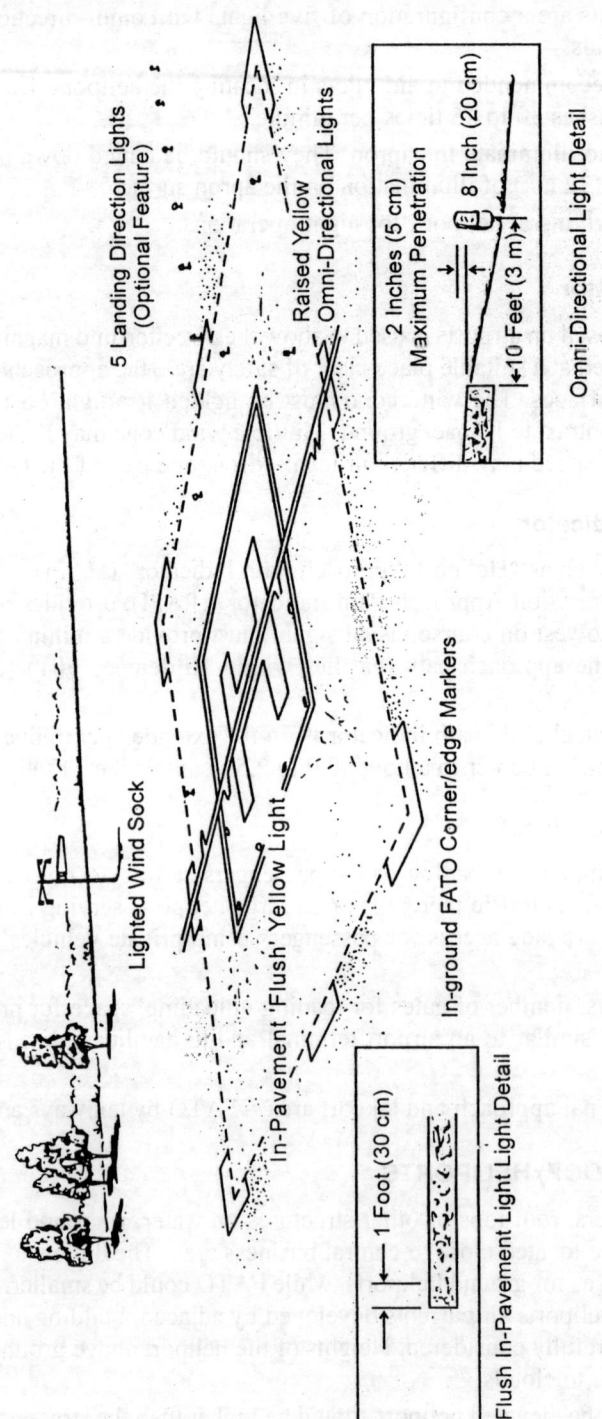

Fig. 17.10. Lighting system at heliport (FAA).

pable of supporting 150 per cent of the maximum takeoff weight of the design helicopter. Roof top heliport TLOF's may be constructed of wood, metal or concrete.

17.9 PAVEMENT DESIGN FOR HELIPORTS

Hard load bearing surfaces should be designed and constructed to support the weight of the design helicopter. Loads are applied through the contact area of the tires for wheel equipped helicopters or the contact area of the skid for skid equipped helicopters. Helicopter landing gear contact area configuration are shown in Fig. 17.1. Helicopter weights are given in Table 17.1.

For design purposes, the static load is equal to the design helicopter's maximum takeoff weight, applied through the total contact area of the wheels or skids. For design purposes, dynamic loadings may be assumed at 150 per cent of the takeoff weight of the design helicopter.

TLOF's, taxiways, parking aprons should usually be paved to improve load carrying capacity, to minimise the erosive effects of rotor wash, and to facilitate surface drainage. Stabilised unpaved portions of FATO and taxi routes are used.

Pavement design principles are the same as for airports. In most of the heliports 15 cm thick portland cement concrete (PCC) pavements are able to support operations by helicopters weighing upto 9216 kg (20000 lbs). Thicker pavements may be required for heavier helicopter.

STOL PORTS

Facilities provided for short takeoff and landing (STOL) aircrafts are called stolports. Demands on the airports on large cities are rapidly increasing. For short air travels, therefore, STOL aircrafts are becoming popular as they provide better service to passengers, because of their high lift capability. STOL aircrafts require shorter runways than conventional aircrafts, they produce lesser noise than VTOL aircrafts and thus best suited for short-haul intercity air travel. By developing STOL ports, the congestion on airport can be reduced. As some airports, separate short runways are provided for STOL aircraft operations for more efficient utilisation of both available air-space and ground facilities.

17.10 CHARACTERISTICS OF STOL AIRCRAFTS

STOL aircrafts presently have capacities of about 25 to 50 passengers. They are operated by turboprop engines. STOL aircraft technology is still developing regarding methods of providing lift, number of passengers in the aircraft and the requirements of runway lengths. The normal runway length requirements of the present day STOL aircrafts is about 1000 to 1500 meters. Table 17.3 shows characteristics of some typical short haul passenger aircrafts. Dimension are given in Ft-lb system. For conversion of meter-kg system, multiply by the conversion factors. STOL airplanes are solution to the increasing airport bottlenecks that are causing deterioration of the quality, increasing operating cost and for short hauls, even the usefulness of air transportation.

17.11 ADVANTAGES OF STOL AIRCRAFTS

STOL aircrafts offer several advantages to the continually growing domestic air travel needs. Some of the main advantages are listed below :

- STOL aircrafts are lesser noisy than VTOL aircrafts.
- STOL aircrafts are best suited to intercity air travel.
- STOL ports can reduce congestion and delays at CTOL airports.

Table 17.3. Characteristics of typical short haul aircrafts

Aircraft type	Fuselage length ft.	Wing span ft.	Wheel base ft	Number and engine type	Number of passengers	Maximum take off weight	Required runway length
1. ATR 42-300	74'-5"	80'-7"	13'-5"	2 TP	42-50	36815	3576
2. ATR-72	89'-2"	88'-9"	13'-5"	2 TP	64-74	47400	4620
3. B Ac 146-100	85'-10"	88'-5"	15'-6"	4 TF	94	84000	4000
4. Canadiar RJ 100ER	87'-10"	69'-7"	25'	2 TF	50	51000	5265
5. Casa C-212-300	53'	66'-7"	10'-2"	2 TP	25	16975	2680
6. Cessana 402C	36'-5"	44'-1"	18'	2 P	10	6850	2195
7. DeHavilland DNC7	80'-8"	93'	23'-6"	4 TP	52	44000	2260
8. Dornier 228-212	54'-4"	55'-8"	10'-10"	2 TP	20	14109	2250
9. Fokker-50	82'-10"	92'-1"	23'-8"	2 TP	50	45900	4450
10. Fokker-100	106'-8"	92'-2"	16'-7"	2 TF	107	98000	5645
11. SAAB 2000	88'-9"	81'-3"	27'	2 TP	50	47000	4920
12. Shorts 330	58'-1"	74'-8"	13'-11"	2 TP	30	22900	3420
13. Shorts 360	70'-10"	74'-10"	13'-11"	2 TP	36-39	27100	4280
14. SToL-BLC	100'-4"	123"	⊞	2 TF	90	101345	–

P : Piston; T : Turbofan; TP : Turboprop.

- Because of high lift capability STOL aircrafts require shorter runways than CTOL aircrafts.
- Facilities closer to origin and destination can be provided by STOL ports.
- STOL aircrafts are more maneouverable and can make efficient use of congested airspace.
- STOL aircrafts offer better economies than VTOL aircrafts.
- Future STOL aircrafts shall be able to accommodate more passengers upto 150 or so with smaller airfields.

17.12 PHYSICAL REQUIREMENTS FOR PLANNING AND DESIGN OF STOL PORT

ICAO and FAA recommendations for design and planning for first STOL aircrafts such as DHC-6, DHC, published in 1988 or so, are given below so that reader may have better understanding of their requirements as compared to CTOL provisions. For STOL aircrafts and ports, research is still ongoing to develop better, lesser noisy aircrafts which may accommodate more passengers, but still require lesser runway lengths and shorter airfield dimensions. Planning criteria for STOL ports are the same as for conventional airports.

- Site for a STOL port should be close to origin and destination of the air traffic.
- Site should be easily accessible by land transportation, e.g., cars, buses, mass transportation systems.
- Adequate parking space should be available at the site.
- The primary surface which is an imaginary plane centered over the runway should be the runway length plus 30 m at either end of runway. Its width should be 90 m.
- The approach and departure surface length should be 3000 m and the inner width should be 100 m. The outer end width should be 1020 m.

- The approach and departure surface slope should be 15 : 1.
- The transitional surface slope is 4 : 1. The width of transitional surface should be 330 m.
- The inner width runway clear zone should be 90 m and outer width should be 160 m. The length of clear zone is 225 m.
- Runway configuration for conventional airports can be adopted for STOL ports also, but for easy approach and faster service, a single runway with parallel taxiway is most common configuration for STOL ports.
- For large STOL aircrafts, the runway widths range from 20 m to 25 m, depending upon the aircraft.
- Runway lengths at mean sea level and 32°C temperatures vary from 450 m to 550 m. Corrections for elevation, temperature and gradient is done like CTOL aircrafts.
- For small STOL aircrafts, taxiway width of 7 m to 7.5 m is adequate. For bigger aircrafts taxiway width may be 18 m or more. A typical layout of STOL port is shown in Fig. 17.11.

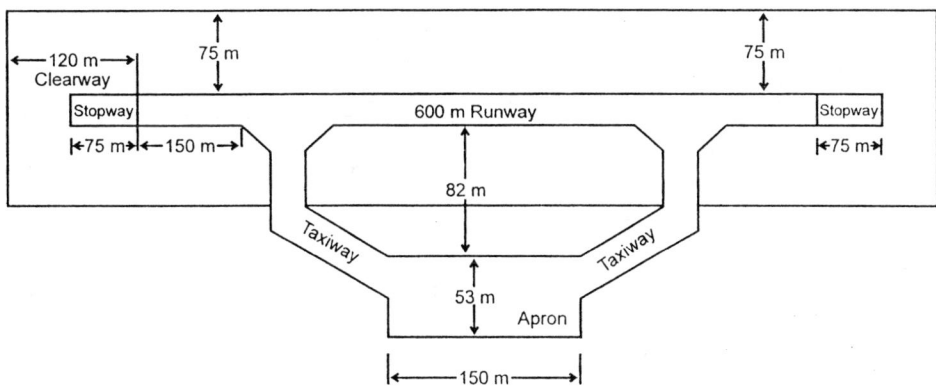

Fig. 17.11. Typical Stolport layout.

17.13 MARKINGS AND LIGHTING OF STOLPORT

17.13.1 Stolport markings

Stolport marking and signing criteria is the same, as for airport marking, signing and lighting already described in earlier chapters. Markings recommended by FAA are shown in Fig. 17.12. White markings are used on taxiways. The basic markings are centreline marking, threshold marking, touchdown zone marking, runway edge marking etc.

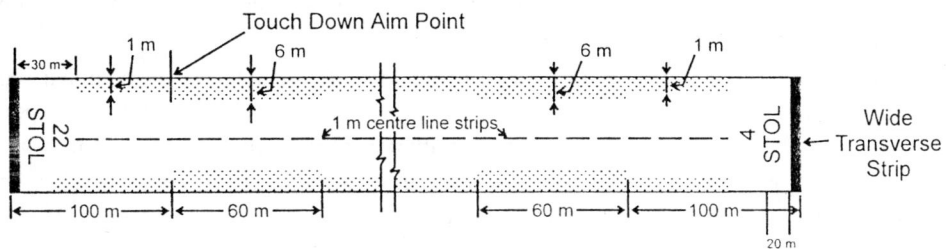

Fig. 17.12. STOLPORT runway marking (FAA)

Threshold marking is shown by the letter STOL and wide transverse strip at the beginning of the runway. STOL letters are 20 m thick as shown in the Fig. Runway direction number is written above the threshold marking. The edges of the touchdown zone are shown by solid lines 6 m wide and 60 m long starting at 100 m from the runway on either side of the centreline of the runway.

17.13.2 STOL port lighting

Lighting mainly consists of threshold lighting, runway edge lighting and approach lights, following the same criteria as described earlier regarding airport runway lighting. Lightings assist landing and take off operation during nights and reduced visibility conditions during the day.

A typical lighting system recommended by FAA is shown in Fig. 17.13.

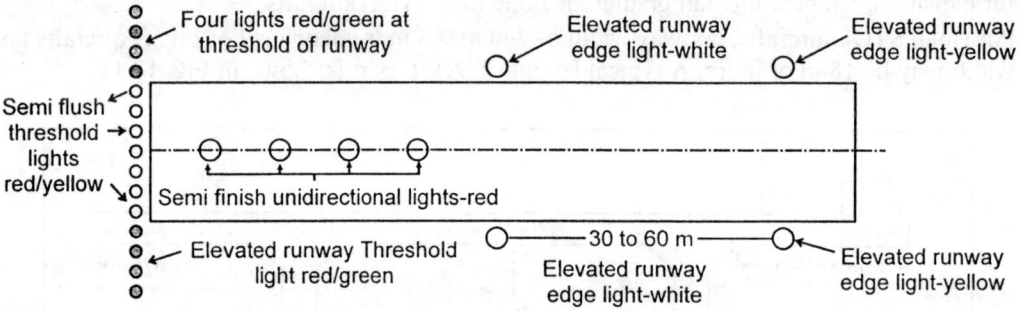

Fig. 17.13. STOL port runway lighting (FAA).

"Threshold lights" consist of a group of 4 lights forming a wing bar on both sides of the centreline of runway.

"Edge lights" are alternate yellow and white in colour spaced at 30 to 60 m interval, 3 m from runway edge.

"Runway end lights" are on line with runway threshold lights spaced 1.5 m, the centre one being on runway centre line extended.

Four lights beginning at 15 m from runway threshold and spaced at 15 m c/c are used as "distance indicator lights".

VERTIPORTS

A vertiport is an identifiable ground or elevated area, used for takeoff and landing of **tiltrotor** aircraft and **rotorcraft**. A vertistop is used/designed solely for takeoff and landing of tilt-rotor aircraft and rotorcraft to drop off or pickup passengers or cargo.

By tiltrotor technology air transportation capability and system capacity will be enhanced. By the tiltrotor aircrafts and vertiports large number of passengers can be provided with air transport facility located near the city as compared to fixed-wing commuter aircrafts using conventional airports.

Vertiports can be on airports, operating without conflict with fixed wing traffic. They can be located near city centres. Tiltrotor aircrafts and other vertical lift aircrafts can serve cities located 600 to 700 kms away and can operate from within the city locations near to origin and destination of passengers (population).

17.14 VERTICAL FLIGHT AIRCRAFTS

Vertical flight aircrafts are those capable of vertical or near vertical takeoffs and landings. Types of vertical flight aircrafts are :

1. *Rotorcraft* : Rotor-winged aircraft that lift vertically (to hover) and sustain forward flight by power driven rotor blades, turning on a vertical axis.

2. *Tilt rotor aircraft* : Tilt rotor aircrafts with the axes of the power-driven proprotor blades capable of pivoting from vertical for vertical takeoff, landing and hover operations to horizontal to drive life from the wing in cruise.

3. *Tilt-wing aircraft* : Rotorcraft with both the wings chord and the axes of the power-driven proprotor blades capable of pivoting from vertical for vertical takeoff, landing, and hover operations to horizontal to drive lift from the wing in cruise.

4. *Fan-in-wing aircrafts* : Fixed wing aircraft with rotor fans in the wings to permit vertical or hover operations.

Table 17.4 and Fig. 17.14 show characteristics of tiltrotor aircrafts.

Table 17.4. Characteristics of tiltrotor aircrafts (FAA)

Aircraft/model	Span tip tip ft.	Rotor diameter ft.	Overall length ft	Height ft.	Landing gear tread ft.	Gear wheel base ft.	Weight lbs	Number of passengers
V-200 (fan in wing)	29	4	32	10	7	14	4500	6
TW-68 (tilt wing)	41	17	39	13	9	15	16500	11-16
CTR-800	58	26	41	15	8	15	15750	8
Eurotor	76	33	52	20	13	28	18000	19
CTR-1900	65	28	47	17	13	20	22800	19
Eurotor	86	36	65	21	13	28	36000	30
CTR-22A/B	85	38	57	18	15	22	45120	31
CTR-22C	85	38	69	21	11	30	46230	39
CTR-22D	86	38	72	23	11	30	49260	52
CTR-7500	109	46	84	28	17	28	79820	75

(all dimensions are approximate)

17.15 SITE SELECTION FOR VERTIPORTS

Following considerations should be kept in mind for location of a vertiport.

- Normally they are located on ground level near population centres, well connected by modes of transportation or public transport.
- Vertiports can also be located on piers, roof of buildings etc.
- Size of vertiport should be enough to accommodate the largest vertical lift aircraft forcasted to use the facility.
- Wind flow around buildings should be studied like a heliport placement.
- Enough parking area should be available near the site.

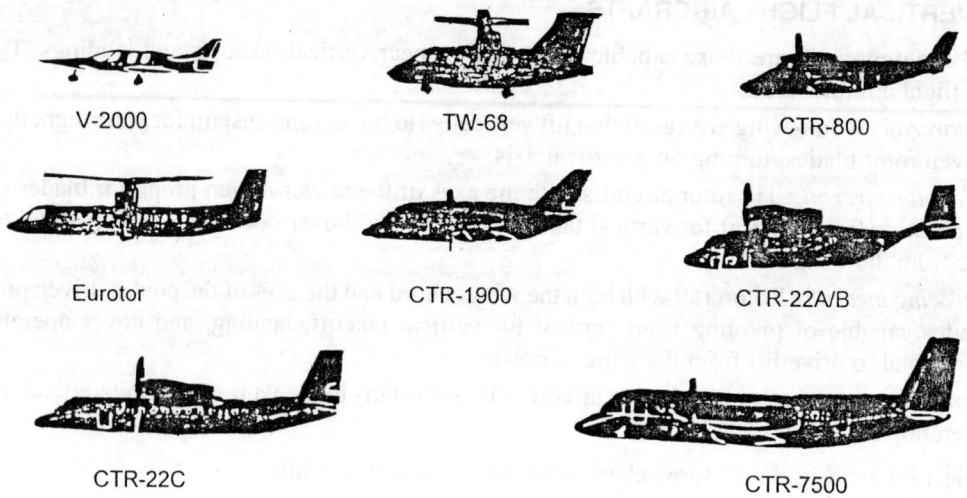

V-2000

TW-68

CTR-800

Eurotor

CTR-1900

CTR-22A/B

CTR-22C

CTR-7500

Fig. 17.14. Some Tiltorotor Aircrafts (FAA).

Rotor span is distance between the extreme edges of the plane(s) generated by spinning rotors or proprotors.

17.16 PHYSICAL REQUIREMENTS FOR PLANNING AND DESIGN OF VERTIPORTS

17.16.1 Final approach and takeoff area (FATO)

- FATO may have any shape, but the size must be such that it will circumscribe a square with 75 m sides. If the elevation of FATO is more than 300 m, the FATO length must be increased by 15 m per 300 m of elevation above mean sea level (msl).
- The surface of FATO must be free from objects that may adversely affect the takeoff or landing of tiltrotor aircrafts and be resistant to rotor downwash.
- The primary axis of FATO should be aligned to provide maximum wind coverage, like airports.
- The slope in any direction in unpaved portion of FATO should be 1.5 to 5%. The maximum slope of paved portions of FATO should not exceed 1%.
- For IFR the FATO is 90 m wide and 373 m long.

17.16.2 Touchdown lifeoff area (TLOF)

- The TLOF, located within FATO is a hard or paved surface capable of supporting the heaviest tilt-rotor aircraft or other vertical-lift aircraft, expected to use the facility.
- The TLOF must be of dimensions such that it is capable of circumscribing a square with 30 m sides for VFR and 45 m side for IFR operations.
- An elongated TLOF upto 120 m offers significant operational and economic benefits and should be used wherever possible.
- Normally the TLOF is centered in the FATO with its primary axis centered on the final approach course. If not centered, there must be at least 22.5 m of clear space between edges or ends of the

TLOF. For IFR operations, the TLOF must be located in FATO with its primary axis centered on the final approach course.

- Longitudinal and transverse gradient of TLOF should not exceed 1 per cent.
- Vertiports with significant number of operations should have paved shoulders, 7.5 m wide.

17.16.3 Taxiways

- A taxiway or hover taxiway must be obstacle free.
- A ground taxiway requires a cleared safety area of 45 m wide. The safety are for hover taxiway is 75 m wide.
- Paved taxiways should be 22.5 m wide.
- For economy, a taxiway may be constructed with a load bearing centre section equal to the aircraft's gear width plus 6 m.
- On a taxiway, a clearance of at least one-half the rotor span but not less than 7.5 m should be provided between the limits of rotor span of an aircraft and a fixed or movable objects, when the undercarriage of the aircraft is located 4.5 m from the edge of the taxiway.
- The gradient of paved taxiways (longitudinal and transverse) should not exceed 2%.
- The minimum distance between the centreline of an elongated TLOF and the centreline of a parallel taxiway or between centrelines of parallel paved taxiways is 60 m.

17.16.4 Aprons and ramps

- Aircraft parking positions should provide a minimum of one-half tip to tip span but not less than 7.5 m, separation between the limits of rotor span of adjacent aircraft or between the limits of the rotor span of an aircraft and fixed or movable object.
- Apron or ramp longitudinal and transverse gradient should not exceed 2% in any direction.

17.16.5 Pavement design

- The TLOF, taxiways and aprons must be designed to support the heaviest tiltrotor or other aircraft expected to use the facility.
- To compensate for the dynamic forces created by the tilt-rotor aircrafts and rotorcrafts, pavement design should be based upon a load equal to 150% of the maximum takeoff weight of the design aircraft with 75% of that total load distributed equally to the main landing gear.
- Light duty paved shoulders 7.5 to 11 m wide around all operational surfaces will help to alleviate dust and debris.
- Materials resistant to high temperatures should be used for pavements and pavement joints.

17.17 AIR SPACE REQUIREMENTS FOR VERTIPORT

17.17.1 VFR : Visual approach

- For VFR operations, the approach surface begins at the edge of the primary surface and extends outward a distance of 1200 m at a slope of 8 : 1 (horizontal to vertical). The inner width of the approach surface is same as that of the primary surface. The outer width is 195 m.
- The primary surface is established in the horizontal plane which has the size and shape of the final approach and takeoff area (FATO).
- The transitional surfaces extend outward and upward from the lateral boundaries of the primary and the approach surface for a distance of 98 m at a slope of 2 to 1.

- Curved approach surfaces are permitted, but the final 360 m segment must be straight and aligned with the centreline of the touchdown and lift area (TLOF).

17.7.2 IFR : Non-precision instrument approach

- The primary surface for non-precision instrument approaches must be at least 100 m long and 100 m wide. The approach surfaces begin at the edge of the primary surface and extend outward a distance of 1500 m at a slope of 20 to 1. The inner width of the approach surface is 150 m and the outer width is 600 m.
- Transitional surfaces extend outward and upward from the lateral boundaries of the primary surfaces and the approach surface. The planes slope upwards at a ratio of 4 : 1 (horizontal to vertical) for distance of 105 m measured horizontally from the boundaries of the primary surface and the distance varies along the boundaries of the approach surface from 105 m at the primary surface end and narrows to 0 at 600 m from the primary surface. Fig. 17.15 and Fig. 17.16 show these surfaces.

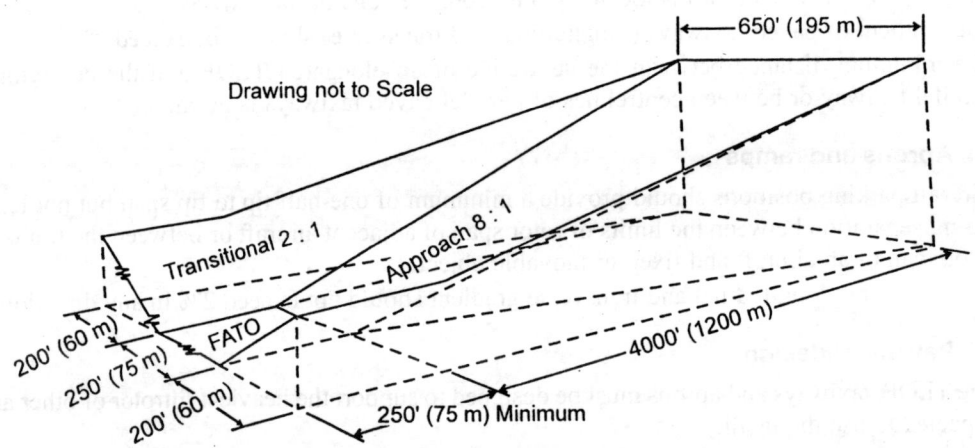

Fig. 17.15. Vertiport visual approach surfaces (FAA).

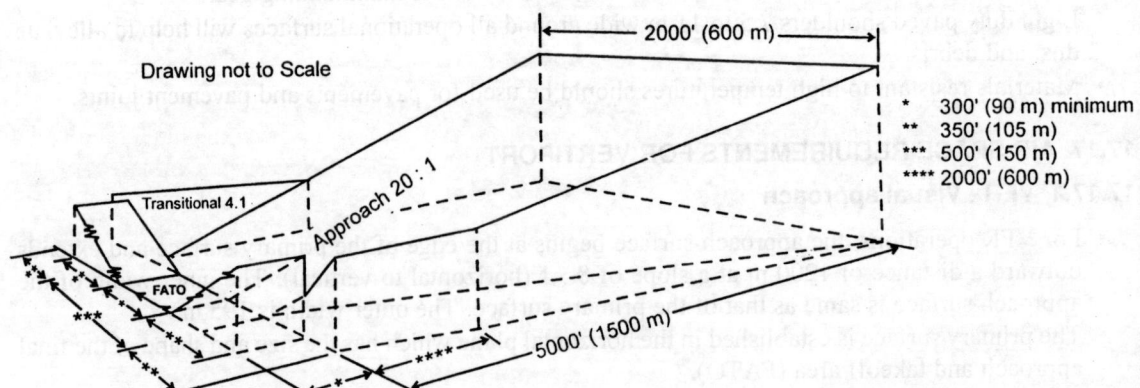

Fig. 17.16. Vertiport non-precision instrument approach surfaces (FAA).

17.17.3 Precision-instrument surfaces for different glide slopes

Precision-instrument operations can be done at vertiports with $4°$, $6°$ or $9°$ glide slopes.

- For $4°$ or $6°$ glide slope, the approach surface is trapezoid in shape. It begins 373 m from the far end of FATO and extends outward a distance of 7500 m at a slope of $17 : 1$ for a $6°$ glide slope and $25 : 1$ for a $4°$ glide slope. The inner width of the approach surface is 300 m and the outer width is 1800 m.
- Transitional surfaces extend outward from the edge of the approach surface for a distance of 450 m at a slope of 7 to 1 (horizontal to vertical) on each side of approach surface.
- The transitional surface provided to the approach surface gradually narrows to 180 m at a distance of 373 m from the beginning of the approach surface. Transitional surfaces also extend outward from the edge of the final approach and takeoff area (FATO) at a distance of 105 m at a slope 7 to 1.
- For a $9°$ glide slope, the approach surface is trapezoid. It begins at the edge and end of the primary surface with a width of 75 m and extends outwards a distance of 3000 m at a slope of $10 : 1$. The inner width of approach surface is 75 m and outer width is 900 m.
- The transitional surfaces extending outward from the edges of the approach surface for a distance of 293 m at slope of $7 : 1$ are required on each side of the approach surface.

The transitional surface gradually narrows to 180 m at the beginning of approach surface. Transitional surfaces also extend outward from the edges of the primary surface area a distance of 180 m at a slope of 7 to 1. Figs. 17.17 and 17.18 show $6°$ and $9°$ approaches.

17.18 MARKING AND LIGHTING OF VERTIPORTS

17.18.1 Vertiport markings

- The limits of the FATO should be defined with in-ground edge marker placed at the corner and at intervals of approximately 37 m along the FATO edges. Alternatively raised markers, no more than 15 cm high may be used in lieu of in-ground marker. Edges of paved FATO are marked with 40 cm wide painted white lines.
- The edges of TLOF are marked with a 40 cm wide painted white lines not more than 30 cm in from the TLOF edge. When the length of the TLOF is more than 3 times its width, the TLOF has a centre line marked with 15 m long, 40 cm wide painted white lines separated by 7.5 m spaces. The centre line begins 7.5 m from the numerals indicating the magnetic direction of the approach.
- A distinctive marking is recommended to identify a vertiport, as shown in Fig. 17.19.
- On a vertiport or vertistop, where the maximum aircraft weight and rotor span are restricted, the restriction is indicated in the near right corner of the TLOF, area by a bar, with number above the bar showing the restricted takeoff weight and number below the bar showing maximum rotor span allowed to land.
- Taxiway centre lines are delineated by solid yellow lines with 15 cm wide line. Taxiway edges are marked with double 15 cm wide lines spaced 15 cm apart.
- The portion of taxiway pavement that is not full strength is marked with yellow 1 m wide bars perpendicular to the taxiway centre line at interval not more than 9 m.
- Cylindrical raised markers are used at the outer edges of the hover taxi route safety area, of 1 m in height, to be retroreflective and provide maximum colour contrast with natural background.
- Painted 15 cm wide yellow lines are used to guide aircrafts into parking positions.

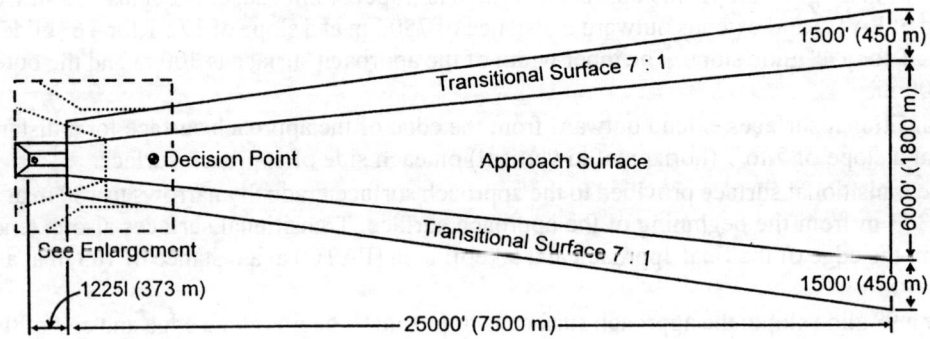

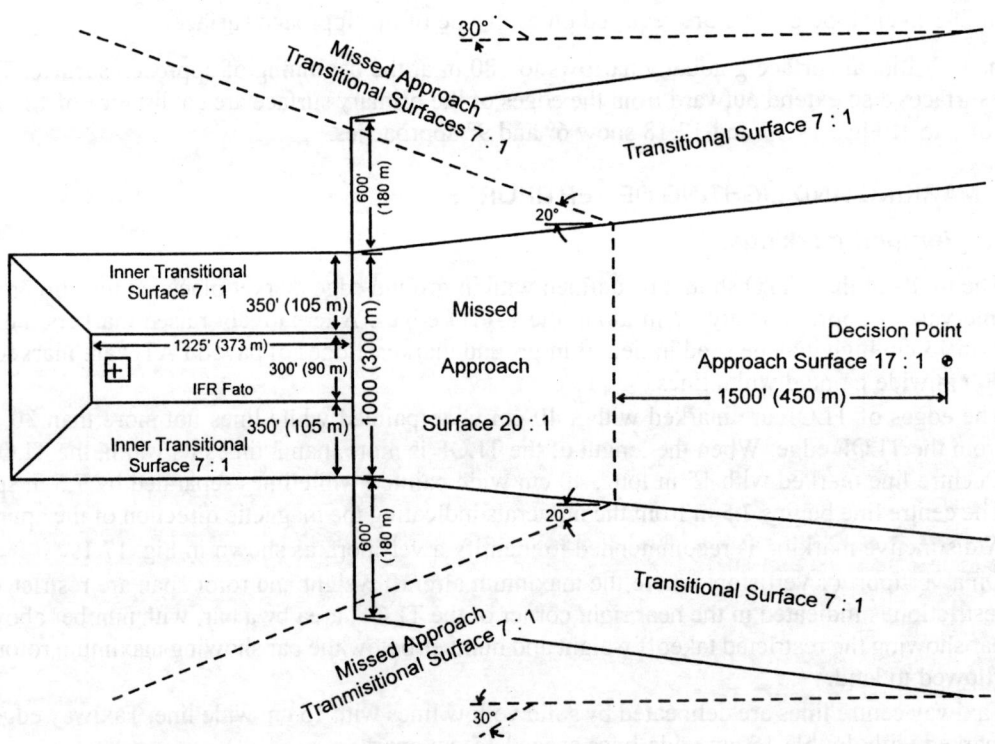

Fig. 17.17. Vertiport 6° precision instrument approach surfaces (FAA).

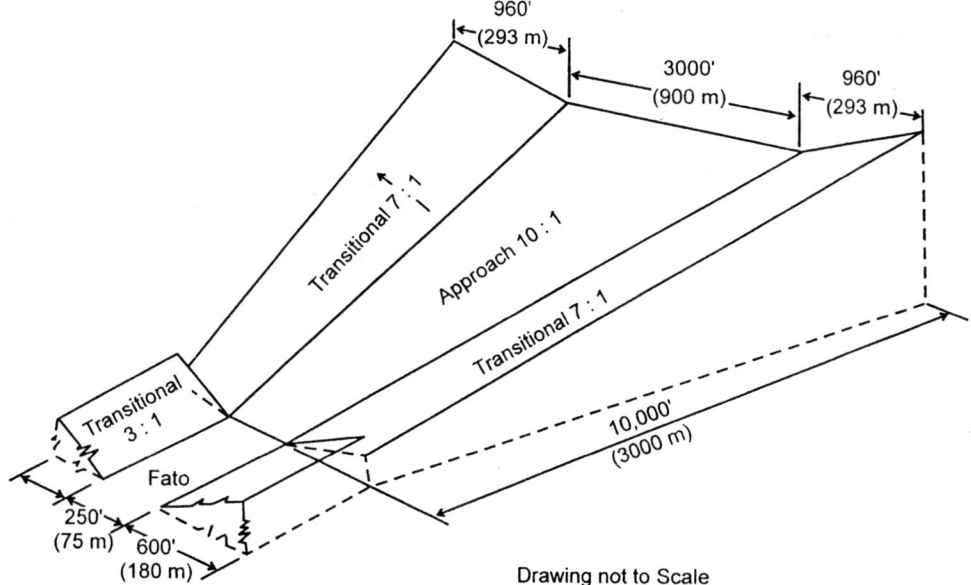

Drawing not to Scale

Fig. 17.18. Vertiport 9° precision instrument approach surfaces (FAA).

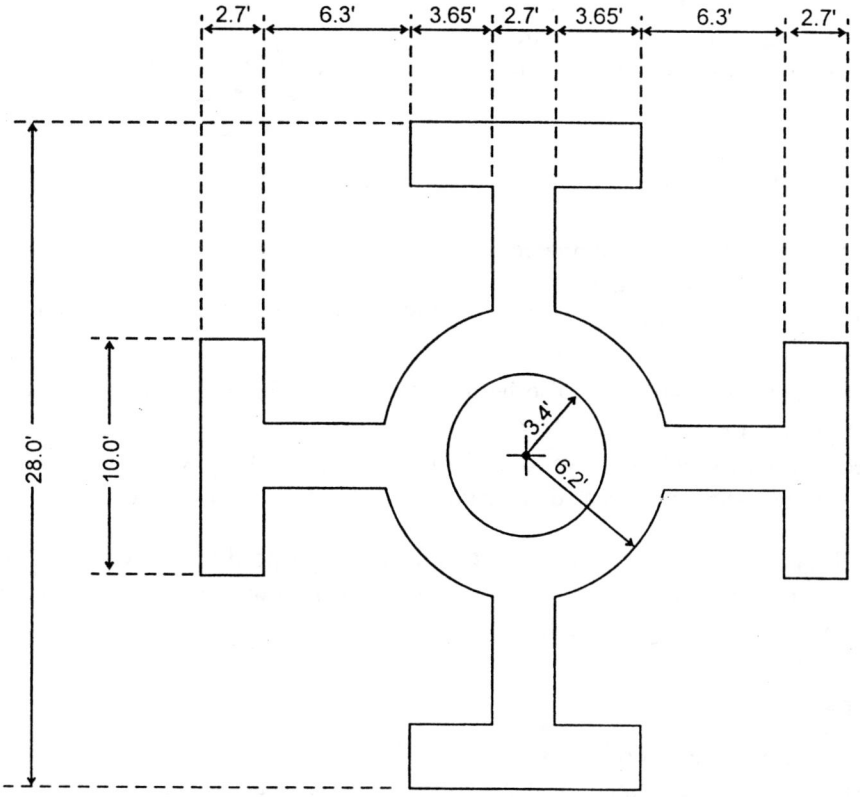

Fig. 17.19. Vertiport marking (FAA).

17.18.2 Vertiport lighting

- Vertiports are lighted with standard aviation lighting fixtures.
- If TLOF is lighted, the limits of a ground level FATO need not be lighted.
- On elevated vertiports or vertistops, an odd number of red double lens obstruction lighting fixture at 15 m interval are used to mark the edges of the building.
- The touchdown and lift-off area (TOLF) is delineated by at least five omnidirectional yellow lights located at least 3 m apart from the outside edge, same lights used for FATO.
- Blue taxiway lights, not more than 20 cm high, define taxiway edges. The lights are located 3 m out from the edge of the full strength pavement.
- Alternatively inset green bi-directional lights may be used to mark the centre line of a paved taxiway, including a taxiway located on a ramp or apron.
- The edges of the hover taxiway are defined by blue taxiway lights spaced not more than 15 m apart.
- Aprons or ramps may be illuminated with flood lights located so that the mounting poles are not hazardous to aircraft operation.

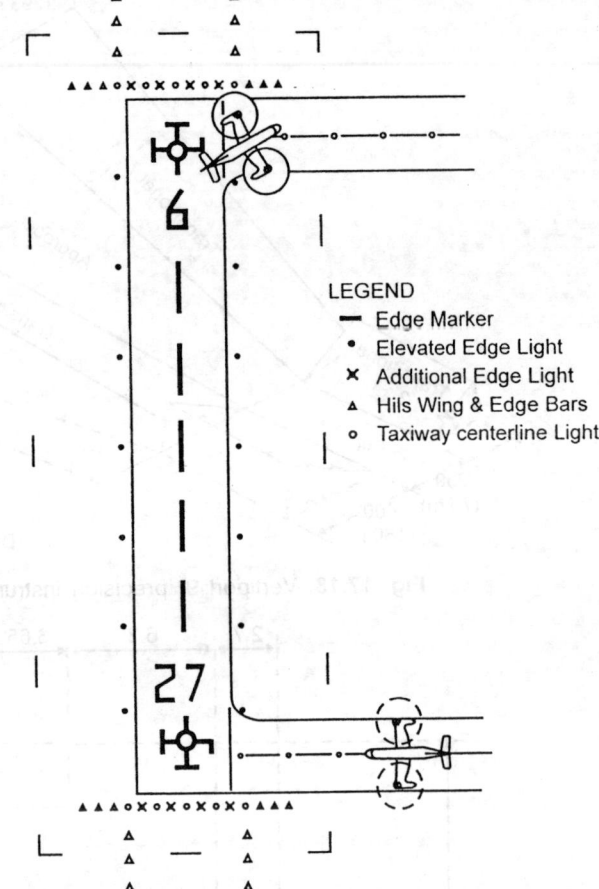

Fig. 17.20. Typical vertiport marking and lighting (FAA).

LEGEND
- Edge Marker
- Elevated Edge Light
- × Additional Edge Light
- ▲ Hils Wing & Edge Bars
- ○ Taxiway centreline Light

- Yellow bi-directional lights may also be used to delineate guidance routes for aprons or ramp parking positions located at 3 m intervals.
- A lighted Vertiport should have an identification beacon flashing white-green-yellow pulses of light. The beacon should be located on or immediately adjacent to vertiport.

The basic FATO and/or TLOF lighting system must be enhanced for the vertiports having precision instrument approach operation, as per special recommendations of FAA. A typical lighting and marking system is shown in Fig. 17.20.

AIR SCOOTER

Based on the principle of a rotorcraft, Newada Air Scooter Manufacturing Company is planning to launch an 'Air-scooter' in near future, which shall be a revolutionary step in aeronautical research.

Developed by one Mr. Woody Novis, this will be an ultralight air vehicle having following characteristics :

Power of the engine	65 HP
Type of the engine	Aero twin, 4 stroke
Diameter of the rotor	14 ft.
Overall height	11 ft.
Overall width	7 ft.
Overall length	12.5 ft.
Total weight of scooter	300 lbs
Fuel capacity	5 gallons
Speed of scooter	55 kms/hr
Expected price	Rs. 50 lacs

CHAPTER 18

Drainage and Environmental Considerations for Airports

DRAINAGE

18.0 INTRODUCTION

Airport drainage is very important for safe movement of the aircrafts. When the runway surface is affected by water to any degree of wetness (i.e. from damp to a flooded state), the friction levels provided by individual runways drop significantly from the dry value. Degradation of available friction can have serious implications on safety, regularity and efficiency of operations. Wet runways could be significant hazard and potential threat to flight operations. Airport drainage characteristics should, therefore, be carefully studied and planned.

Standing or excessive water on the pavement or below the pavement reduces the structural ability of the airport pavement, and may result in early failure/deterioration. Natural drainage at site is rarely sufficient to satisfy the needs of airports and artificial drainage system must therefore, be installed.

Adequate drainage system is required for :

  (i) Removal of surface water.

 (ii) Diversion of surface and ground water from adjacent areas.

(iii) Removal of sub-surface flow of water from the airport.

The drainage system should have sufficient capacity to handle present and future drainage problems. The drainage demand is determined by local precipitation rates. Proper grading of the site with adequate longitudinal and cross slopes, helps in quick drainage of storm water. The topographic map of the area helps in identifying the directional of natural surface flow of water, from the airport area.

Surface, sub-surface drains, intercepting drains, properly designed and located, consist the adequate drainage system for an airport.

18.1 CONSIDERATIONS FOR AIRPORT DRAINAGE

For designing the drainage system for an airport, it is necessary to know the design storm – the quantity of flow, which the system should accommodate, the intensity and duration of storm, and the amount of runoff.

18.1.1 Design storm

Selection of the design storm is an economic consideration. A severe storm occurring infrequently, will result in interruptions if system is designed for storm of lesser capacity, but if higher capacity is provided, it may not be economically justified.

FAA therefore, recommends that for civil airports, the drainage system should be designed for a storm whose probability of occurrence is once in 5 years. The design should, however, be checked with a storm of lesser frequency (10-15 years) to make sure if serious damage or interruption of traffic would result from such a storm. Drainage for military airport is based on a 2-year storm frequency.

18.1.2 Intensity-duration for design storm

The amount of rainfall at the site of airport is determined for designing the drainage system. How often a rainfall of particular intensity shall occur is also important. Rainfall intensity is expressed in mm/hour or inches/hour. Sufficient rainfall data concerning intensity-duration, frequency of storm should be developed/collected for the airport site, through meteorological studies. The rainfall intensity-duration curves are then prepared for the site.

18.1.3 Determination of amount of runoff

The analysis of airport surface drainage and amount of runoff depends upon several factors such as porosity and type of soil, vegetation cover over soil, slopes, moisture content of soil, etc.

The runoff, for airport drainage is calculated from the following expression :

$$Q = CIA$$

where

  Q = runoff from given drainage area cub m/sec.
  C = ratio of runoff to rainfall
  I = rainfall intensity for time of concentration of runoff mm/sec.
  A = drainage area in 100 sq. metres.

18.1.4 Time of concentration

Time of concentration is the time taken by water to reach the drain inlet from the most remote point in the tributary area. There are two components of the time of concentration :

  (i) *The inlet time* : It is the time required for water to flow overland from the most remote point in the drainage area to the inlet, and

  (ii) *The flow time* : Which is the time taken by the water to flow from the drain inlet, to the point in the system under consideration.

The time of flow given by empirically derived formula :

$$D = KT^2$$

where

  D = distance in metres
  K = a dimensionless empirical factor which is dependent upon slope, roughness of terrain, vegetation cover and distance to the drain inlet
  T = time in minutes

The inlet times can be estimated from Fig. 18.1 given by FAA.

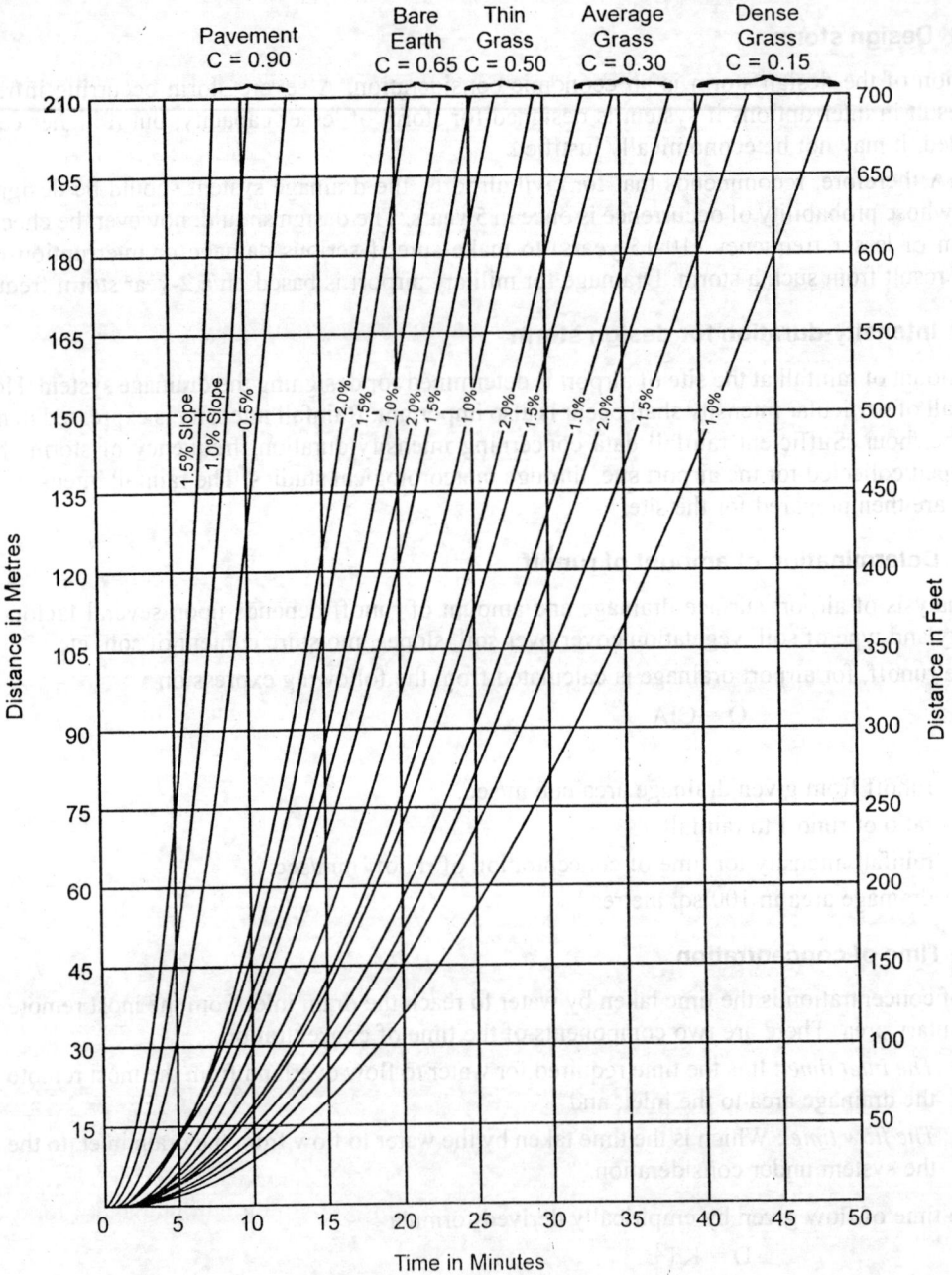

Fig. 18.1. Inlet time curve (FAA).

18.1.5 Coefficient of runoff

The runoff rate is variable from storm to storm and even during a single period of rainfall. The coefficient of runoff depends upon :

- slope and type of surface
- type of soil and its porosity, and
- size of the drainage area.

The range of values suggested by FAA are given in Table 18.1.

Table 18.1. Coefficient of runoff C (FAA)

Types of surface	Factor C
All water tight roof surfaces	0.75–0.95
Asphalt runway pavements	0.80–0.95
Concrete runway pavements	0.70–0.90
Gravel or mecadam pavements	0.35–0.70
For impervious soils (heavy)*	0.40–0.65
Impervious soils with turf*	0.30–0.55
Slightly pervious soils*	0.15–0.40
Slightly pervious soils with turf*	0.10–0.30
Moderately pervious soils*	0.05–0.20
Moderately pervious soils with turf*	0.00–0.10

* For slopes from 1 to 2 per cent.

For the drainage area consisting of different types of surfaces with different infiltration characteristics, the runoff coefficient is computed by following expression :

$$C = \frac{A_1 C_1 + A_2 C_2 + A_3 C_3}{A_1 + A_2 + A_3}$$

where
A_1, A_2, A_3 = areas of different types of surfaces
C_1, C_2, C_3 = respective coefficient of runoff

18.1.6 Intensity-duration rainfall pattern

To accommodate the worst condition, in designing the pipe system a separate duration of storm is selected for each sub-drainage area tributary to a drain inlet. The duration of the storm is made equal to the sum of the inlet time and time to flow. A typical intensity-duration rainfall pattern for a five-year frequency storm, at a site is shown in Fig. 18.2.

18.1.7 Ponding

In design of the airport system, ponding (temporary storage of runoff before its entry to underground system) is used to reduce the cost. For purposes of design computation the ponded volume is assumed to be an inverted pyramid or a truncated pyramid, the height of which is depth of water above the inlet at any stage. The area of the base of pyramid is taken as the area of the pond.

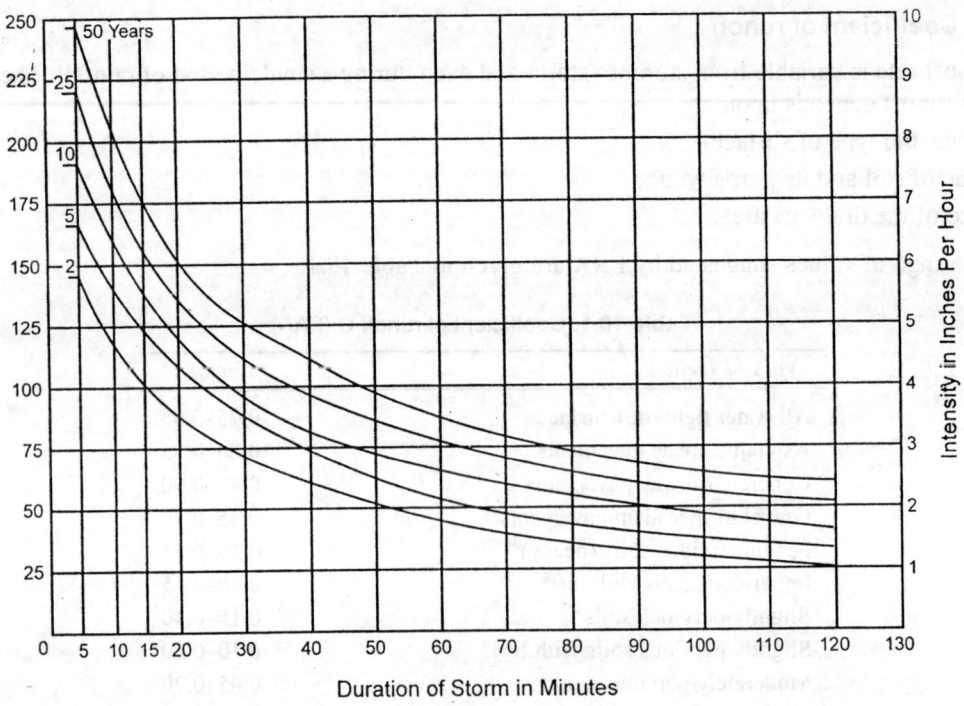

Fig. 18.2. Typical rainfall intensity-duration curve.

Ponding is not allowed on runways, taxiways, parking aprons, runway and taxiway shoulders and on any area where standing water may damage subgrade, sub-base due to saturation. It is permitted on open airfield areas, where damage is not likely to occur. Ponding area edges are kept at least 22.5 m away from edges of the pavements, to prevent damage due to saturation of ground near the pavement.

The volume of the possible storage is calculated by the formula :

$$V = \frac{1}{3} b \left(A + B + \sqrt{AB} \right)$$

where

V = volume of pond in cubic metre

b = height of pond in metre

A = area of upper surface of the pond in sq.m.

B = area of lower surface of the pond in sq.m.

18.2 DESIGN PROCEDURE FOR SURFACE DRAINAGE

Surface drainage is a basic requirement of importance for airport. It serves to minimise water depth on the surface. Adequate surface drainage is provided by an appropriately sloped surface (in both the longitudinal and transverse direction) and surface unevenness. The capacity of the drainage system should be adequate.

The procedure of surface drainage layout is described below in short.

(a) No ponding provided

- Contour map of the taxiway, runway, apron and airport site is studied and several drainage layouts are worked out for selecting the most economical one.
- The grades of the storm drain should have a mean velocity of 0.75 m/sec to provide sufficient scouring action to avoid silting and to main adequate cross-section for flow all the times. The diameter of the storm drain should not be less than 30 cms.
- Water from the drainage area is collected into storm drain by inlet structures like a concrete box with a covered grate of cast iron, cast steel or RCC. It should be strong enough to support aircraft wheel loads and tire pressures.
- The drain inlets may be placed at an interval of 60 m to 120 m on straight lengths, at suitable places as per configuration of the airport and grading plan, when ponding is allowed.
- Inlet should be located not less than 22.5 m from the edge of the pavement. If no ponding is provided inlets are closer at 52 m to 53 m interval.
- Adequate depths of cover should be provided over the pipes so that the pipes can support traffic.
- Having selected the structure (inlet) and calculated the length of pipe, the time of flow to the inlet from most distant point is established, using suitable value of 'C' and inlet time curves.
- Quantity of runoff is then calculated, using rainfall intensity and concentration data, by the formula $Q = CIA$. This water is drained off through pipe.
- Size of the pipe is calculated by Mannings formula given below :

$$Q = \frac{AR^{2/3}S^{1/2}}{n} \text{ cubic m/second}$$

where :

Q = quantity of water to be carried by the (pipe) drain cub.m/sec.
A = cross-sectional area of the drain in sq.m.
R = mean hydraulic radius in metre, which is area of cross-section of drain pipe divided by wetted perimeter of drain pipe
n = coefficient of roughness, which depends upon type of pipe and channel (Table 18.2).

Table 18.2. Coefficient of roughness n (FAA)

Type of drain	n
(a) Clay and concrete pipes	
Good alignment, smooth joint, smooth transitions	0.013
Less flow favourable conditions	0.015
(b) Corrugated metal pipes	
100% of periphery smoothly lined	0.013
Paved invert, 50% of periphery paved	0.018
Paved invert, 25% of periphery paved	0.021
Unpaved, bituminous-coated or uncoated	0.024
(c) Open channels paved	0.015–0.020
(d) Unpaved channels	
Bare earth, shallow flow	0.020–0.025
Bare earth depth of flow over 3 m	0.015–0.020
Turf depth of flow over 0.3 m	0.040–0.060

(b) Ponding provided

- The pipe of the diameter, as calculated above, is laid at the suitable slope, to carryout the quantity of run-off.
- When ponding is provided the volume of possible storage is calculated by the formula :

$$V = \frac{1}{3} b \left(A + B + \sqrt{AB} \right)$$

where : V = volume of pond in cubic metre; b = height of pond in metre
A = area of upper surface of the pond in sq.m.
B = area of lower surface of the pond in sq.m.

- Runoff volumes are then calculated for rainfall intensities of 5, 10, 15, 20, 30, 60, 90 and 120 minutes duration, using the figure and cumulative runoff verses time curve is plotted.
- Using Mannings formula, discharge capacity of pipe is determined.
- Discharge capacity versus time curve when superimposed over cumulative runoff versus time curve, shows the ponding after beginning of storm. For safety reasons, the maximum ponding should be less than the calculated volume of the pond.

Figs. 18.3 and 18.4 show typical drainage plan and section for runway pavement and runway and parallel taxiway pavement, recommended by Federal Aviation Agency, USA.

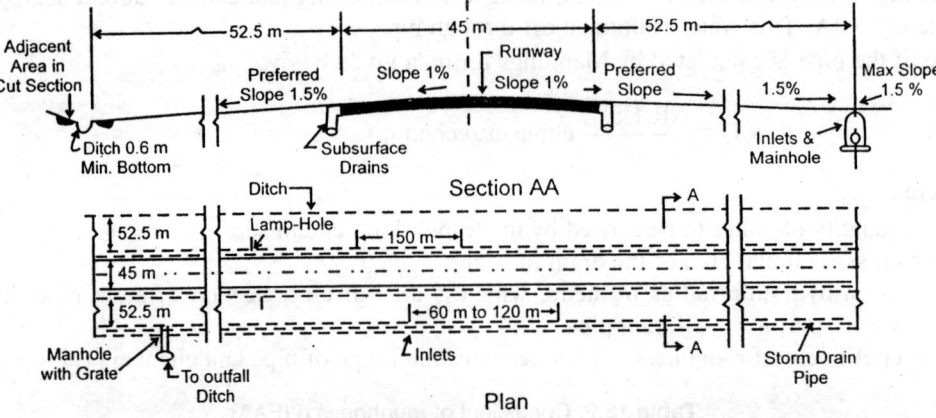

Fig. 18.3. Recommended pavement drainage section and plan (FAA).

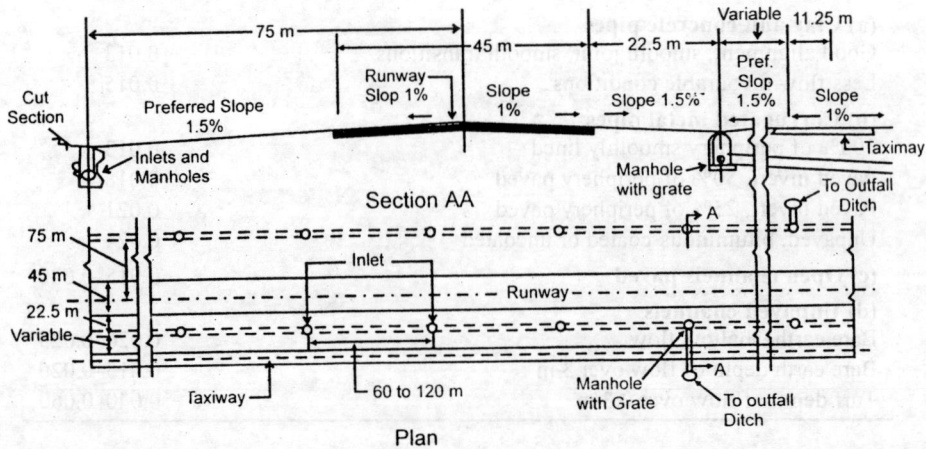

Fig. 18.4. Drainage layout section (plan and section) for a runway and parallel taxiway (FAA).

18.3 SUB-SURFACE DRAINAGE

Sub-surface drainage involves keeping the water away from the subgrade, sub-base and base by intercepting it, and removing it from base course and subgrade.

Base drainage is mainly required where :

   (i) Frost action takes place in subgrade, beneath a pavement.
  (ii) Ground water is likely to rise to the base course level.
 (iii) Subgrade is highly impervious (low permeability).
 (iv) Pavement is subjected to frequent inundation.

Subgrade drainage is needed at locations where water may rise beneath the pavement to less than 30 cms below the bottom of base course.

Intercepting drains are desirable where sub-surface waters from adjacent areas are seeping towards the airport pavements.

Certain soils like gravelly sands, sand and sandy loam etc. are self draining and require very little if any subsurface drainage. Clays, loams etc. require good drainage.

18.3.1 Methods of draining subsurface water

Base-course drainage

Base courses are generally drained by installing subsurface drains adjacent to and parallel to the edges of the pavement. The pervious material in the trench should be up to the bottom of the base course. The drainage pipe is placed at least 30 cms below the bottom of the base course.

The filter material should be put around the drain pipe. Drain pipe is put at a sufficient depth below to provide adequate cover for wheel loading. Diameter, slope of the drain pipe and spacing of drain pipe are determined, after calculating the maximum discharge, coefficient of permeability etc. by hydraulic formulae.

Sub-grade drainage

Subgrade drainage is provided through drainage pipes installed along the edges of pavements and sometimes where ground water table is high, underneath the pavements. A typical subgrade drainage is shown in Fig. 18.5. The centre of the sub-surface drain should be placed not less than 30 cms below the level of the ground water. When subgrade drains are placed along the edges of the pavement, they may also serve the purpose of base course drainage. After determining the maximum rate of discharge the slope and diameter of subdrain pipe are determined using Manning's formula.

The pipe diameter is usually 15 cms. The depth below the base course is not less than 30 cms. Adequate cover is provided for protection under the wheel loads.

Intercepting drainage

Intercepting drainage is provided by means of open ditches well beyond the pavement areas, to prevent water flowing towards pavement. If this is not possible, then subdrains are used. Fig. 18.6 shows a typical intercepting drain.

18.3.2 Sub drainage pipes

Pipes which are used for subdrainage are :

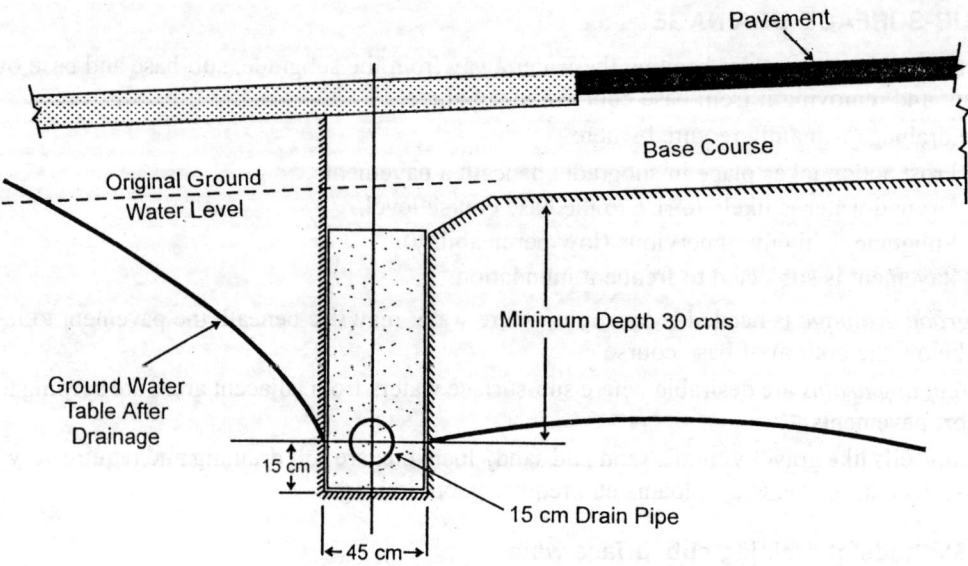

Fig. 18.5. Subgrade drainage details.

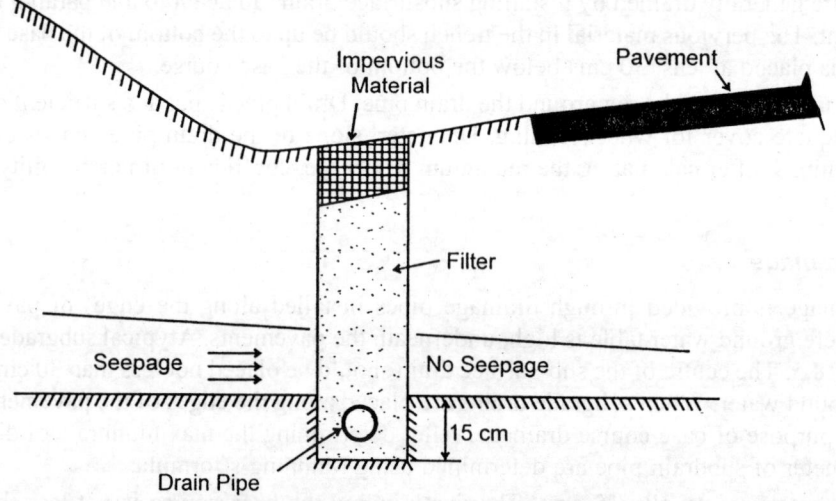

Fig. 18.6. Typical intercepting drain.

- Performed metal, concrete or vetrified clay pipes, their joints are sealed. Perforated areas placed adjacent to the soil. Perforated area normally extends over one third of the circumference of pipe.
- Porous concrete pipes are used. They collect water through seepage by concrete wall of the pipe. Joints of pipes are sealed.
- Special type of bell-and-spigot pipes with slots at the bells made to vitrified clay and cast iron are also used.

Normally 15 cm diameter pipes are sufficient unless extreme ground water problem is faced. To ascertain the size of the pipe, flow is estimated through well known formulae of soil drainage, found in soil mechanic/hydraulic books.

The recommended minimum slope of the pipe is 1 in 650. A minimum thickness of 15 cm of filter material should surround the drain.

For cleaning and inspection, utility holes should be installed along the drain at an interval not more than 300 m. With one riser approximately midway between holes to be able to insert a hose for flushing the system.

18.3.3 Filter materials and gradation

"Filter material" is the ganular material which is used as a backfill in the trenches where subdrains are placed. To allow free water to reach the drain, the filter material must be much more pervious than the protected soil. If the filter is too pervious, then soil particles to be drained will move into the filter and clog it. It is, therefore, necessary that filter material and its gradation should be selected on the rational basis. Based on studies and experience, the following criteria is used for selection of the gradation of the filter material.

(i) *To prevent clogging of a perforated pipe with filter material*, the following requirements must be satisfied.

$$\frac{85 \text{ per cent size of the material*}}{\text{Diameter of perforation}} > 1$$

* (85 per cent [by weight] is finer than the specified size).

(ii) *To prevent the movement of particles from the protected soil into the filter material*, following conditions must be satisfied.

$$\frac{15 \text{ per cent size of filter material}}{85 \text{ per cent size of protected soil}} \leq 5$$

and

$$\frac{50 \text{ per cent size of filter material}}{50 \text{ per cent size of protected soil}} \leq 25$$

(iii) *To permit free water to reach the pipe*, following condition must be satisfied.

$$\frac{15 \text{ per cent size of filter material}}{15 \text{ per cent size of protected soil}} \geq 5$$

(iv) *To prevent segregation of filter material*, the following requirement should be met.

$$\frac{60 \text{ per cent size of filter material}}{10 \text{ per cent size of filter soil}} \leq 20$$

A single gradation of filter material is preferred for simplicity of construction. Filter material should always be placed in the moist state, to reduce segregation.

ENVIRONMENTAL IMPACTS

18.4 IMPORTANCE OF ENVIRONMENT

Many problems are created to people, ecological systems, water resources, air quality, noise, by the existing airports in many cities, due to lack of planned approach in their location, size and configuration. Airports must be planned with utmost considerations to environment, and community concerns on a long term basis, well coordinated with comprehensive regional plan of development and expansion of the city.

Environmental impacts of the airport must be assessed and suitable mitigation and management measures should be implemented, as an integral part of the total airport project. Site selected for an airport should have minimum adverse effects on human activities and environmental impacts on air and water quality, ecological and biological resources, fisheries, forests, noise etc. Alternative-sites must be evaluated and compared for above effects along with quality of life, economic resources and cultural values, before selecting the final location of an airport. A holistic approach to location, planning and development of an airport is required, with complete awareness of its impacts on users and non-user from economical, soil, environmental and ecological view points. Adverse environmental effects should be minimised, while quality of the environment has to be enhanced and restored to the possible extent. An interdisciplinary approach is required for assessing and preserving the environment.

18.5 ENVIRONMENTAL IMPACTS FROM AIRPORT PROJECTS

Construction, maintenance and operation of an airport project has its impacts on environmental, social and ecological factors, which must be properly evolved and understood to reduce detrimental effects and long and short term implications of these projects, rendering efficient and economic plans and designs.

Various impacts/effects of an airport project can be grouped in following categories :

1. *Pollution impacts* :
   - Air pollution
   - Water pollution
   - Noise impacts
   - Pollution during construction process

2. *Social impacts* :
   - Land development
   - Displacement and re-location
   - Displacement of recreational, historical and natural assets
   - Consistency with planning

3. *Ecological impacts* :
   - Wild and water life
   - Flora and fauna
   - Endangered species and birds
   - Contamination of lakes and streams

4. *Engineering factors* :
   - Flood hazards
   - Energy and natural resources

5. *Economic factors*
   - Cost-benefit analysis

18.5.1 Air Pollution

Airports with high volumes of commercial jet aircrafts can contribute significantlly in polluting the air affecting public life, health, damaging vegetation, wildlife, animals and soil, resulting in reduced visibility which may create several other transportation and traffic hazards.

The amount of air pollution created by aircrafts depends upon type of engine and mode of operation of the aircraft. Study of the air quality impacts requires the determination of ambient air quality, local meteorological conditions, number and path of aircrafts using the airport, their emission rates in different operating modes. A study of other ground transportation vehicles which provide services to airport is also undertaken.

Before and after concentrations of pollutants in the area of the airport are studied for comparison with the allowable air quality standards, and their impacts on plant, animal and human life. Corrective measures are then undertaken to reduce or eliminate the harmful effects due to air pollution.

The six pollutants of air are carbon monooxide (Co), Hydrocarbons, nitrogen oxides, sulphur dioxide, suspended particles and photo chemical oxidants. Airport projects may result in increase percentage of these pollutants, than prescribed limits, resulting in detrimental effects. Prescribed standards of ambient air quality should not be violated.

Continuous monitoring of air quality, near the airport should be undertaken, on hourly, 8 hourly and daily basis.

18.5.2 Water pollution

Construction and operation of airports can degrade the quality of water, reduce the quantity of ground water and surface water.

A water quality study in the vicinity of the airport should be undertaken to study the direct and indirect effects of the airport project, such as runoff from facility, infiltration, turbidity, sewage disposal, disruption of nutrient cycle. Removal of material for construction purposes alters the natural system of filtration, storage etc. water quality study identifies the sources of pollutants as well as the amount of degradation of water quality and supply. Water and waste water management plans are then worked out to preserve water quantity and quality.

Dust and chemical in the air, finally get deposited on the land surface, and are mixed with surface rain water, polluting the water and degrading its quality.

Continuous monitoring of water quality should be done, to keep water free from harmful ingredients.

18.5.3 Noise Impacts

The effects of noise on the communities living near an airport, are major environmental problems as jet aircrafts are very noisy. Airport planners should estimate the noise produced by airport operation and its impact on community. Magnitude of sound, quality of sound, duration of sound, frequency of sound during the seasons and time of the day, affect the community. The distance from noise, landuse, type of the building construction, noise level are the factors which show the communities altitude towards airport projects and tolerance of noise.

The human sensitivity to noise tolerance varies, depending upon pitch, frequency, loudness of the noise, time of the day and number of time it occurs. The airport noise is quantified and related to human assessment of loudness and tolerance. Many models and metrices are developed for noise analysis.

Sounds created by the source is radiated and transmitted through the air, fluctuating in form of sound pressures which impinge on the ear, producing audible sound. Human auditory system is sensitive to a very wide range of sound pressures.

Pitch of the sound is an important factor. Frequency or pitch is the number of times per second that the second pressure oscillates about atmospheric pressure. The units of frequency are Hertz (Hz). All sounds have wide range of frequencies. The human auditory system has normal range of frequencies of hearing varying from 16 Hz to about 16000 Hz. The system is not equally sensitive to all ranges of frequency. For this reason A-weighted sound level (A level) was developed and adopted as the basic unit for analysis of environmental impact.

A-weighted sound levels are measured in decibles. Some typical A-weighted sound levels in decibles (dB) are :

747-100 aircraft take-off 6500 m away = 100 dB
727-200 aircraft 6500 m away from takeoff = 90 dB
757-200 aircraft 6500 m from start of take off = 70 dB
Threshold of hearing – 0-10 dB.

Another important factor in environmental sound impact is its variation overtime. During an aircraft pass-by, the sound level fluctuates. It continues to increase until an aircraft passes the hearer and then decreases as aircraft goes away. Maximum sound level measures on a sound-level meter is reported for comparison sounds of different aircrafts.

The "duration", how long sound of a particular level continues also effects the noisiness and human tolerance. Sound level measurements are also necessary for longer period of times like an hour, several hours, or a day. These measurements are necessary to define cumulative effects of the noise. Day and night, and day to day variability of noise is measured near the airport. The averaging must be done on a sound-energy basis. It is represented as Day and Night Levels (DNL) of sound.

The perceived noise level (PNL) and effective perceived noise levels (EPNL) are quantities which also represent A sound exposure level. They are developed specifically to correlate with aircraft sound. The perceived noise level represented by the unit PNdB is a quantity which varies from moment to moment. Determination of PNL and EPNL is a complex procedure for measurement, requiring frequency analysis and non-linear amplitude adjustments.

People get affected by noise in two ways :
  (a) Day to day life effects
  (b) Long term effects.

Day to day effects interfere with activities, like rest, sleep, communication, mental disturbances etc. while long term effects are on health like hearing loss, nonauditory effects, such as hypertension etc. Guidelines are introduced for assessing the noise with landuses and to protect public health and welfare, by measures of controlling noise.

Research has shown that it is almost unlikely that aircraft noise around airport could ever produce hearing loss. More than 1000 flights per day with a sound level of 100 dB each, produce a time-weighted average sound level of 85 dB.

FAA has established DNL as cumulative noise exposure measure, for use in airport noise analysis and developed guidelines for noise and landuse compatibility evaluation. People often get used to a level of noise exposure and may not accept guidelines. Guidelines indicate that all landuses are considered compatible with aircraft day-night average levels below 65 dB.

Noise problems developed by airport activities are predicted through computer based simulation models and quantified through portable and permanent measurement and monitoring systems.

18.5.4 Control of noise

Aircraft related noise problems, can be reduced by increasing distance between aircraft and noise sensitive community, and by reducing the noise emission levels and their frequencies. Some of the techniques, which could provide some relief to noise problems are :

(a) Noise barriers

Noise barriers are used to reduce the noise, when aircraft is on the ground. When aircraft is the barrier airborne, the barriers are not effective. Barrier must be put close to the source or close to the receivers, for best effects. Barrier could be in form of walls, earth berms etc. of modest height.

(b) Sound proof structures

Sound insulation is is provided in structures like hospitals, schools, auditoriums and residences. Airgaps between windows and doors are kept minimum, by glass panes, and cost-effective sound insulation techniques, and use of sound absorbing materials.

(c) Selective runway use

On airports where number of runways is adequate, the runway whose use affects the least population in neighbourhood is widely used, when weather permits. Preferential runway optimises the runway use, to minimise noise impacts on the population as much as possible.

(d) Noise abating flight paths

Aircraft paths arc so designed that it flies over least populated area, thus abating the noise, from noise sensitive areas. Air traffic is thus routed over less sensitive land uses. Noise abatement departure procedures uses specific headings and turns to avoid flying over residences and sensitive areas.

(e) Restrictive use of airports

Those aircrafts which are more noisy, may not be allowed to use a particular airport to reduce the noise problem. Such a restriction is adopted, in view of local circumstances, community expectations and support.

Such restrictions on use of an airport may involve complete elimination of certain aircrafts or night use restrictions, during late hours of night when people are more sensitive to noise while sleeping.

Detailed analysis of noise reduction and benefits to the community should be undertaking before implementing restrictive uses.

(f) Noise regulations

Federal aviation agency of USA (FAA) has framed several regulations, which must be followed by airport authorities and airlines, for obtaining airworthiness certificates and use of airports. Permanent noise monitoring units are used to see that noise is within prescribed units.

18.5.5 Pollution during construction process

During the construction of an airport, several types of adverse impacts may be for community and travellers for short time or long time. Such impacts include :

- Soil erosion, and excavations encouraging erosion of soil.
- Noise due to construction equipment.
- Reduced freshness in air, due to particulate matter and dust.
- Water pollution and reducing of water table.
- Reduction in quantity of construction material.
- Dislocation of business, residences and other activities.
- Disruption of traffic, detours creating new problems.
- Drainage problems due to cut or fills, water logging etc.
- Loss of trees.
- Effects on other construction projects.
- Restriction on access to land uses.

A detailed study must be undertaken to identify the likely impacts which construction activity of an airport will develop and adequate measures should be undertaken to safeguard the community from such impacts, throughout the entire construction period.

Soil erosion can be reduced by replanting or turfing, and adopting erosion protection measures.

18.5.6 Social impacts

Several social impacts are created on the community by an airport project, mainly concerning land development, displacement and relocation.

(a) Land development

Several type of land development impacts may be there such as :

- change in pattern of landuse near airport.
- changes in economic activity and landuse values.
- increased residential, commercial, retail and other activities.
- changed pattern of transportation network for access to the airport.
- increased industrial/commercial activities in the region near the airport due to better accessibility.
- New facilities, hotels etc. for tourists.

A detailed study of land development impacts, considering factors that influence industrial, commercial and residential landuse, must be undertaken. The analysis should examine historical trends, for similar area, conducting surveys etc. to have realistic prediction of land development. The overall land development should be planned in accordance with regional policy and existing regulations.

(b) Displacement and relocation

Construction of an airport, requiring large areas of land or expansion of an airport, demanding additional land results in removing of existing residences, business and commercial activities. This results in public annoyance and problems. Airport planners should undertake a comprehensive study and measures to minimise adverse consequences to the community.

Relocation sites and resources should be identified and rationally selected as per needs and wishes of the displaced population. Changes and demands of public services should be identified.

(c) Displacement of recreational, historical, and natural assets

Impact of construction of airports on parks, historical places, recreational areas, cultural and archaeological resources, etc. is to be studied and analysed. Detailed studies must be conducted and measures undertaken to preserve these facilities, their character and compatibility. Acquisition of such lands should be avoided as far as possible. When it is required to displace any of the above activity, efforts should be made to develop similar or better facility in the vicinity with easy and adequate accessibility and conformity to the existing environment.

(d) Consistency with local planning

All efforts should be made for design, planning, construction and operation of an airport towards its location in an environment which is compatible and well coordinated in the regional plan of the area. When fitted in the overall planning of area, it will eliminate and reduce community conflicts, and shall fit in public inspiration and objectives.

A multi-disciplinary approach involving all the concerned agencies, clearly stating the goal and objectives should be undertaken so that airport plans are in integration with economic, landuse and transportation planning of the region. Planning should involve public participation so that community concerns and adverse impacts are properly and adequately addressed.

18.6 ECOLOGICAL IMPACTS FROM AIRPORT PROJECTS

Ecosystem gets affected by airport construction and operation changing the natural ground and water flowing system. It is necessary to protect the ecosystem through proper measures.

(a) Wild and water life

For supporting the life of various species, plants and animals living on land or in water, it is essential to maintain conducive ecosystem without any degradation due to airport activities. Cutting of trees and clearing shrubs affects ecology of plants and animals. Turbidity of water, formation of dust, creation of noise during construction of airport is harmful for life.

(b) Flora and fauna

Disruptions to flora and fauna, due to development of an airport must be minimised. Changes in topography of the land, clearing of bushes, plants, trees and shrubs, eliminate or reduce the ecological balance due to loss of flora and fauna.

(c) Endangered species and birds

Airport construction and its activities destroy the natural habitat and feeding grounds for wild life. Aircrafts strike the birds in the air. Migration routes of birds, wild animals, should be avoided for use of the aircraft to provide protection to birds and wild life. Life support system of different species in the area should be identified and impact of airports on their requirements for food, water and life support requirement of vegetation etc. should be studied for providing long term protection. Bird scaring devices may be used for safety of aircrafts in areas where birds are large and too many in number.

(d) Contamination of lakes and streams

Contamination of streams, lakes and waterways is likely to take place by aircrafts, ground vehicles, washing, servicing and fuelling of aircrafts and vehicles, maintenance operations of airports and terminals, mainly in case drainage system is not properly designed or inadequate. During construction and site clearing operations, occurring of contamination is possible.

Water quality and its preservation are important to preserve natural resources before implementation of airport development project.

Ecological environment should be periodically assessed.

18.7 ENGINEERING IMPACTS FROM AIRPORT PROJECTS

(a) Floods

Alteration in ground levels, topography, soil cover, may affect storing capacity of nearby rivers, waterbodies, streams etc. Potential of flooding in such cases should be examined and analysed and measures should be undertaken to avoid hazards (if any), as the part of the overall project.

(b) Energy and natural resources

The impact of airport design elements on energy consumption such as navigational aids, layout of apron and taxiways, runway systems etc. should be examined in detail. Alternative designs and layout must be evaluated.

Extensive use of construction materials can also affect natural resources and affect the cost. Water demands for the airport project increase stressing the quantities of existing surface and ground water sources.

(c) Economic factors

Engineering costs for an airport project mainly consist of :
- land acquisition costs and environmental quality costs.
- construction costs of facilities, including man, material and machinery.
- operation, maintenance and management costs.

For purposes of evaluation or studying economic impact of developing an airport, these costs are related to passengers and aircraft traffic, to arrive at cost per enplaned passengers or cost per air carrier operation. For justification of the project, costs are also allotted to various other users and non-users of the projects such as tenants, airlines, aviation cargo, business, and local private and governmental agencies involved with the project.

A cost-benefit analysis including both direct and indirect benefits to both users and non-users is undertaken for justification of the project.

Revenue generation and requirements are then worked out.

18.8 ENVIRONMENTAL IMPACT ASSESSMENT (EIA)

Before undertaking an airport project, it is essential that an environmental impact study (EIA) is conducted at the planning stage, considering all the above mentioned impacts. Costs involved in maintaining/building the environment and protecting it, should be included in the total costs of the airport project. Measures should be undertaken to avoid detrimental impacts of the existing airports on the environment.

The basic contents of an environmental study report are :
- need of the new or improvement project, its description.
- an inventory of problems and associated issues.
- identification of constraints and opportunities.
- various improvement components.
- measures to increase benefits and reduce harms.
- an analysis of alternatives and their impacts on environment, short as well as long term.
- protection and mitigation measures for the environment and resources.
- selection of the particular alternative and its justification from environmental and economic point of view.

18.9 ENVIRONMENTAL MANAGEMENT PLAN (EMP)

For mitigation of adverse impacts and protection of the environment, different measures to be under-taken are listed and described in an environment management plan report. Recommendations made in EMP are to be incorporated and implemented as the airport project proceeds. Right from planning, design and implementation/operational stage, the environmental protection and enhancing measures are to be implemented.

EMP is a teamwork of many involved agencies which are responsible for maintaining, enhancing and continuous monitoring of the environment, through observations, testing, sampling and regular analysis.

Typical Questions for Practice

1. How air transportation helps in economic development of a nation? What are specific advantages of air transportation.

2. What was the necessity of establishing an international organisation for air transportation? Mention in brief the functions of International Civil Aviation Organisation (ICAO).

3. With what objectives National Airport Authority of India (NAA) was created? What functions it undertakes?

4. Write a brief note on development of air transportation in India and abroad.

5. Draw typical layout of an airport, naming its basic components. Mention functions of each component.

6. What are the different gear configurations used in aircrafts to support the load? How the load is distributed?

7. Explain briefly how the pilot controls the vertical and horizontal movement of an aircraft from his cabin?

8. What are different characteristics of an aircraft which affect planning and design of an airport?

9. What factors affect planning of an airport? Why regional planning is necessary?

10. For location of a large airport and for selecting a suitable site what factors should be taken into account.

11. What is a master plan for airports? What factors are considered in developing a master plan?

12. What do you understand by imaginary surfaces for an aircraft? Name and describe different imaginary surfaces.

13. What is an approach zone? Draw a typical sketch showing height limitation in approach zones for safety of the aircraft.

14. Write a short note on the following :
    (i) Transitional surfaces
    (ii) Instrumental runway
    (iii) Clear zones
    (iv) Horizontal and transitional surfaces

15. Why wind rose diagram is prepared for orientation of a runway? How it helps?

16. Mention different types of wind-rose diagrams and their utility.

17. How runways are oriented? Explain the term 'coverage' and 'cross wind component'.

18. What is basic runway length? How actual runway length is determined?

19. What corrections are applied to basic runway length for obtaining actual runway length at a particular site?

20. Why it is necessary to classify airports? Explain different ways of classifying airports.

21. What are the various elements of runway geometrics? Mention ICAO recommendations for the same.

22. Draw a sketch to show various runway geometrics for an instrumental runway.

23. Write short notes on :
    (i) Cross wind component
    (ii) Runway sight distance
    (iii) Corrections to basic runway length
    (iv) Airport classification.

24. Describe need for 'Runway orientations' and 'preparation of wind rose diagram'.

25. What factors affect capacity of an airport? How practical capacity is determined?

26. What are different runway patterns? Mention merits and applicability of various patterns.

27. What are various elements of a taxiway? What are ICAO recommendations for each one of them?

28. Draw a sketch of a taxiway configuration at an international airport, showing various geometric elements.

29. What are the main differences between a runway and a taxiway? Draw a cross-section of a typical runway and taxiway.

30. What factors are considered in structural design of airport pavement?

31. How equivalent single wheel load (ESWL) is determined? Why it is necessary to determine ESWL?

32. What is a flexible pavement? How it differs from rigid pavement? What are their advantages and disadvantages?

33. Describe the 'CBR' method of flexible pavement design of airport in brief.

34. Draw a sketch to show typical joints in rigid airport pavements. Write functions of each type of joint.

35. Compare 'Rigid' and 'Flexible Pavements'.

36. What is an overlay? Name various types of overlays.

37. Explain different types of distresses caused in flexible pavements. How they can be prevented?

38. Why pavement evaluation is done? What remedial measures are taken to rehabilitate distressed pavements?

39. What is a load classification number (LCN)? What is its utility for airports?

40. What is meant by strengthening of runway pavements? Explain the procedure of designing a flexible overlay over a flexible pavement.

41. Describe in brief the procedure for designing rigid overlay over rigid pavements.

42. What is a terminal for an airport? Draw a typical layout plan for terminal area showing its important components and their functions.

43. Draw sketches to show common types of runway layout. What are the requirements of an ideal runway layout.

44. Why visual aids are necessary for an airport? Draw a neat sketch to show :
    (a) Typical markings
    (b) Typical lighting arrangement for airports

45. What are the functions of runway threshold lights and beacon light?

46. With the help of suitable sketches describe how an aircraft lands under instrumental landing system?

47. Explain the following terms as used in airport engineering :
    (i) Precision approach radar
    (ii) Air traffic control system
    (iii) Control tower
    (iv) Localiser and glide slope antenna

48. Explain the need for an adequate airport drainage system. How subsurface drainage is done?

49. What is meant by ponding? When ponding should be permitted on an airport drainage system.

50. Draw typical sketches to show surface and subsurface drainage arrangement at an airport.

Typical Questions for Practice

1. For long distance travel the most convenient mode of transportation is :
   - (a) Railway
   - (b) Highway
   - (c) Waterways
   - (d) Airways

2. The earliest mode of transportation was :
   - (a) Water transportation
   - (b) Air transportation
   - (c) Highway transportation
   - (d) Railway transportation

3. In India first air flight was started by :
   - (a) G.D. Birla
   - (b) J.R.D. Tata
   - (c) B.C. Roy
   - (d) Government of India

4. The international body which provides standards and regulations for air flights and airports is :
   - (a) FAA
   - (b) ICAO
   - (c) AAI
   - (d) ICO

5. First international flight was started in :
   - (a) 1901
   - (b) 1918
   - (c) 1950
   - (d) 1952

6. ICAO headquarter is located in :
   - (a) Canada
   - (b) United States of America
   - (c) Japan
   - (d) France

7. Air India was originally known as :
   - (a) Imperial Airways
   - (b) Hans Airways
   - (c) Tata Airlines
   - (d) Pawan Airlines

8. 'Air bus' is the name for :
   - (a) a bus which has high speed
   - (b) an aircraft
   - (c) rail transport on elevated track
   - (d) a bus whose shape is like an-aircraft

9. The first Jumbo jet plane of Air India was named :
   - (a) Kanishka
   - (b) Shehanshah
   - (c) Emperor Ashok
   - (d) Vayudoot

10. The first airport was built in India in 1932 at :
    - (a) Ahmedabad
    - (b) Mumbai
    - (c) Delhi
    - (d) Kolkata

11. The lateral control of the aeroplane is provided by :
    - (a) Wings
    - (b) Rudder
    - (c) Aileron
    - (d) None of the above

12. The facility provided for emergency landing is called :
    - (a) Heliport
    - (b) Air strip
    - (c) Air field
    - (d) Landing strip

13. The approach surfaces at the end of runways have a shape which is :
    - (a) Rectangular
    - (b) Circular
    - (c) Irregular
    - (d) Trapezoidal

14. The wind velocity at airport is known by :
    - (a) Anemometer
    - (b) Mach scale
    - (c) Beaufort scale
    - (d) Barometer

15. Which number on Beaufort scale represents a "Hurricane".
    - (a) 1
    - (b) 6
    - (c) 8
    - (d) 12

16. Wind velocity is usually expressed in following unit :
    - (a) Knots
    - (b) Mach
    - (c) Beaufort
    - (d) None of the above

17. The air control provided at the tail end of the aeroplane is called :
    - (a) Rudder
    - (b) Flap
    - (c) Propeller
    - (d) Air screw

18. The minimum separation spacing between two airports operating jet-aircrafts should be atleast :
    - (a) 40 kms
    - (b) 80 kms
    - (c) 160 kms
    - (d) 600 kms

19. Obstruction clearance surface for safe landing of aircrafts at airports is :
    - (a) Approach surface
    - (b) Conical surface
    - (c) Horizontal surface
    - (d) All the above

20. 'Calm period' on a wind rose diagram, is when wind intensity is :
    - (a) Less than 4 kmph
    - (b) Less than 6.4 kmph
    - (c) Less than 10 km kmph
    - (d) Zero

21. A wind rose diagram shows :
    - (a) Direction of wind
    - (b) Intensity of wind
    - (c) Duration of wind
    - (d) All the above

22. The maximum longitudinal gradient for an 'A' type airport as specified by ICAO is :
    (a) 1.5%   (b) 2.0%
    (c) 2.5%   (d) 3.0%

23. An 'Hanger' at airport is meant for :
    (a) Safety of aircraft   (b) Overhauling of aircraft
    (c) Parking of aircraft   (d) Aircraft nosing

24. Jet exhaust is deflected by :
    (a) Blast fences   (b) Pitching
    (c) Turfing   (d) Non of the above

25. 'STOL' represents :
    (a) Safe landing   (b) Short landing
    (c) Surface landing   (d) Sea landing

26. During the poor visibility flights are operated by :
    (a) VFR   (b) ILS
    (c) IFR   (d) CLS

27. "Standard temperature" at the airport is considered as :
    (a) 10°C   (b) 12.5°C
    (c) 15°C   (d) 25°C

28. Outer horizontal surface may not be required at an airport having runway length less than :
    (a) 800 m   (b) 1000 m
    (c) 1100 m   (d) 900 m

29. Minimum value of radius of curvature for airports serving jet aircraft is :
    (a) 100 m   (b) 120 m
    (c) 150 m   (d) 160 m

30. Runway length at an airoport depends upon :
    (a) Elevation of site   (b) Rainfall at the site
    (c) Direction of the wind   (d) All the above

31. The imaginary conical surface extends upwards and outwards from the periphery of :
    (a) Approach surface   (b) Inner horizontal surface
    (c) Boundary of the airport   (d) Runway ends

32. As per ICAO recommendations basic runway length is to be increased for every 300 m rise in elevation above mean sea level, at the rate of :
    (a) 5%   (b) 8%
    (c) 9%   (d) 11%

33. As per ICAO total corrections applied to basic runway length for a runway for elevation and temperature should not exceed :
    (a) 25% basic length   (b) 30% basic length
    (c) 35% basic length   (d) 40% basic length

34. Structural design of airport pavement involves :
    (a) Finding the pavement thickness   (b) Determining total space requirement
    (c) Designing terminal building   (d) Providing best suited geometric layout

35. Runways are generally oriented in the direction of :
    (a) Minimum ground slope   (b) Magnetic north
    (c) Least wind   (d) Prevailing winds
36. The noise level of a jet-aircraft while taking off is about :
    (a) 120 db   (b) 200 db
    (c) 180 db   (d) 60 db
37. Efforts are made to limit cross wind velocity at an airport to :
    (a) Less than 30 kmph   (b) Less than 5 kmph
    (c) Less than 25 kmph   (d) Less than 1 kmph
38. Fuselage of an aircraft does not include :
    (a) Space for passengers   (b) Pilot cabin
    (c) Toilets   (d) Wings
39. Speed of an aircraft relative to ground is called :
    (a) Wind speed   (b) Cruising speed
    (c) Sonic speed   (d) Air speed
40. Colour of the lights at runway threshold is :
    (a) Red   (b) Yellow
    (c) Blue   (d) Green
41. ICAO classifies runway length according to codes in :
    (a) Alphabets   (b) Numerals
    (c) Combination of alphabet and numerals   (d) None of the above
42. The imaginary conical surface of the approach area has a slope upward of :
    (a) 1 in 10   (b) 1 in 15
    (c) 1 in 20   (d) 1 in 25
43. As per ICAO recommendations the maximum longitudinal gradient along taxiway should not exceed :
    (a) 1.5%   (b) 1.25%
    (c) 2%   (d) 3%
44. All markings on a taxiway should be painted in which colour as per ICAO recommendations :
    (a) Blue   (b) Black
    (c) Yellow   (d) White
45. Threshold markings have a width of :
    (a) 3 m   (b) 4 m
    (c) 5 m   (d) 6 m
46. Taxiway lighting colour is :
    (a) Blue   (b) Green
    (c) Red   (d) Yellow
47. Obstruction lighting colour is :
    (a) Red   (b) Green
    (c) Yellow   (d) Voilet

48. The radius of airport control area is :
    (a) 50 kms
    (b) 100 kms
    (c) 160 kms
    (d) 200 kms
49. The pioneering work in rigid pavement design for airport pavements is based on theories of :
    (a) Westeergard
    (b) Macadam
    (c) Tresguet
    (d) Burmister
50. ILS stands for :
    (a) International light system
    (b) Intended landing site
    (c) Indian lighting standard
    (d) Instrument landing system
51. TORA is an abbreviation for :
    (a) Total operating runway available
    (b) Take off run available
    (c) Time of return availed
    (d) Towards other runway approach
52. PAPI stands for
    (a) Precision approach path indicator
    (b) Project area path indicator
    (c) Precise airspace planning idea
    (d) Perfect airport planning inventory
53. Airport master plan is :
    (a) Concept for ultimate development of an airport
    (b) Plan for short term improvements
    (c) Providing immediate remedial measures to improve capacity
    (d) Arranging for finances for developing an airport
54. Site selection for an airport involves determination of :
    (a) Atmospheric conditions
    (b) Surrounding obstructions
    (c) None of the above
    (d) All of the above
55. Rigid airport pavement consists of :
    (a) Hard stone base
    (b) Portland cement concrete
    (c) Two or more layers of high quality material
    (d) Full depth bituminous pavement
56. International Civil Aviation Organisation (ICAO) came into existence in the year :
    (a) 1947
    (b) 1940
    (c) 1952
    (d) 1948
57. An organisation under Central Government which looks after the airports in India :
    (a) Planning Commission
    (b) Central Airport Authority
    (c) Airport Authority of India
    (d) C.P.W.D.
58. Airport Authority of India was setup in the year :
    (a) 1990
    (b) 1992
    (c) 1995
    (d) 1998
59. As per ICAO, the maximum specified width of runway is :
    (a) 30 m
    (b) 45 m
    (c) 60 m
    (d) 70 m

60. The rate of rise or fall of a runway along its length is called :
    (a) Ramp
    (b) Gradient
    (c) Slope
    (d) Camber
61. The minimum width of a runway as recommended by ICAO is :
    (a) 10 m
    (b) 15 m
    (c) 18 m
    (d) 21 m
62. A test used to determine modulus of subgrade is called :
    (a) CBR test
    (b) Plate bearing test
    (c) Consolidation test
    (d) Flexural strength test
63. The base course in flexible runway pavement is :
    (a) A concrete slab
    (b) A prepared material of hard core whose function is to act as a weight distribution layer
    (c) Protective layer to withstand wear and tear
    (d) A layer to provide smooth and water proof surface
64. The purpose of providing blast fences is :
    (a) To prevent damage to buildings and equipment from blast
    (b) To protect aircrafts from fast winds
    (c) To provide safety from other vehicles
    (d) To prevent passengers crossing a runway
65. Wind direction indicator should have minimum length of :
    (a) 4 m
    (b) 4.53 m
    (c) 3 m
    (d) 3.6 m
66. A drain provided in the slope of cutting to intercept the water flowing down the cut slope is called:
    (a) Catch drain
    (b) Side ditch
    (c) Slope drain
    (d) Gradient
67. A sign showing taxiway/runway intersection is a :
    (a) Regulatory sign
    (b) Warning sign
    (c) Informatory sign
    (d) Mandatory sign
68. Which of the following aircrafts will have maximum weight :
    (a) DC-3
    (b) Airbus 310
    (c) Boeing 747
    (d) Concorde
69. The piston engine aircraft among the following is :
    (a) DC-8
    (b) Viscount 802
    (c) Boeing 720
    (d) DC-3
70. The height of outer horizontal surface above airport reference point (ARP) is :
    (a) 250 m
    (b) 200 m
    (c) 150 m
    (d) 100 m
71. In CBR method of flexible pavement design of airport fo'lowing information is not required :
    (a) CBR value of subgrade and sub-base material
    (b) Modulus of subgrade reaction

    (c) Gross weight of design aircraft

    (d) Number of annual departures of design aircraft

72. In rigid pavement design, the radius of relative stifness is given by formula :

    (a) $4\sqrt{\dfrac{1000\ Eh^3}{12\,(1-\mu^2)\,K}}$
          (b) $4\sqrt{\dfrac{3000\ Eh^3}{12\,(1-\mu^2)\,K}}$

    (c) $3\sqrt{\dfrac{1000\ Eh^3}{12\,(1-\mu^2)\,K}}$
          (d) $2\sqrt{\dfrac{1000\ Eh^3}{12\,(1-\mu^2)\,K}}$

73. The purpose of a pavement overlay is :

    (a) To strengthen the existing pavement

    (b) To build a new pavement replacing the old one

    (c) Change the flexible pavement to rigid

    (d) Change the rigid pavement to flexible

74. The term "Ponding" in airport drainage is used for :

    (a) Temporary storage of runoff
          (b) A pond in the vicinity

    (c) A pond made to collect rain water
     (d) Where all drains meet

75. Jumbo jet is the name for :

    (a) Airbus
               (b) DC-8

    (c) F-16
                 (d) Boeing 747

CORRECT ANSWERS TO QUIZ TYPE QUESTIONS

1. (d)	2. (a)	3. (b)	4. (b)	5. (b)	6. (a)	7. (c)	8. (b)	9. (c)	10. (b)
11. (c)	12. (b)	13. (d)	14. (c)	15. (d)	16. (a)	17. (a)	18. (c)	19. (d)	20. (b)
21. (d)	22. (a)	23. (b)	24. (a)	25. (b)	26. (c)	27. (c)	28. (d)	29. (b)	30. (a)
31. (b)	32. (b)	33. (c)	34. (a)	35. (d)	36. (a)	37. (c)	38. (d)	39. (b)	40. (c)
41. (a)	42. (c)	43. (d)	44. (c)	45. (b)	46. (a)	47. (a)	48. (c)	49. (a)	50. (d)
51. (b)	52. (a)	53. (a)	54. (c)	55. (b)	56. (a)	57. (c)	58. (c)	59. (b)	60. (b)
61. (c)	62. (b)	63. (b)	64. (a)	65. (d)	66. (a)	67. (d)	68. (c)	69. (d)	70. (a)
71. (b)	72. (a)	73. (a)	74. (a)	75. (a)					

A-380 Double Deck Aircraft

Incorporating the latest technologies, Airbus – a French manufacturing company located in Toulouse, France has produced the world's largest aircraft having high capacity, long range, two decks and wide body. A-380 will be a successor to the Boeing-747 aircrafts.

A-380 has 49% more floor space and 35% more seating capacity (number of seats : 555) than existing largest aircraft. It has 10-15% more range, lower fuel burn and emission and less noise. Several A-380 models are planned. The basic model is A380-800. The freighter aircraft model A380-800F will be able to carry a 150 tonne payload and is due to enter in service in 2008. Potential future models are: A380-700 (480 seats) and A380-900 (656 seats).

The first successful Airbus 380 flight was undertaken by Qantas Airways in February 2007 to Los Angeles, USA. It will soon be used for commercial flights by Singapore Airlines, Emirates Airlines, Kingfisher Airlines etc.

A380 aircraft landed at New Delhi airport for a demonstration flight, from Toulouse, France, on May 6, 2007 after a 9 hours flight, and flew to Mumbai airport on May 8, 2007. It will be used by Kingfisher airlines in India, for the first time in near future.

Characteristics of A380, Jumbo Jet, Double Deck Aircraft

(a) Aircraft dimensions

Length	73.00 m
Width	79.80 m
Height	24.10 m
Fuselage diameter	7.14 m
Cabin length	50.68 m
Wing area	845 m^2
Wheel base	30.4 m
Wheel track	14.3 m
Number of wheels	22

(b) Design weights

Maximum take-off weight	155 tonnes
Maximum landing weight	140 tonnes
Maximum zero fuel weight	133.5 tonnes
Maximum fuel capacity	34,430 litres

(c) Basic operating data

Engine thrust range	119-120 KN
Maximum operating Mach number	0.82
Speed	1049 km/hour
Engine type	Royal Royce Trent 900, Four Engines
Passenger capacity	Economy class : 853
	or
	Others : 555
Approximate cost	Rs. 1485 crores

References

1. Air Transport Association of America, "Runway Capacity Criteria for Airport Planning Purposes", Air Navigation Traffic Control report No. 118, 1976.
2. Air Transport Association of America, "Airport Geometrics", AD/SC, Report No. 5, July, 1977.
3. Aaron Henry, "Performance of Airport Pavements", American Road Builder Association, Bulletin No. '75.
4. Bruce, G.H., "Canadian Air Travel Demand", Air Transport Division of ASCE, 1993.
5. Dutt and Narayanan, "Speedy Construction of Aerodromes", Paper No. 243, Indian Road Congress, New Delhi.
6. Donald N. Brown, "Flexible Pavement for Tomorrow's Major Airports", Paper No. 6874, American Society of Civil Engrs., 1970.
7. David K. Witheford, "Airports and Accessibility", Transportation Quarterly, Eno Foundation, 1967.
8. David S. Lawrence, "Airport Again Access or Avoidance", Transportation Quarterly, Eno Foundation, 1970.
9. David S. Lawrence, "Helicopters and Today's Air Transportation System", Traffic Quarterly, Eno Foundation, USA.
10. Douglas K. Fleming, "Competition in the US Airline Industry", Traffic Quarterly, Eno Foundation, USA.
11. Eno Foundation for Transportation, "International Freight Transportation" by John H. Muhoney, 1995.
12. Edward N. Hall, "The Aircity", Traffic Quarterly, Eno Foundation, USA.

Federal Aviation Administration (FAA), Washington, USA

13. "Airport Design", Advisory Circulation, AC/150/5300-13, 1989.
14. "Runway Length Requirements for Airport Design", Advisory Circular, AC150/5325, 1990.
15. "General Operating and Flight Rules", 1992.
16. "Airport Master Plans", Advisory Circular, 150/5070-6A, 1985.
17. "Terminal Area Forecasts", 1991.
18. "Airport Capacity and Delay", Advs. Cir., AC150/1300-13, 1989.
19. "The Apron-Terminal Building Planning report", 1975.
20. "Airport Ground Access", 1978.
21. "Planning and Design Guidelines for Airport Terminal Facilities", Advisory Circular AC150/5360-13, 1994.
22. "Heliport Design", Advisory Circular AC 150/5390 2A, 1992.
23. "Planning and Design Criteria for Metropolitan SToL Ports", Advisory Circular, 150/5300-8.

24. "Vertiports", Advisory Circular, AC 150/5390-3, 1991.
25. "Airport Pavement Design and Evaluation", Advisory Circular No. 150/5320-6C, 1988.
26. "Use of Non-Destructive Testing Devices in the Evaluation of Airport Pavements", Advisory Circular No. AC150/5370-11, 1990.
27. "Marking on Paved Areas on Airports", Advisory Circular, 150/5340, 1987.
28. "Obstruction, Marking and Lighting", Advisory Circular, AC 70/7460-1H, 1991.
29. "Precision Approach Path Indicator (PAPI) System", Advisory Circular, AC 150/5345-28D, 1985.
30. "Runway and Taxiway Edge Lighting System", Advisory Circulation No. 150/5340-24, 1975.
31. "Runway Visual Range (RVR)", Advisory Circular, 97-1A, 1977.
32. "Taxiway Centreline Lighting System", Advisory Circular, AC150/5340-19, 1968.
33. "Airport Drainage", Advisory Circular, AC150/5320-5B, 1970.
34. "Airport Environmental Handbook", 1985.
35. "Aviation Noise E/effects", 1985.
36. "Noise Control and Compatibility Planning for Airport", Advisory Circular, AC 150/5020, 1983.
37. "Planning the Metropolitan Airport System", Advisory Circular, AC 150/5070-5, 1970.
38. "Airport Design Standard Site Requirement for Terminal Navigation Facilities", Advisory Circular, AC150/5300-2, 1987.
39. "Recommended Standard Traffic Pattern and Practices for Aeronautical Operating at Airports Without Operating Control Towers", AC No. 90-66A, 1993.
40. "FAR Guidelines Material Airport Radar Services Area", ARSA AC No. 90-88, 1986.
41. "Maintenance of Airport Visual Aid Facilities", Advisory Circular, AC 150/5340-26, 1982.
42. "Plane-Sense", FAA-H 8083-19, 2000.
43. "Airport Pavement Design for the Boeing 777 Airplane", Advisory Circulation, AC 150/5320-6.
44. "FAR Guidance Material", AC No. 90-88, 1986.
45. "Safety in and Around Helicopter", Advisory Circular, 91-32B, 1997.
46. "Pavement Management System", Advisory Circular, AC 150/5380-7.
47. "Aircraft Weight and Balance Control", AC-120-27B, 1990.
48. "Approval of Offshore Helicopter Approaches", AC No. 90-80A, 1988.
49. Gilbert, G.A., "Air Traffic Control", Ziff-Davis Publishing Company, Chicago.
50. Harper and Madda, "Airport Towers", A new generation Civil Engineering, American Society of Civil Engrs., Nov., 1993.
51. Horonjeff, R. and McKelvey, "Planning and Design of Airports", McGraw Hill Co., USA.
52. Henry, H.K., "Balanced Highway Airport Design", Paper No. 6420, ASCE, Feb., 1969.
53. "Institution of Transportation Engineers, USA", Transportation and Traffic Engineering Handbook, 2nd ed., Prentice Hall, USA.
54. Institute of Civil Engineers, "Airports and Automation", edited by G.M. Crook.

International Civil Aviation Organisation (ICAO), Montreal, Canada

55. "Aerodromes - Annex 14 to the Convention of International Civil Aviation", 1990, Vol. I and II.
56. "Aerodrome Design Manual Part I Runways", Doc 9157-AN/901, 1984.
57. "Aerodrome Design Manual Part II Taxiways", Aprons and Holding Bay, 1984.
58. "Aerodrome Design Manual Part III Pavements", Doc 9157-AN/901, 1983.
59. "Airport Planning Manual", Doc 9184 - AN/902, Part I and II, 1985.
60. "Standards for Airport Sign Systems", Advisory Circular, 150/5340, 18C, 1991.
61. "Specification for Wind Cone Assemblies", Advisory Circular, AC 150/5345, 27C, 1985.

62. John R. Wiley, "Airport Administration and Management", Eno Foundation for Transportation, 1986.

63. John M. Reibe, "SToL Aircraft Flight and Landing Area Considerations", Transportation Engineering Journal of American Society of Civil Engineers (ASCE), 1973.

64. Kenneth, K.W., "International Challenges to Airports", Paper 7455, ASCE, 1970.

65. Macatee, W.R., "Flexible Pavements - Requirement for Airports", American Road Builder Association, 1947.

66. McGrath, D.C., "Aircraft Noise", Fugitive factor in landuse planning, Journal of Urban Planning and Development, American Society of Civil Engineers, 1969.

67. Ministry of Works, Tanzania, "A Review of Aerodrome Statistics", Govt. of Tanzania, 1984.

68. Neufville, R., "Airport Construction : The Japanese Way", Civil Engineering, ASCE, Dec., 1991.

69. Robert R. Piper, "Runway Length Optimisation and Airport Pricing", Traffic Quarterly, Eno Foundation, USA.

70. Robert L. Mitchel, "Pavement Design for Airfields with Infrequent Traffic", Ministry of Road and Works, Rhodesia, 1975.

71. Reiss, S.M., "Down to Earth Terminal Design", Civil Engineering, ASCE, Feb., 1995.

72. Richard Tilles, "Curb Space at Airport Terminals", Traffic Quarterly, Eno Foundation, USA.

73. Transportation Research Board, USA, "Airport Capacity Planning", Transport Research Record No. 655, 1977.

74. Stuart, G.T., "Role of Air Travel in Transportation Systems", Traffic Quarterly, Eno Foundation, USA.

75. Transport Research Board, USA, "Airport and Air Transport Planning", Transport Research Record No. 588, 1976.

76. Vallerga and Lee, "Airport Challenges of the Future", American Society of Civil Engineers, 1973.

77. Vickil Golich, "Airline Deregulation : Economic Boom or Safety Bust", Traffic Quarterly, Eno Foundation, USA, 1988.

78. Venus Jamal, "Erbil Airport", University of Sulaimania Project Report, 1979.

79. Walter, E.C., "Airport Approach Runway and Taxiway Lighting System", American Society of Civil Engineers, Journal and Transportation Engineers, 1968.

Index